D0947240

# MEDICAL IMAGING

# MEDICAL IMAGING

## Principles, Detectors, and Electronics

*Edited by*

**Krzysztof Iniewski**

*Redlen Technologies Inc.*

**WILEY**

A JOHN WILEY & SONS, INC., PUBLICATION

Copyright © 2009 by John Wiley & Sons, Inc. All rights reserved

Published by John Wiley & Sons, Inc., Hoboken, New Jersey

Published simultaneously in Canada

No part of this publication may be reproduced, stored in a retrieval system, or transmitted in any form or by any means, electronic, mechanical, photocopying, recording, scanning, or otherwise, except as permitted under Section 107 or 108 of the 1976 United States Copyright Act, without either the prior written permission of the Publisher, or authorization through payment of the appropriate per-copy fee to the Copyright Clearance Center, Inc., 222 Rosewood Drive, Danvers, MA 01923, (978) 750-8400, fax (978) 750-4470, or on the web at www.copyright.com. Requests to the Publisher for permission should be addressed to the Permissions Department, John Wiley & Sons, Inc., 111 River Street, Hoboken, NJ 07030, (201) 748-6011, fax (201) 748-6008, or online at www.wiley.com/go/permission.

Limit of Liability/Disclaimer of Warranty: While the publisher and author have used their best efforts in preparing this book, they make no representations or warranties with respect to the accuracy or completeness of the contents of this book and specifically disclaim any implied warranties of merchantability or fitness for a particular purpose. No warranty may be created or extended by sales representatives or written sales materials. The advice and strategies contained herein may not be suitable for your situation. You should consult with a professional where appropriate. Neither the publisher nor author shall be liable for any loss of profit or any other commercial damages, including but not limited to special, incidental, consequential, or other damages.

For general information on our other products and services or for technical support, please contact our Customer Care Department within the United States at (800) 762-2974, outside the United States at (317) 572-3993 or fax (317) 572-4002.

Wiley also publishes its books in variety of electronic formats. Some content that appears in print may not be available in electronic format. For more information about Wiley products, visit our web site at www.wiley.com.

*Library of Congress Cataloging-in-Publication Data:*

Iniewski, Krzysztof
  Medical imaging electronics/Krzysztof Iniewski.
    p. cm.
  Includes bibliographical references and index.
  ISBN 978-0-470-39164-8 (cloth)
1.  Imaging systems in medicine.   2.   Medical electronics.   I.   Title.
  R857.O6I55 2009
  616.07′54—dc22

                                                                                    2008045490

Printed in the United States of America

10  9  8  7  6  5  4  3  2  1

# CONTENTS

**3   Circuits for Digital X-Ray Imaging: Counting and Integration**   **59**

*Edgar Kraft and Ivan Peric*

# PREFACE

The ability to peer into the human body is an essential diagnostic tool in medicine and is one of the key issues in health care. Our population is aging globally; for example, over 20% of the population in Japan is already over 65 years old. Older people require many more imaging investigations than do younger ones. Cancer and heart disease are the number-one killers, approaching 40% of all deaths. Improved image quality becomes essential for effective diagnostics in these cases. Shorter examination times, shift to outpatient testing, and noninvasive imaging are rapidly needed. The challenges to contain health-care costs are enormous, and technology solutions are needed to address them.

This book addresses the state-of-the-art in hardware design in the context of medical imaging of the human body. There are new exciting opportunities in ultrasound, magnetic resonance imaging (MRI), X-ray, computed tomography (CT), and nuclear medicine (PET/SPECT). Emerging detector technologies, circuit design techniques, new materials, and innovative system approaches are explored. This book is a must for anyone serious about electronics in a health-care sector.

There are four major imaging modalities described in this book. Their effective signal positions on the electromagnetic spectrum vary from kilohertz (kHz) for ultrasound, through gigahertz (GHz) for magnetic resonance imaging (MRI), to $10^{18}$ Hz for X-ray/computed tomography (CT) and nuclear medicine, over 15 orders of magnitude variation! Despite their vastly different frequencies and principles of operation, there are numerous commonalities in signal processing of signals received by these imaging detectors, such as signal amplification, filtering, multiplexing, and analog-to-digital conversion. These hardware commonalities among imaging techniques merit putting all related knowledge and know-how into one publication. I sincerely hope that this book will help improve the understanding of medical imaging electronics and stimulate further interest in the development and use of this equipment to benefit us all.

KRZYSZTOF (KRIS) INIEWSKI

*Vancouver 2008*

# ABOUT THE EDITOR

**Krzysztof (Kris) Iniewski** is managing R&D chip development at Redlen Technologies Inc., a start-up company in British Columbia. His research interests are in VLSI circuits for medical and security applications. From 2004 to 2006 he was an Associate Professor in the Electrical Engineering and Computer Engineering Department of University of Alberta where he conducted research on low power wireless circuits and systems. During his tenure in Edmonton, he wrote a book for CRC Press, entitled *Wireless Technologies: Circuits, Systems, and Devices*.

From 1995 to 2003, he was with PMC-Sierra and held various technical and management positions. Prior to joining PMC-Sierra, from 1990 to 1994 he was an Assistant Professor at the University of Toronto's Department of Electrical Engineering and Computer Engineering. Dr. Iniewski has published over 100 research papers in international journals and conferences. He holds 18 international patents granted in the United States, Canada, France, Germany, and Japan. He received his Ph.D. degree in electronics (honors) from the Warsaw University of Technology (Warsaw, Poland) in 1988. Together with Carl McCrosky and Dan Minoli, he is an author of *Network Infrastructure and Architecture: Design High-Availability Networks*, John Wiley & Sons, 2008. He is also an editor of *VLSI Circuits for Biomedical Applications*, Artech House, 2008 and *Circuits at the Nanoscale: Communications, Imaging, and Sensing*, CRC Press, 2008. Kris can be reached at kris.iniewski@gmail.com.

# CONTRIBUTORS

ANNA CELLER, UBC, VGH Research Pavilion, Vancouver, BC, Canada

GIANLUIGI DE GERONIMO, Instrumentation Division, Brookhaven National Laboratory, Upton, NY

BRUNO HAIDER, Imaging Technologies, GE Global Research, Niskayuna, NY

KRZYSZTOF (KRIS) INIEWSKI, CMOS ET, Coquitlam, BC, Canada

KARIM KARIM, Department of Electrical and Computer Engineering, University of Waterloo West, Waterloo, Ontario, Canada

PIOTR KOZLOWSKI, UBC, Vancouver, BC, Canada

EDGAR KRAFT, Siemens AG, Forchheim, Germany

BÖRJE NORLIN, Department of Information Technology and Media, Mid Sweden University Holmgatan, Sundsvall, Sweden

IVAN PERIC, Institut für Technische Informatik (ZITI) der Universität Heidelberg, Lehrstuhl für Schaltungstechnik und Simulation, Mannheim, Germany

JAN THIM, Department of Information Technology and Media, Mid Sweden University Holmgatan, Sundsvall, Sweden

KAI E. THOMENIUS, Imaging Technologies, GE Global Research, Niskayuna, NY

ROBERT WODNICKI, Imaging Technologies, GE Global Research, Niskayuna, NY

NICOLA DE ZANCHE, Cross Cancer Institute, Edmonton, Alberta, Canada

# PART I
# X-Ray Imaging and Computed Tomography

# 1 X-Ray and Computed Tomography Imaging Principles

KRZYSZTOF INIEWSKI

## 1.1. INTRODUCTION TO X-RAY IMAGING

X-ray imaging is a well-known imaging modality that has been used for over 100 years since Röntgen discovered X-rays based on his observations of fluorescence. His initial results were published in 1885, and reports of diagnoses of identified fractures shortly followed. A year later, equipment manufacturers started selling X-ray equipment. Today, X-ray and its three-dimensional (3D) extension, computed tomography (CT), are used commonly in medical diagnosis.

X-rays are high-energy photons. Their generation creates incoherent beams that experience insignificant scatter when passing through various media. As a result, X-ray imaging is based on through transmission and analysis of the resulting X-ray absorption data. Typically, X-rays are detected through a combination of a phosphor screen and a light-sensitive film, as shown in Fig. 1.1. The current system, which has been used for mammography and radiography for many years, provides a good-quality analog image that is not compatible with digital storage and transmission requirements of the modern digital era. A slight variation of this common technique is used in fluoroscopy where image intensifier is used as transition stage to supply signals to CMOS cameras producing an analog image directly on a TV screen. Multiple conversions steps in this case from X-rays to electrons to light to camera display lead to poor image quality.

An alternative to the conventional detection technique, also shown in Fig. 1.1, uses a digital detector that converts X-ray photons directly into an electrical signal of digital nature. Chapter 2 in this book discusses an example of this direct detection technology using a large-area active matrix flat panel based on the amorphous silicon (a-Si). Having a digital image leads to lower storage cost and ease of electronic transmission in a future e-Healthcare era.

*Medical Imaging: Principles, Detectors, and Electronics,* edited by Krzysztof Iniewski
Copyright © 2009 John Wiley & Sons, Inc.

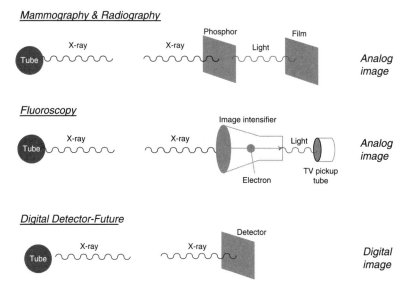

**Fig. 1.1.** Typical X-ray detection methods used today that provide an analog image (**top** and **middle**) and that in the future will use a digital detector (**bottom**). (From http://www.ecse. rpi.edu/censsis/.)

There are some important differences in characterization of film-based imaging and digital imaging. In this chapter and the rest of the book we will focus on digital imaging because it is a more modern technique which with time is expected to completely replace film-based imaging the same way that digital cameras have displaced analog films in consumer cameras. In digital imaging we use terms such as brightness, dynamic range, linearity, or signal-to-noise ratio instead of density, latitude, film speed, or image sharpness, the terms associated with film-based technology.

Digital X-ray detectors can operate in two regimes: photon counting and integration. In the photon counting mode, each individual photon is detected; and if its energy is higher than the set threshold, the photon is counted with its corresponding energy registered if desired. In the integrating mode the charge generated by the incoming photon is integrated in a selected time interval. Due to this principle of operation, a count rate in the photon counting mode is limited, typically to $10^6$ counts per second (c/s), while it is virtually unlimited for the integrating mode. The photon counting mode can detect smaller signals, down to an individual photon, and offers a higher dynamic range (typically $10^6$) compared to the integrating mode ($10^4$). Advantages of the photon counting mode include higher detector quantum efficiency (DQE), lower electronic noise, no need for signal digitization, and possibility of energy discrimination. The integrating mode, in turn, can operate with high count rates, and it is simple and inexpensive to implement. Chapter 3 in this book discusses differences in both photon counting and integration modes of operation in more detail.

While operation of modern X-ray based scanners can be quite complex, a basic principle behind X-ray imaging is quite simple. The technique relies on analyzing attenuation data of the object (patient) that undergoes X-ray exposure. Because different materials (internal organs) experience different levels of X-ray intensity attenuation, an image corresponding to these properties can be readily created. The attenuation characteristics are governed by the so-called Beer–Lambert law, which is expressed as follows:

$$I(z) = I_0 * \exp(-\mu z) \qquad (1.1)$$

where $I(z)$ is the X-ray intensity at the detector, $I_0$ is the X-ray intensity at the source, $z$ is the distance between the source and the detector planes, and $\mu$ is the attenuation coefficient that has a different value for different materials. By measuring $I(z)$ for a set of detectors, one can establish the corresponding value of the attenuation coefficients that give a representation of the image. X-ray imaging is particularly good for providing a contrast between soft and hard tissues, because the attenuation coefficient has a quite different value in both media; hence one of the first applications was to identify fractured bones.

To operate as a diagnostic technique, X-ray imaging needs a radiation source, a means of interactions between the X-ray beam and the object to be imaged, ways of registration of the radiation carrying information about the object, and finally the ability to convert that information into an electrical signal. Although widely used and inexpensive, standard X-ray technique have quite severe limitations. First, 3D structures are collapsed into 2D images, leading to highly reduced image contrast. Second, it is difficult to image soft tissues due to small differences in attenuation coefficients. Finally, standard film-based technology does not provide quantitative data and requires specialized training for accurate image assessment.

Fortunately, a 3D extension of 2D X-ray technology, called computed tomography (CT), was invented in 1972 and is in widespread use today. A basic principle behind CT is to take a large number of X-ray images at multiple angles and, based on that information, calculate the 3D image of the imaged object. CT hardware used for this application is typically called a CT scanner and is similar to an ordinary X-ray machine, albeit with much more computational power. With today's multiple-row detector helical CT scanners, 3D images can be obtained with spatial resolution approaching that of conventional radiographic images in all three dimensions.

This chapter is organized as follows. The radiation source, a well-known X-ray tube, is discussed briefly in Section 1.2. Details of interaction between photons and the object, which include absorption, reflection, scattering, and diffraction, are considered in Section 1.3. Detectors used to register the radiation events are discussed in Section 1.4, while conversion of electrical signals is mentioned in Section 1.5. Principles of computed tomography (CT) are introduced in Section 1.6, while CT scanner design is described in Section 1.7. Extension of X-ray imaging that takes into account photon energy, referred to as "color" X-ray imaging, is discussed in Section 1.8 followed by summary of future trends in Section 1.9. For more details on X-ray imaging modalities, the reader is referred to numerous books on this subject [1–7].

## 1.2. X-RAY GENERATION

A typical X-ray tube is shown in Fig. 1.2. Generation of X-rays depends on thermionic emission and acceleration of electrons from a heater filament. During that process, electrons emitted from cathode are accelerated by anode voltage. Kinetic energy loss at an anode is converted to X-rays. The relative position of an electron with respect to the nucleus determines the frequency and energy of the emitted X-ray.

X-rays produced in an X-ray tube contain two types of radiation: *Bremsstrahlung* and characteristic radiation. The word *Bremsstrahlung* is retained from the German language to describe the radiation that is emitted when electrons are decelerated. It is characterized by a continuous distribution of X-ray intensity and shifts toward higher frequencies when the energy of the bombarding electrons is increased. Characteristic X-rays, on the other hand, produce peaks of intensity at particular photon energies as shown in Fig. 1.3. In practice, emitted radiation is filtered, intentionally or not, producing high-pass filter response as low-energy radiation is completely attenuated. As a result, the final X-ray spectrum has band-pass type characteristics with several local peaks superimposed on it (Fig. 1.4).

The filtering effect shown in Fig. 1.4 is intentional, used to cut off X-ray energies below 20 keV in the shown example. A similar effect can be achieved unintentionally if the gap between the source and the detector is large. Figure 1.5 shows transmission characteristics through air. While 40-keV radiation is not affected by the air gap, 10-keV rays are severely attenuated, and the degree of their attenuation is dependent on the distance.

X-ray generation is a fairly inefficient process because most of the electrical power ends up as heat at the anode. Therefore, an X-ray tube is also a heater, and heat

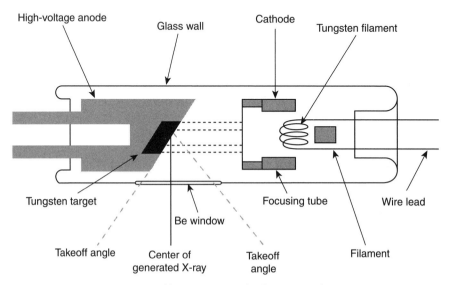

**Fig. 1.2.** X-ray tube. From http://www.siint.com/en/technology/xrf_descriptions1_e.html.

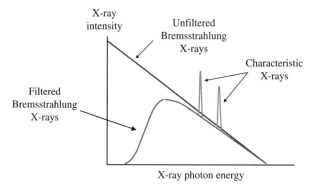

**Fig. 1.3.** Schematic representation of X-ray intensity frequency characteristics.

extraction problems are primary problems in the equipment design and manufacturing. In addition, only a few percent of the generated X-rays end up being absorbed at the detector because the X-ray beam is not collimated and photons are radiated in all possible directions. X-ray photon energy is related to acceleration voltage; so if the acceleration voltage is 20 kV, it will produce 20-keV photons. Clearly, X-ray-based equipment is clearly not suitable for home use! The total number of photons generated is proportional to the cathode current, which typically is several milliamperes. A typical X-ray system uses step-up transformers to produce high-voltage (HV) as schematically shown in Fig. 1.6.

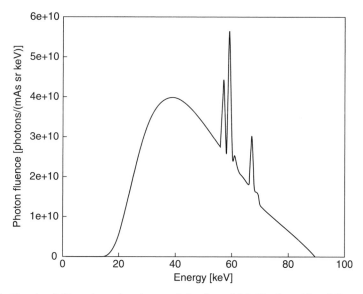

**Fig. 1.4.** Simulated X-ray intensity characteristics for a 90-keV tube with a 1.5-mm Be and 2.7-mm Al filter. (From Roessl and Proksa [8], with permission.)

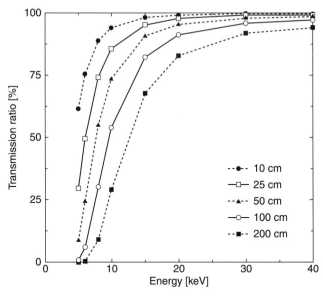

**Fig. 1.5.** X-rays transmission characteristics for the air path as a function of X-ray energy. (From Miyajima and Imagawa [9], with permission.)

X-ray tubes used in computed tomography (CT) are subjected to higher thermal loads in than in any other diagnostic X-ray application. In early CT scanners, stationary anode X-ray tubes were used, since the long scan times meant that the instantaneous power level was low. Long scan times also allowed significant heat dissipation. Shorter scan times in later versions of CT scanners required high-power

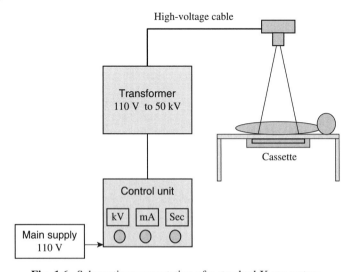

**Fig. 1.6.** Schematic representation of a standard X-ray system.

X-ray tubes and use of liquid-cooled rotating anodes for efficient thermal dissipation. The recent introduction of helical CT with continuous scanner rotation placed even more demands on X-ray tubes; this is clearly a challenging engineering problem because the dissipated power is in the kilowatt range.

X-rays represent ionizing radiation that at significant dose will cause tissue damage. The traditional unit of absorbed dose is the rad. 1 rad is defined as the amount of X-ray radiation that imparts 100 ergs of energy per gram of tissue or, as re-stated in SI units, causes 0.01 joule of energy to be absorbed per kilogram of matter. As a frame of reference, a typical chest X-ray exposure is about 50 mrads, while exposure of 50 rads causes radiation sickness. In the SI system, rad is now superseded by gray, with the following simple relationship between the two: 1 gray equals 0.01 rad.

## 1.3. X-RAY INTERACTION WITH MATTER

X-rays interact with matter in several ways that can be divided into absorption and scattering effects. Primary effects at energies of interest in medical applications are photoelectric effect, Compton scatter, and coherent scatter. In the photoelectric effect, the energy of an X-ray photon is absorbed by an orbital electron, which in turn is ejected from an atom. During this process, X-rays are converted into electric charges, a process very useful for radiation detection. Scattering effects can be of Compton nature, where some energy loss is involved, and coherent, without any energy loss. In Compton scatter, some of the X-ray energy is transferred to an electron, and the X-ray photon travels on with an altered direction and less energy. The Compton process might sometime be utilized in medical imaging in so-called Compton cameras, but frequently it is an undesired effect. As opposed to the Compton effect in coherent scatter, all X-ray energy interacts with the atom, but is later re-radiated with same energy in an arbitrary direction. As a result, the photon changes direction but still carriers the same energy, a process quite detrimental to medical imaging because the original path of photon from a source to a detector is altered.

The relative probability of above processes is dependent on the photon energy and characteristics of the matter with which it interacts. In order to focus our discussion here, we will discuss some details of photon interaction with a semiconductor material called CZT. CZT stands for cadium zinc telluride and is currently considered as the most promising detector material for X-ray and γ-ray direct detection in medical imaging for reasons that are explained later in this chapter. The relative probability of absorption/effect is plotted in Fig. 1.7. The photoelectric effect is a dominant one in the considered energy range of 20–300 keV; although at higher energies, Compton scattering becomes equally probable. At the energy of 122 keV, which represents a characteristic cobalt radiation line, the photoelectric effect has 82% probability of happening, Rayleigh scattering 7%, and Compton scattering 11%.

Note that the photoelectric line shows an interesting behavior in the 20- to 40-keV range due atomic structure. The corresponding attenuation length, shown in Fig. 1.8, varies from 0.05 to 0.17 mm. This indicates that even a thin CZT detector will effectively absorb all radiation in that energy range.

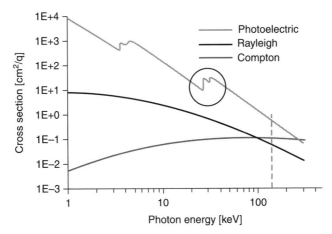

**Fig. 1.7.** Effective photon cross sections for the photoelectric effect, Rayleigh scattering, and Compton scattering in CZT. The dashed vertical line indicates the 122-keV cobalt line.

The photoelectric effect is one of the energy loss processes where the photon effectively disappears after the interaction. A complete absorption of the photon energy is the desired effect for X-ray detection. The name photoelectron comes from a process of ejecting an electron from one of the atomic shells of the media. After the ejection of the photoelectron, the atom is ionized. The vacancy in the bound shell is refilled with

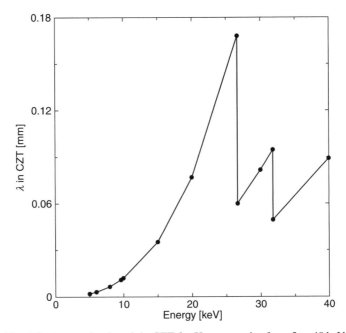

**Fig. 1.8.** Attenuation length in CZT for X-ray energies from 5 to 40 keV.

an electron from the surrounding medium or from an upper atom shell. This may lead either to the emission of one or more characteristic fluorescence X-rays or to the ejection of an electron from one of the outer shells, called an Auger electron. Depending whether tellurium, cadmium, or zinc atoms are involved, the resulting soft X-rays energies might be in an 8- to 31-keV range (Te 27–31 keV; Cd 23–26 keV; Zn 8–10 keV).

Having addressed absorption effects, let us change our focus to scattering events: Rayleigh and Compton. Rayleigh scattering describes photon scattering by atoms as a whole, frequently also called coherent scattering because the electrons of the atom contribute to the interaction in a coherent manner and there is no energy transferred to the CZT material. The elastic scattering process changes only the direction of the incoming photon. For these reasons, Rayleigh scattering is detrimental to medical imaging as the original photon trajectory is changed.

Unlike Rayleigh scattering, the Compton effect deals with photons that are scattered by free electrons and as a result lose some of their primary energy. This mechanism can contribute significantly to the measured energy spectrum. The change in the photon energy increases with increasing scattering angle. The energy and momentum lost by the photon is transferred to one electron, called the recoil electron, which is emitted under a certain angle with respect to the direction of the incoming photon and can have a maximum kinetic energy defining the so-called Compton edge. The Compton edge can frequently be visible in the measured spectrum as an abrupt end to the energy tail caused by Compton scattering (Fig. 1.9).

The Compton scattering equation describes the change in photon energy and its corresponding wavelength as

$$\lambda' - \lambda = \frac{h}{m_e c}(1 - \cos\theta)$$

where $\lambda$ is the wavelength of the photon before scattering, $\lambda'$ is the wavelength of the photon after scattering, $m_e$ is the mass of the electron, $\theta$ is the angle by which the

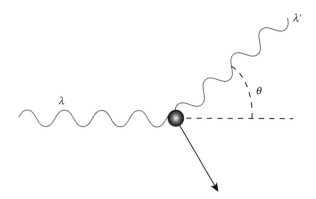

**Fig. 1.9.** Illustration of Compton scattering event.

photon's trajectory changes, $h$ is Planck's constant, and $c$ is the speed of light. Substituting textbook values for $m_e$, $c$, and $h$, we can obtain a characteristic Compton wavelength defined as $h/(m_e c)$ to be equal to 2.4 picometers (pm).

The Compton equation has two interesting properties. First, the characteristic Compton wavelength value is small compared to the typical X-ray wavelengths used in medical imaging (10-keV wavelength is about 120 pm). As a result, the maximum wavelength change is only a fraction of the original wavelength value. Second, the largest change in the photon energy can only be expected for angles $\theta$ close to 180 degrees. The maximum wavelength change is twice the Compton wavelength change.

## 1.4. X-RAY DETECTION

To detect X-rays, one needs a material that is capable of absorption of the incoming radiation and transformation of the corresponding photon energy into an electrical signal. There are two ways of accomplishing this result: indirectly and directly. The indirect process uses materials called scintillators that convert X-rays photons to visible light. The generated light is in turn detected using conventional photodetectors. The direct conversion process uses semiconductor detectors that convert X-ray photons directly into electrical signals.

In typical X-ray indirect detection applications, either thallium-doped sodium ionide NaI(Tl) or thallium-doped cesium iodide CsI(Tl) is used as a scintillator because thallium-doped crystals are one of the best-known materials for scintillation processes. One of the fundamental problems with scintillators is a low efficiency of the indirect process. Once converted to visible light, the signal is no longer proceeding along the direction of the incoming photon; instead light is emitted in all directions. As a result, very few generated photons reach the subsequent photosensitive detector.

Other manufacturers use direct conversion techniques with amorphous selenium, amorphous silicon, or cadmium zinc telluride (CZT) semiconductor detectors. Chapter 2 of this book deals with circuit, process, and device development for large-area, digital imaging applications using amorphous silicon (a-Si), and it discusses in detail the implementation of this new technology. For low-energy X-ray detection, standard silicon is frequently used, mainly because of good homogeneity and very well stabilized technology. A 300-$\mu$m-thick silicon detector converts nearly all 8-keV X-rays, but only 2% of 60-keV X-rays. High-purity germanium (HPGe) and compound semiconductors, like gallium arsenide (GaAs), cadmium telluride (CdTe), cadmium zinc telluride (CdZnTe), and mercuric iodide, show much better radiation stopping power compared to silicon due to their higher atomic numbers.

In the indirect detection mode, an additional photomultiplication is required to compensate for very low signal magnitude. The process can utilize devices called photomultiplier tubes, which provide very high gain ($10^6$), low noise, fast response, and capability of detecting single photoelectrons. Unfortunately, these devices are sensitive to magnetic field and bulky. Their large size makes some medical imaging equipment look like vacuum tube technology used in electronics before transistors was invented.

More miniaturized equipment can be built using avalanche photodiodes (APD). Silicon APDs have high quantum efficiency (QE over 60%) and broad spectral response, but their gain is not that high ($10^3$ at best) and their additional noise degradates system performance. Another promising approach is based on silicon photomultipliers (SiPM). These devices can be manufactured in a standard CMOS process and offer very high gain as PMTs. They are, however, sensitive to voltage and temperature fluctuations and present several operation problems that include cross-talk and afterpulsing degradation issues.

PMTs are still the main technology in today's clinical nuclear medicine. Ongoing improvements in QE and timing performance while reducing already low cost for large sizes promise this incumbent technology to be dominant for a number of years at least in low-end, low-performance equipment. APDs have matured and a number of APD-based scanners have been built. SiPM is a new technology with a promising future, and it remains to be seen whether it is applied in commercial equipment. However, ultimate gain in performance can only be obtained with the direct detection. CZT-detector-based scanners offer the highest performance currently available in X-ray and $\gamma$-ray medical imaging, albeit at high cost of manufacture at the moment.

CZT detectors have high potential as X-ray spectrometers due to relatively high atomic number (Cd 48, Zn 30, and Te 52) and high density (approximately 5.9 g cm$^{-3}$). The wide CZT bandgap (about 1.6 eV) enables the detectors to be operated at room temperature. The main drawback of CZT detectors is their poor hole charge transport properties that create polarization problems in high-count rate systems. Another disadvantage is high cost of manufacturing defect-free CZT crystals. With recent improvements in CZT volume production, these problems are expected to be overcome soon [10–12].

## 1.5. ELECTRONICS FOR X-RAY DETECTION

Advanced Very Large Scale of Integration (VLSI) electronics and techniques of designing and prototyping integrated circuits (ICs) allow integration of a large number of channels into a single silicon chip, so each imaging pixel of the detector array can be read out by an individual electronic channel. Such hardware offers post-processing capability and good spatial resolution of an image, as well as large dynamic range. If, in addition, digital imaging systems are able to extract some energy information about X-ray radiation, this information can be used to extract additional media information in radiology, material screening, or diffractometry as discussed later in the color X-ray section of this chapter (Section 1.8).

The fast front-end electronics for a large array of X-ray sensors should amplify and filter small signals from the each sensor element, perform analog-to-digital conversion, and then store the data in digital form. Because of the complexity of the multi-channel mixed-mode integrated circuit, the important problems such as power limitation, low level of noise, good matching performance, and crosstalk effects must be solved simultaneously. Chapters 3, 4, and 6 in this book, as well as references 13–16, discuss architectures and related circuit solutions for the readout electronics for the digital X-ray imaging systems.

A primary example of VLSI electronics used for X-ray detection is Medipix consortium and related chip development (see http://medipix.web.cern.ch/MEDIPIX/ for details). A third-generation Medipix3 chip is a CMOS pixel detector readout chip designed to be connected to a segmented semiconductor sensor such as CZT. Like its predecessor, Medipix2, it acts as a camera taking images based on the number of X-ray photons which hit the pixels when the electronic shutter is open. Medipix3 aims to allow for color imaging and dead time free operation. It also seeks to mitigate the effect of charge sharing by summing charge between neighboring pixels and allocating the sum or hit to the individual pixel with the highest collected charge.

## 1.6. CT IMAGING PRINCIPLE

Standard X-ray imaging poses fundamental limitations, making it only useful for clinical diagnosis in cases of initial screening or simple bone fractures. The primary limitation comes from the fact that three-dimensional (3D) objects are collapsed into 2D images. The technique offers only low soft-tissue contrast and is not very quantitative. As mentioned earlier X-ray CT tomography relies on taking a large number of X-rays at multiple angles. The Greek roots of the term "tomography" are "tomo"—act of cutting and "graphos"—image. Based on the X-ray tomographical information a 3D image is calculated although the computational load can be very heavy. In the era of powerful Intel processors this is no longer a limitation, a conventional personal computer (PC) is sufficient in most cases.

As shown in Fig. 1.10, a CT scanner makes many measurements of attenuation through the plane of a finite-thickness cross section of the body. The system uses these data to reconstruct a digital image of the cross section, with each pixel in the image representing a measurement of the mean attenuation of a boxlike element (called a voxel) that extends through the thickness of the section. An attenuation measurement quantifies the fraction of radiation removed in passing through a given amount of a specific material of thickness. Each attenuation measurement is called a *ray sum* because attenuation along a specific straight-line path through the patient from the tube focal spot to a detector is the sum of the individual attenuations of all materials along the path. Rays are collected in sets called *projections*, which are made across the patient in a particular direction in the section plane. There may be about a thousand rays in a single projection. To reconstruct the image from the ray measurements, each voxel must be viewed from multiple different directions. A complete data set requires many projections at rotational intervals of $1°$ or less around the cross section. Back-projection effectively reverses the attenuation process by adding the attenuation value of each ray in each projection back through the reconstruction matrix.

The attenuation coefficient of water is obtained during calibration of the CT machine. Voxels containing materials that attenuate more than water (for example, muscle tissue or bones) have positive CT numbers, whereas materials with less attenuation than water (for example, lungs) have negative CT numbers.

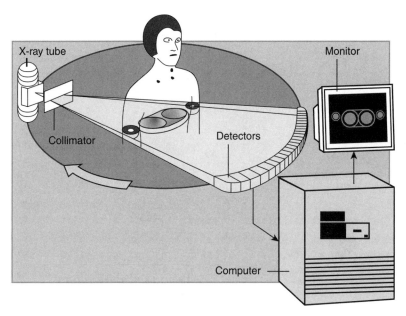

**Fig. 1.10.** Schematic illustration of CT imaging operation.

The CT concept was invented in 1972 by Godfrey Hounsfield, who introduced computed axial transverse scanning. As a result, the first CT machines were known as CAT scanners. The scanner is quite similar to other pieces of X-ray imaging equipment with additional requirements for movements and signal processing. In the next section we will review architecture of CT scanners as they evolved from 1972 until now.

## 1.7.  CT  SCANNERS

Computed tomography (CT) is a method of acquiring and reconstructing the image of a thin cross section on the basis of measurements of X-ray attenuation. In comparison with conventional radiographs, CT images are free of superimposing tissues and are capable of much higher contrast due to elimination of scatter. Most of the developments in CT since its introduction can be considered as attempts to provide faster acquisition times, better spatial resolution, and shorter computer reconstruction times. From the early designs, the technology progressed with faster scanning times and higher scanning plane resolution, but true three-dimensional (3D) imaging became practical only recently with helical/spiral scanning capabilities.

Despite its internal complexity and hefty price, a CT scanner is basically a set of rotating X-ray tubes and radiation detectors. Figure 1.11 shows a schematic representation of a typical CT system. In order to achieve a 3D image, lots of computing power is required so every scanner has a built-in personal computer (PC). In addition to

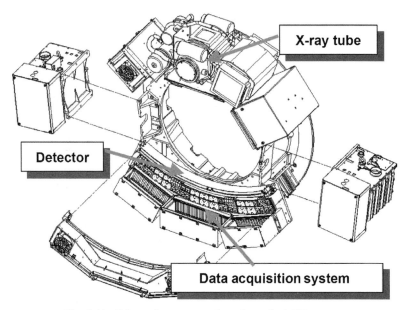

**Fig. 1.11.** Schematic representation of a typical CT scanner.

medical imaging CT scanners are being used for nondestructive evaluation of materials and luggage inspection.

First generation of scanners used parallel beam design with one or two detectors used in parallel. The early versions required about 5 minutes for a single scan and thus were restricted to regions where patient motion could be controlled (head, for example). Since procedures consisted of a series of scans, procedure time was reduced somewhat by using two detectors so that two parallel sections were acquired in one scan. Although the contrast resolution of internal structures was unprecedented at the time, images had poor spatial resolution of the order of a few millimeters.

Second-generation scanners introduced a small fan beam, rotation, and a larger number of detectors. Later designers realized that if a pure rotational scanning motion could be used, then it would be possible to use higher-power, rotating anode X-ray tubes and thus improve scan speeds in thicker body parts. One of the first designs to do so was the so-called third-generation or rotate–rotate geometry. In these scanners, the X-ray tube is collimated to a wide, fan-shaped X-ray beam and directed toward an arc-shaped row of detectors. The number of detectors varied from few hundred in early versions to nearly a thousand in modern scanners.

Fourth-generation scanners made additional improvements and introduced even more detectors (up to 5000 of them) in one machine placing them in a concentric ring. It was later realized that if multiple sections could be acquired in a single breath hold, a considerable improvement in the ability to image structures in regions susceptible to physiologic motion could result. However, this required some technological advances, which led to the development of helical/spiral CT scanners. In conventional CT, 360 degrees of data are acquired for one image. To get another image,

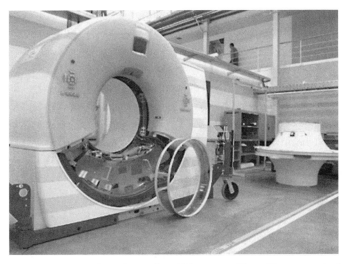

**Fig. 1.12.** Manufacturing of a modern CT scanner. (From Siemens web site.)

the gantry is moved to next location. Helical CT, on other hand, covers a nonplanar geometry. In this technique a table with a patient moves while the image is being taken. This movement is slow: A typical speed is a few millimeters per second. Most algorithms used in helical scanners are the same as with conventional CT, but an added interpolation step in the direction of the table movement is required. The development of helical or spiral CT around 1990 was a truly revolutionary advancement in CT scanning that finally allowed true 3D image acquisition within a single breath hold.

CT scanners for full-body analysis are large industrial machines. Figure 1.12 shows a manufacturing step of a dual-source CT scanner, which in this example is primarily used for heart examinations. The device contains two X-ray tubes and two set of detectors which allow taking clinical images at two different X-ray energies simultaneously. This dual-energy scanning offers potential of differentiating materials beyond the visualization of morphology—for example, direct subtraction of either vessels or bone during scanning. Using multiple energy scanning leads eventually to color X-ray imaging, a technique described in the next section. Before we conclude the CT scanning section, let us note that the hardware shown in Fig. 1.12 has 78-cm gantry bore, 200-cm scan range, and 160-kW generator power. One can easily imagine special physical accommodation conditions required for these machines to be deployed in hospitals.

## 1.8. COLOR X-RAY IMAGING

A traditional X-ray imaging relies on analysis of attenuation data without paying any attention to particular energy of transferred X-ray photons. However, some additional, useful information can be obtained if energy information is retained, even in a crude fashion by binning the entire energy range into few energy bins. This technique is called "color" X-ray imaging and is getting increasing attention in the imaging field [13].

As discussed previously, attenuation of X-rays by matter is dependent on the energy of the photons. The Bremsstrahlung sources of radiation in clinical X-ray computed tomography (CT) systems are characterized by a broad-energy spectrum. Therefore, different spectral components are attenuated with different strengths, a process called beam hardening. In conjunction with integrating X-ray detectors, beam hardening causes characteristic artifacts in the reconstructed images when left uncorrected.

Imaging modalities based on X-ray transmission measurements are usually regarded as rather insensitive with respect to the detection of specific drugs, contrast materials, and other substances present in the object of interest. This is due to the fact that in the transmission case the signal coming from these samples is always super-imposed on the signal resulting from the anatomic background. Conventional X-ray transmission systems based on current integration do not allow us to separate the two signals. The spectral decomposition allows the selective separation of the ana-tomic background and contributions coming from heavy elements. In connection with photon counting detectors with reasonable energy resolution, X-ray transmission modalities—for example, X-ray CT—could allow the selective, high-resolution imaging of targeted contrast materials in addition to the conventional imaging of the anatomy.

Conventional X-ray/CT imaging relies on measuring total absorption of an object regardless of the particular energy of the incoming radiation. Different combinations of objects can produce equal absorption, while a color X-ray can differentiate between these different objects. The goal of a color X-ray is therefore to improve detectability of details, improve signal difference-to-noise ratio (SDNR), discriminate between different materials and different types of tissue, and enhance visibility of contrast media.

Color X-ray imaging is already used in some baggage inspection systems, as an automatic exposure control in mammography, and for bone mineral density measure-ment discrimination of bones and soft tissue using dual-energy contrast-enhanced digital subtraction techniques. VLSI chips that can discriminate against X-ray photon energies are already available in production volumes. The use of a color X-ray in medi-cal X-ray imaging is expected to increase dramatically in the next 5–10 years.

## 1.9. FUTURE OF X-RAY AND CT IMAGING

The most basic task of the diagnostician is to separate abnormal subjects from normal subjects. In many cases there is significant overlap in terms of the appearance of the image. Some abnormal patients have normal-looking films, while some normal patients have abnormal-looking films. Technology optimization can be used to select an abnormality threshold; however, true medical diagnosis must be determined independently, based on biopsy confirmation, long-term patient follow-up, and other medical factors.

Medical X-ray imaging can be viewed from the physics, application, or medical points of view. The physical point of view is originating in technological aspects of X-ray imaging. What wavelength and photon energy are used? What are available sources of radiation? What electronics is used for processing? While this point of view is a dominant one in this book, we have to briefly address the other two.

The application point of view addresses the use of the technology. Is technology used for diagnosis or intervention? Morphological or functional imaging? What spatial resolution can be achieved? The medical point of view, on the other hand, relates heavily toward a patient. Is this screening procedure performed on a presumably healthy person? Or is it used for a diagnosis if patient is suspected of having a disease already? Is the image used for diagnostic purposes or as a helping tool in image-guided therapeutic interventions?

Standard X-ray screen-film technology is in widespread use but it needs to be replaced/complemented with much more accurate CT imaging, which, as discussed, is a method for acquiring and reconstructing a 3D image of an object. CT differs from conventional projection in two significant ways. First, CT forms a cross-sectional image, eliminating the superimposition of structures that occurs in plane film imaging because of compression of 3D body structures onto the two-dimensional recording system. Second, the sensitivity of CT to subtle differences in X-ray attenuation is at least a factor of 10 higher than normally achieved by screen-film recording systems because of the virtual elimination of scatter.

One obvious further development in X-ray-based imaging is digitization. As discussed in this chapter, modern CT equipment is capable of producing digital images of underlying organs by directly transforming photon energy information into electrical signals. Availability of digitally captured images throughout the whole health-care system provides enormous advantages and is instrumental in so-called telemedicine. Needless to say, there are also important security and privacy issues related to electronic patient records. In fact, one might argue that the largest obstacles to widespread implementation of this new digital imaging technology are of nontechnical nature.

While this introductory chapter is primary devoted to transmission-based X-ray modalities, there are alternative medical imaging techniques called nuclear medicine that can be used as complementary techniques that rely on a somewhat different principle. Nuclear medicine contains two major modalities: positron emission tomography (PET) and single particle emission computed tomography (SPECT). In both techniques, no external radiation is used to scan the object as in X-ray imaging. Instead, the patient is subjected to injection or oral administration of tracers with radioactive isotopes. During the procedure the clinician waits for distribution and accumulation of these isotopes in the region of interest—for example, in tumor. Subsequently, radiation is detected to determine the source distribution of the isotopes. Mapping of emitted radiation is performed either by collimator (SPECT) or by coincidence using two detectors (PET). 2D or 3D images can be created using iterative reconstruction and filtered back projection. As opposed to CT, nuclear medicine does not rely on attenuation data; in fact it is desirable to correct for attenuation effects if possible. Nuclear medicine techniques are described extensively in Chapter 5.

X-ray and γ-ray detection is performed in a very similar manner in both nuclear medicine and CT. There are, however, some important differences in how the electronics is being to used to process the collected information. SPECT imaging requires that the detector channels count individual photons (γ-ray events), which produce small quantities of charge in proportion to their energy. These charge pulses range from 0 to 20 fC and occur at rates in the range of 10–1000 photons/s/pixel. The charge (energy) from each γ-ray must be measured with a resolution better than of 0.1 fC for low image noise and good contrast. In contrast to SPECT, CT detector channels measure X-ray events at a relatively high flux (tens of millions of photons per second per pixel) producing a continuous current due the overlapping of individual charge pulses. This photon flux can vary over a 140-dB dynamic range, producing detector currents ranging from 1 pA to 10 μA, and must be measured with better than 74-dB SNR (signal-to-noise ratio) at a kilohertz sampling rate.

Standard X-ray and CT provide an anatomic image of the underlying organ. Other modalities discussed later in this book (SPECT, PET, functional MRI) are used to detect an organ function. It would be extremely advantageous if both the anatomic image and the organ function were combined in one. The area of multimodality image scanning is emerging rapidly as a main direction toward improvements in medical imaging. Combining anatomical imaging (CT, MRI, ultrasound, diffuse optical tomography) and functional imaging (SPECT, PET) promises to lead to dramatic improvements in medical diagnosis. The early multimode X-ray-based implementations (PET with CT, SPECT with CT) simply join two separated apparatus; this offers limited advantages. The true integration would involve using the same data acquisition electronics and detector technologies to obtain a perfect fusion of anatomical and metabolical images.

To accomplish the goal of multimode operation with high spatial resolution and sensitivity, several breakthrough technologies have to be used. First, novel semiconductor detectors such as CZT offer unprecedented energy resolution that is currently being exploited in SPECT gamma cameras. Avalanche photodiode and silicon multiplier coupled individually to crystal arrays of LSO scintillators would be another potential detector technology. Second, innovative circuit techniques that utilize subthreshold operation can be implemented in front-end ASICs to provide very low power systems (reduction of power dissipation by an order of magnitude compared to the state-of-the-art). Third, creative use of FPGA resources is required to provide efficient signal processing of hundreds of channels simultaneously. Finally, proven signal processing algorithms in FPGA can be implemented in medical imaging signal processor that would serve as the "Pentium" equivalent used for personal computing.

CT imaging systems create high-resolution three-dimensional X-ray images that provide anatomical definition with a millimeter spatial resolution, but lack direct physiological information. In the medical literature, it has been recognized that combining CT and SPECT in a single machine would improve the correlation of these two data sets and improve diagnostic process. Let us hope that the evolution from film-based 2D

X-ray screening toward precise imaging using 3D techniques such as CT and/or SPECT will happen in the not-too-distant future in order to improve quality of life for all of us.

## REFERENCES

1. H. H. Barrett and W. Swindel, *Radiological Imaging*, Vols. 1 and 2, Academic Press, New York, 1981.

2. S. R. Deans, *The Radon Transform and Some of Its Applications*, John Wiley & Sons, New York, 1983.

3. G. T. Herman, *Image Reconstruction from Projections*, Academic Press, New York, 1980.

4. A. C. Kak and M. Slaney, *Principles of Computerized Tomographic Imaging*, IEEE Press, New York, 1988.

5. W. A. Kalender, *Computed Tomography*, Publicis MCD Verlag, Munich, 2000.

6. E. Krestel, *Imaging Systems for Medical Diagnostics*, Publicis MCD Verlag, Munich, 1990.

7. A. Webb, *Introduction to Biomedical Imaging*, IEEE Press, New York, 2003.

8. E. Roessl and R. Proksa, K-edge imaging in x-ray computed tomography using multi-bin photon counting detectors, *Phys. Med. Biol.* **52**, 4679–4696, 2007.

9. S. Miyajima and K. Imagawa, CdZnTe detector in mammographic x-ray spectroscopy, *Phys. Med. Biol.* **47**, 3959–3972, 2002.

10. H. Chen et al., Characterization of large cadmium zinc telluride crystals grown by traveling heater method, *J. Appl. Phys.* **103**, 014903, 2008.

11. K. Iniewski et al., Charge sharing modeling in pixellated CZT detectors, *IEEE NSS-MIC*, Honolulu, 2007.

12. K. Iniewski et al., CZT pixel scaling for improved spatial resolution in medical imaging, *IEEE NSS-MIC*, Dresden, 2008.

13. J. Thim, CMOS for Color X-Rays, *2008 CMOS Emerging Technologies Workshop, Vancouver*, Canada, August 5–7, 2008, www.cmoset.com.

14. E. Charbon, Single-Photon Imaging: The Next Big Challenges, *2008 CMOS Emerging Technologies Workshop*, Vancouver, Canada, August 5–7, 2008, www.cmoset.com.

15. P. Grybos, Detector interface circuits for X-ray imaging, in *Circuits at the NanoScale: Communications, Imaging and Sensing.* K. Iniewski, ed., CRC Press, Boca Raton, FL, 2008, pp. 575–600.

16. D. Pribat and C. Cojocaru, Active matrix back planes for the control of large area X-ray imagers, *IWORID '05.* 4–7 July 2005, ESRF Grenoble.

# 2 Active Matrix Flat Panel Imagers (AMFPI) for Diagnostic Medical Imaging Applications

KARIM KARIM

## 2.1. INTRODUCTION

Pixelated arrays of electronics in amorphous silicon (a-Si) technology, routinely used for liquid crystal displays, are now being extended to several new and significant application areas in large-area digital imaging [1, 2]. Interest in a-Si technology stems from its desirable material and technological attributes. Some of the benefits of large-area a-Si technology are: the low-temperature deposition ($<300°C$); high uniformity over a large area; few constraints on substrate size, material, or topology; standard integrated circuit (IC) lithography processes; and low capital equipment cost associated with the a-Si material. Notable application areas include: contact imaging for document scanning, digital copiers, and fax machines; color sensors/imaging; position/ motion detection; radiation detection/imaging of X-rays in biomedical applications, gamma-ray space telescopes, airport security systems; and nondestructive testing of mechanical integrity of materials or structures [1, 3, 4]. This chapter deals with circuit, process, and device development for large-area, digital imaging applications using amorphous silicon (a-Si) technology for diagnostic medical imaging.

### 2.1.1. Digital Imaging

The most commonly used method to detect X-rays is to convert the radiation to visible light through a phosphor screen, which is subsequently detected by a light-sensitive film. Modern screen-film systems typically employ one or a pair of phosphor screens in combination with light-sensitive film packaged in a light-tight cassette.

*Medical Imaging: Principles, Detectors, and Electronics*, edited by Krzysztof Iniewski
Copyright © 2009 John Wiley & Sons, Inc.

This system, which has been accepted by radiologists for nearly a century, provides for good-quality X-ray images. The main drawbacks of the screen-film system are storage costs and digital incompatibility. Another method that is widely used particularly in real-time fluoroscopic imaging [5] is the image-intensifier tube, which is integrated with a CCD or CMOS camera. Here, X-rays first interact with the input phosphor to produce optical photons, which in turn eject electrons from the photocathode. The electrons are accelerated across the vacuum tube and are focused on an output phosphor to produce optical photons, which are then viewed by a camera. This method employs multiple transduction steps to convert X-rays to visible light, which compromises image quality. Moreover, the complex lens and mechanical systems, along with the assembled arrays of CCD or CMOS, lead to cost and size issues.

As an alternative to these conventional detection methods, large-area active matrix flat panel digital X-ray imagers (AMFPIs) based on the a-Si technology have been developed over the past decade. The motivation for developing AMFPIs targeted toward medical imaging applications includes improved image quality, large-area X-ray imaging capability, a compact flat panel structure, low storage cost, better waste management since photographic film is obviated, and computerized handling/ storage of sensory information.

There are two architectures currently employed in large area AMFPIs: the linear architecture, used in photocopiers, fax machines, and scanners, and the two-dimensional array architecture, employed in digital (including video) lens less cameras as well as X-ray imaging systems. In both linear and array architectures, the basic imaging unit is the pixel, which consists of an image sensor and a switch. The pixel is accessed by a matrix of gate and data lines, and it is operated in storage (or integration) mode. Here, during the off-period of the switch, the sensor charge is integrated, and when the switch is turned on, the charge in the sensor is transferred to the data line where it is detected by a charge sensitive amplifier. There are various metallic lines used to control the readout of the imaging information from the array of pixels. The imaging system is completed with (a) peripheral circuitry that amplifies, digitizes, and synchronizes the readout of the image and (b) a computer that manipulates and distributes the final image to the appropriate soft- or hard-copy device. A schematic diagram of a complete flat-panel X-ray imaging system is shown in Fig. 2.1. Recently, a-Si AMFPIs using the sensor-and-switch pixel architecture with areas of 30 cm × 40 cm were demonstrated for diagnostic medical chest radiography and digital mammography [6].

### 2.1.2. Detection Schemes

The detection schemes for X-rays are divided into two categories: direct detection, where the X-rays are directly absorbed and converted to electrical charge in the detector; and indirect detection, where the X-rays are first converted into visible light by a phosphor layer, which in turn is converted to electrical charge in the photodetector. The electrical charge is read out by means of an a-Si active matrix thin-film transistor (TFT) array.

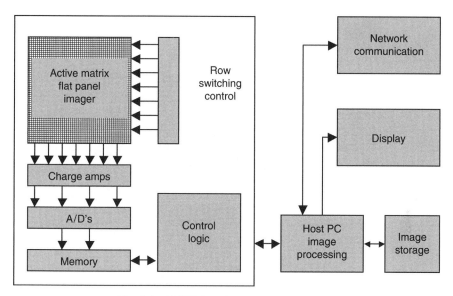

**Fig. 2.1.** AMFPI imaging system diagram [5].

In the direct detection scheme, a thick photoconductor, such as amorphous selenium (a-Se), is deposited over the a-Si active matrix [5], as depicted in Fig. 2.2a. Incident X-rays are absorbed in the a-Se film and directly converted to electron–hole (e–h) pairs, which are then collected by an electric field that is applied across the photoconductor layer. A high electric field ($\sim$10 V/$\mu$m) is required to separate the e–h pairs. Normally, a 500-$\mu$m-thick a-Se layer is required to absorb all of the X-rays, and a bias voltage as high as 5–10 kV is applied across the a-Se film for efficient charge collection. The indirect detection scheme, as shown in Fig. 2.2b, relies on a phosphor layer that is either assembled or integrated with the photodetector. The phosphor film absorbs the incident X-ray photons and produces high-energy electrons, which in turn generate many e–h pairs in the phosphor. Visible light is emitted from the phosphor when the electrons and holes recombine, and it is detected by a-Si image sensor arrays located beneath the phosphor film.

To be able to replace and improve upon current medical imaging techniques, the digital counterparts must meet the same performance criteria as existing X-ray imaging technologies. For example, the digital imager must cover the same field of view as conventional techniques as well as meet the appropriate resolution requirements for each modality. Also, the digital imager must meet the requirements on X-ray spectrum and mean exposure to the detector as determined from current clinical practices. The constraints on detector and pixel size are summarized in Table 2.1.

Attention is drawn to three specific items in Table 2.1 that comprise the major challenges posed by diagnostic medical imaging applications to the current generation of large-area, a-Si flat panel digital imagers. First, fluoroscopy is a real-time modality with very low patient exposure. The minimum X-ray input signal yields merely 1000 signal electrons (for both direct and indirect X-ray detection schemes),

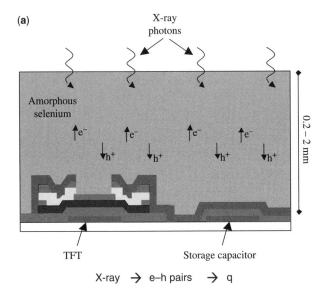

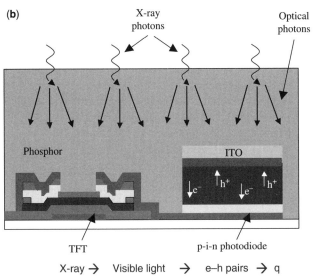

**Fig. 2.2.** (a) Direct X-ray detection method using a-Se and (b) indirect X-ray detection method based on ITO/a-Si Schottky photodiode integrated with phosphor [5].

making the input vulnerable to noise added by the external readout electronics. Next, the pixel size for mammography must be less than or equal to $50 \times 50$ μm while maintaining high fill factor for adequate resolution in breast imaging, although recent developments indicate that 70- to 100-μm pitch pixels are acceptable in practice. Lastly, the pixel count for all modalities is greater than $1000 \times 1000$, making

**TABLE 2.1. Design Requirements for Digital X-Ray Detectors [5]**

|  | Radiography | Mammography | Fluoroscopy |
|---|---|---|---|
| Imager size (cm) | $35 \times 43$ | $18 \times 24$ | $25 \times 25$ |
| Pixel size ($\mu$m) | 100–150 | 50 | 150–200 |
| Readout time (s) | $<5$ | $<5$ | 0.033/frame |
| X-ray energy (kVp) | 80–130 | 25–50 | 80 |
| Exposure (mR) | 0.03–3 | 0.6–240 | 0.0001–0.01 |
| Patient thickness (cm) | 20 | 5 | 20 |
| Object size | 0.5 mm | 50–100 $\mu$m | 2 mm |
|  | (bone detail) | ($\mu$ calcifications) |  |

the AMFPI interface to off-panel external electronics bulky unless some form of on-panel multiplexing is implemented. Of the three modalities presented in Table 2.1, fluoroscopy has the most stringent requirements of low noise and real-time operation while mammography needs small-area pixels for high-resolution imaging.

### 2.1.3. Chapter Organization

First, the traditional a-Si TFT switch-based pixel readout circuit architecture is examined. Then, the process to achieve implement these pixel architectures including high fill factor pixels is introduced. Amorphous silicon TFTs suffer from a bias stress-related characteristic threshold voltage shift ($\Delta V_T$) that is described and modeled next for TFT switch applications. The noise due to the a-Si on-pixel amplifier readout circuit can potentially degrade the noise performance. The a-Si TFT thermal and flicker noise models discussed enable the determination of noise performance for a large variety of a-Si TFT imaging pixel readout circuits. Lastly, recent developments in a-Si TFT-based amplified pixels are presented.

## 2.2. PIXEL TECHNOLOGY

### 2.2.1. Operation

*2.2.1.1. Introduction.* The passive pixel, first introduced by Gene Weckler in 1967, is the workhorse of the digital X-ray imaging industry. A PPS (see Fig. 2.3) consists of a sensing element (e.g., a-Se photoconductor or a-Si photodiode biased in the integration mode) and a switch. The data line is connected to a column charge amplifier to convert the accumulated signal on the charge detection node, $V_x$, into a stable voltage for image processing by external electronics.

For the PPS architecture, the TFT gate is connected to a common row gate signal line and a scanning clock generator addresses all the rows. During readout, a row of TFT switches is activated via a gate line by the clock generator sequentially. The signal charge from each pixel is transferred through common data lines into column charge amplifiers. The amplified signals are then digitized by an A/D converter and sent for image processing.

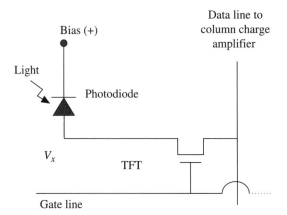

**Fig. 2.3.** Passive pixel sensor architecture.

***2.2.1.2. Operation.*** There are three modes of operation for the PPS:

- *Integration Mode*: Signal charge accumulates on $C_{PIX}$ (sum of sensor and parasitic capacitances at node $V_x$) proportional to the amount of X-ray radiation.
- *Readout/Reset Mode*: Following integration, the accumulated signal charge is transferred from $C_{PIX}$ to the column charge amplifier, simultaneously causing pixel readout and reset to occur. The pixel capacitance, $C_{PIX}$, is reset to some steady state value determined by the charge integrator positive input before the next signal integration begins.
- Charge accumulated on $C_{PIX}$ is read out through the TFT ON resistance and converted to a voltage via the column charge amplifier. During readout, the TFT is biased in the linear region to provide a low ON resistance for quick readout. When the TFT is OFF, it is biased to be nonconducting. However, a small leakage current (on the order of fA for a-Si TFTs) flows from TFT drain to source and decreases the charge on $C_{PIXEL}$. Choosing an optimal TFT OFF voltage can reduce the leakage current.

An a-Se photoconductor-based PPS pixel with associated parasitic capacitances is shown in Fig. 2.4 [7]. Both a large sensor bias and an additional capacitance, $C_{st}$, between the node $V_x$ and ground are required for a-Se based X-ray detection technology. These requirements stem from the large a-Se film thickness (0.2 to 1 mm) necessary for detecting X-rays. The large sensor thickness reduces the sensor capacitance, and a larger detector bias (in kilovolts) is necessary in order to have a sufficiently strong electric field across the sensor for charge collection.

One shortcoming of the thick a-Se film and the consequently small a-Se sensor capacitance and higher detector bias is that a large voltage can build up on the integration node, $V_x$, as the X-ray input photons generate signal electrons. Since $V_x$ is connected to the TFT drain, a large voltage can cause TFT breakdown. For very small sensor capacitances (such as tens of femtofarads), the voltage can become fatal to

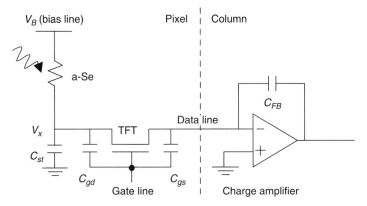

**Fig. 2.4.** Analysis model for PPS architecture.

the TFT even at low-intensity inputs. Thus, an additional capacitance, $C_{st}$ of about 1 pF, is usually designed into the pixel at the charge detection node in order to prevent the $V_x$ voltage from rising up to a value fatal to TFT operation. Since $C_{st}$ comprises the majority of capacitance at the integration node, the pixel capacitance $C_{PIX}$ is approximately $C_{st}$.

The output voltage on the charge amplifier is obtained directly from the charge ($Q_P$) accumulated on $C_{PIX}$ as

$$V_{OUT} = \frac{Q_P}{C_{FB}} \tag{2.1}$$

Here, $C_{FB}$ is the integrator's feedback capacitance. If $C_{FB}$ equal to $C_{PIX}$, $V_{OUT}$ equals $V_x$.

***2.2.1.3. Charge Sensing or Voltage Sensing?*** With a PPS architecture, either charge can be read out or a voltage can be sensed. From past studies done in literature [8], voltage-sensitive amplifiers shown in Fig. 2.5 are not suited for very small input signals such as the ones encountered in diagnostic medical X-ray imaging applications. In the voltage amplifier configuration, the signal charge from $C_{PIX}$ is transferred to the data line capacitance, $C_{LINE}$, at the input of the amplifier, creating a tradeoff between the charge sensitivity of the amplifier and the charge transfer efficiency (CTE) from $C_{PIX}$ to $C_{LINE}$.

If the $C_{LINE}$ is small compared to $C_{PIX}$, then insufficient charge, $Q_{LINE}$, will be transferred from $C_{PIX}$ to $C_{LINE}$ and the CTE will be degraded.

$$Q_{LINE} = Q_P \times \left[ \frac{C_{LINE}}{C_{LINE} + C_{PIX}} \right] \tag{2.2}$$

From Eq. (2.2), to move half of $Q_P$ onto $C_{LINE}$, $C_{LINE}$ must be equal to $C_{PIX}$. Thus, $C_{LINE} \gg C_{PIX}$ is needed for almost complete charge transfer. Conversely, if $C_{LINE}$

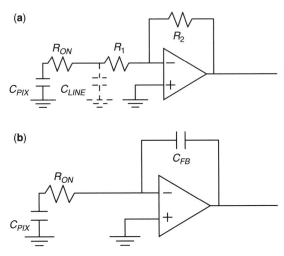

**Fig. 2.5.** (a) Voltage- and (b) charge-sensing amplifiers.

is large, most of $Q_P$ will be transferred but the voltage, $V_{LINE}$, developed at the input of the amplifier will be smaller than $V_x$ since $C_{LINE}$ needs to be at least 10 times greater than $C_{PIX}$ for good CTE and, hence, charge sensitivity is degraded. A better solution is to connect the integrator directly to the pixel output in order to collect the entire signal charge on the integrator's feedback capacitor as is implemented in current PPS architectures.

***2.2.1.4. Gain and Linearity.*** Since the passive pixel readout circuit is a simple switch, it does not provide any inherent gain to the signal on the charge detection node. However, voltage gain can be achieved by using a charge amplifier with a $C_{FB}$ less than the $C_{PIX}$, although there is a corresponding increase in noise too. Also, as long as the sensing element behaves linearly, the pixel output is linear.

***2.2.1.5. Readout Rate.*** During image readout, the pixel is discharged exponentially with a time constant $\tau_{ON}$ given by

$$\tau_{ON} = R_{ON}C_{PIX} \tag{2.3}$$

where $R_{ON}$ is the ON resistance of the switching TFT and $C_{PIX}$ is the pixel capacitance. $R_{ON}$ can be approximated by the following equation for $V_{DS} \ll V_{GS} - V_T$:

$$R_{ON} = \left[ \frac{W}{L} \mu_{EFF} C_G (V_{ON} - V_T) \right]^{-1} \tag{2.4}$$

Here $\mu_{EFF}$ is the effective carrier mobility, $V_{ON}$ is the TFT ON voltage, $V_T$ is the threshold voltage, and $C_G$ is the gate capacitance per unit area of the TFT. For a $W/L = 160/20$, $\mu_{EFF} = 0.5 \, \text{cm}^2/\text{V} \cdot \text{s}$, $V_{ON} = 12 \, \text{V}$, $V_T = 2 \, \text{V}$, $C_G = 25 \, \text{nF}/\text{cm}^2$, and $R_{ON} \sim 1.1 \, \text{M}\Omega$. Next, the intrinsic photodetector capacitance, $C_{a\text{-}Se}$ is given as

$$C_{a\text{-}Se} = \frac{\varepsilon_0 \times \varepsilon_{a\text{-}Se} \times A_{PD}}{t_{a\text{-}Se}} \tag{2.5}$$

For $\varepsilon_0 = 8.85 \times 10^{-12} \, \text{F}/\text{m}$, $\varepsilon_{a\text{-}Si:H} = 6.5$, $A_{PD} = 250 \times 250 \, \mu\text{m}^2$, and $t_{a\text{-}Se} = 500 \, \mu\text{m}$, $C_{a\text{-}Se}$ is around 7 fF. Hence, $C_{PIX} \sim C_{st} = 1 \, \text{pF}$ as noted previously where the TFT parasitic capacitances have been neglected for now.

$\tau_{ON}$ can then be calculated to be 1.1 μs. For complete charge readout (95–99.3%), about 3–5 time constants are usually sufficient, which gives a total pixel readout time of 5.5 μs (for 5 time constants). The result implies that for a PPS-based $1000 \times 1000$ X-ray imaging array with column parallel readout, a frame rate of up to 181 frames/second can be achieved, which is sufficient for real-time operation (30 frames/second).

### 2.2.2. Fabrication

***2.2.2.1. TFT Structure and Process.*** For a-Si TFTs, the most widely used structure is the bottom gate, inverted staggered TFT, which uses amorphous silicon nitride as the gate insulator. In contrast, the staggered TFT structures have the drain/source and gate contacts on opposite sides of the semiconductor layer. The inverted staggered approach with a top passivation nitride is preferred because it minimizes the defect states at the top interface and improves device performance [9].

The fabrication steps, at 260°C, can be described using Fig. 2.6 as an example. First, a 120-nm molybdenum (Mo) layer is DC magnetron sputter deposited and patterned to form the gate metallization. This is followed by PECVD deposition of a 250-nm gate nitride (a-SiN), a 50-nm a-Si semiconductor layer, and a second 250-nm a-SiN layer all within the same vacuum pump down. The source and drain contact layers are defined by photolithography and etched using a buffered HF acid solution. The etching is immediately followed by deposition of a 50-nm $n^+$ a-Si contact layer and a masking a-SiN layer in the same vacuum pump down. The masking nitride is required for the proper etching of the contact layer. The third step involves patterning the masking nitride via photolithography and HF etching. The patterned

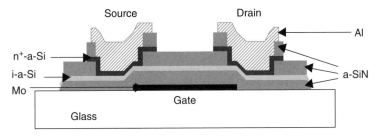

**Fig. 2.6.** Conventional inverted staggered bottom gate TFT structure.

a-SiN nitride layer then becomes the mask (for KOH etching) used to pattern the silicon-based contact layer (for separation of the TFT drain and source) as well as the a-Si film (for TFT to TFT isolation). A fourth photolithographic step is used to define contacts in the masking nitride layer, and aluminum (Al) is deposited as the final metal layer. The fifth and final photolithographic mask is used to pattern the TFT source/drain Al metal. Further details of this TFT process can be found in reference 10. a-Si inverted staggered TFTs can also be fabricated using a four-mask process where the $n^+$ a-Si contact layer and the top source/drain Al metal are patterned via a combination of dry and wet etching using only one photolithographic mask. Introduction of the masking nitride layer into the TFT process increases the mask count to five and is a requirement of a wet-etch fabrication process used to fabricate in-house TFTs.

***2.2.2.2. Nonoverlapped Electrode Process.*** A traditional coplanar imaging pixel process places the sensor and TFT beside each other on the imaging panel. While this gives process simplicity, it also tends to reduce the pixel fill factor, which is determined by the ratio of the X-ray or light-sensitive area of each pixel to the overall pixel pitch. A cross section for a conventional nonoverlapped pixel process using an a-Se photoconductor appears in Fig. 2.7. Here, the sensor back electrode and a TFT shield are included on the same metallization. The Al shield is necessary to prevent charge from accumulating at the bottom of the a-Se film over the TFT channel, thus giving rise to a parasitic back channel that would degrade TFT leakage characteristics, and is discussed more in the next section.

While coplanar pixel architectures may be suitable for imaging pixels where the readout circuit is a simple TFT switch (e.g., PPS), further improvements in sensitivity can be achieved by a move toward vertically stacked pixel architectures.

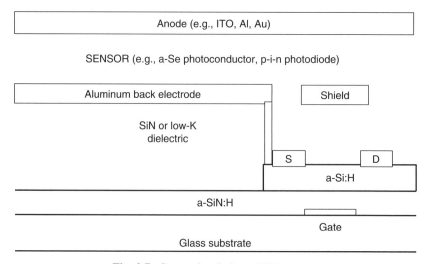

**Fig. 2.7.** Conventional planar TFT process.

Also, since a-Si TFTs consume larger areas than do their CMOS counterparts, a diagnostic medical imaging modality such as large-area mammography that requires small pixel pitch (50 μm × 50 μm) is challenging to implement in a coplanar architecture.

***2.2.2.3. Fully Overlapped Process.*** In large-area flat-panel X-ray imagers, it is important to ensure that the fill factor should be high enough to ensure sufficient charge collection. Conventional imagers are based on a coplanar architecture where the sensor and readout circuitry are placed adjacent to one another. Increasing the on-pixel density of TFTs (such as for an amplified pixel readout circuit) or scaling down pixel sizes reduces the fill factor. Using a fully overlapped architecture [11] with a continuous back electrode over the pixel can circumvent this. However, the continuous back electrode can give rise to parasitic capacitance, whose effect becomes significant when the electrode runs over the various TFTs. One possible effect of the back electrode is to induce a parasitic channel in the a-Si layer of an overlapped TFT during its OFF state, giving rise to a larger $I_{DS}$ leakage current [12]. This leakage is particularly severe if the overlapping back electrode is separated from the TFT by a thin insulating layer. To make the $I_{DS}$ leakage independent of insulator thickness, an alternate TFT structure based on a dual-gate architecture [13] is employed, in which the voltage on the top gate (metal shield) can be chosen to minimize the charge induced in the (parasitic) top channel of the TFT.

Figure 2.8a illustrates the structure of a dual-gate TFT in a fully overlapped sensor architecture where a second gate is inserted between the sensor back plate and the TFT. Figure 2.8b shows static current measurements made on a single dual-gate TFT structure ($W/L = 250$ μm/25 μm) embedded underneath a continuous back electrode using amorphous silicon nitride as the insulator. Modeling details of the dual-gate TFT structure are outside the scope of this chapter, and the reader is referred to reference 13 for further information. For the Al back electrode potential, $V_{DIODE} = 10$ V, the lowest values of leakage current are obtained when the TFT second gate bias, $V_{SHIELD}$, is set to 0 V or to the TFT bottom gate voltage, $V_{GATE}$. In the case where $V_{SHIELD}$ is floated, the leakage current is significantly larger.

### 2.2.3. TFT Metastability

***2.2.3.1. Physical Mechanisms.*** Studies of a-Si TFTs have concluded that two mechanisms can explain the electrical instability [14, 15]. The first is carrier trapping in the gate insulator, which occurs primarily in PECVD a-Si:N gate insulator TFTs where the high density of defects in the insulator can trap charge when the TFT gate undergoes bias stress. Injection of electrons into the insulator layer of a-Si TFTs has long been known to cause $\Delta V_T$. Electrons are first trapped within the localized interfacial states at the a-Si/a-SiN boundary and then thermalize to deeper energy states inside the a-SiN layer either by variable-range hopping [14] or through a multiple-trapping and emission process [16–18]. The broad distribution of energies for the trapping levels within the a-SiN layer leads to two kinds of electron trapping behavior [17].

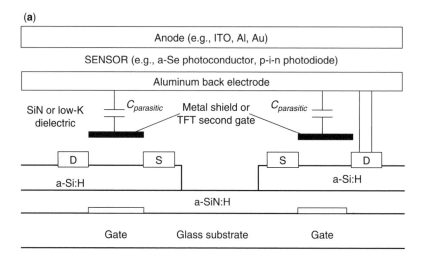

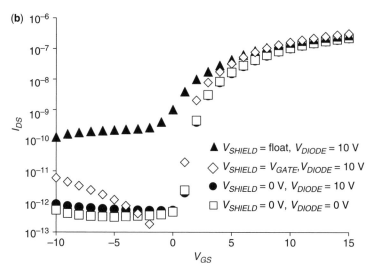

**Fig. 2.8.** (a) Fully overlapped architecture (b) TFT transfer characteristics ($V_{DS} = 1$ V).

The first is due to interfacial traps with fast re-emission times (or fast states). These traps were found to cause a hysteresis in the $I_{DS} - V_{GS}$ behavior of TFTs [16, 19, 20]. The density of these fast states decreases when the optical bandgap of the nitride layer is increased. Moreover, the optical bandgap of the nitride was found to increase when the nitrogen content of the layer was increased [19, 20]. Nitrogen-rich nitrides are generally obtained by using a higher ratio of $NH_3/SiH_4$ ($\sim$10) during PECVD deposition.

The second nitride related degradation effect is due to interfacial and bulk traps with slow re-emission times (slow states). These traps cause a metastable increase in the threshold voltage [14, 17, 18, 20, 21]. However, the number of these deep traps was found to decrease for wider-bandgap, nitrogen-rich, a-SiN layers [19]. As such, using nitrogen-rich a-SiN has become prevalent for TFTs since it improves the reliability and reduces degradation associated with both slow and fast states in the gate insulator. The in-house TFT fabrication process used a ratio of $NH_3/SiH_4 \sim 20$ for the a-SiN films to achieve good metastability characteristics.

The second mechanism related to metastability is point defect creation in the a-Si layer or at the a-Si/SiN interface that increases the density of deep-gap states [22]. When a positive gate bias causes electrons to accumulate and form a channel at the a-SiN/a-Si interface, a large number of these induced electrons reside in conduction band tail states. These tail states have been identified as weak silicon–silicon bonds which, when occupied by electrons, can break to form silicon dangling bonds (deep state defects) [23, 24]. Deep state defect creation forms the basis of the defect pool model [22], which has some similarities to the Staebler–Wronski effect [25] in which photogenerated carriers result in the generation of dangling bonds [26]. In the defect pool model, the rate of creation of dangling bonds is a function of $\phi$, the barrier to defect formation, the number of electrons in the tail states, and the density of the weak bond sites [22]. The metastable increase in the density of deep defect states is shown in Fig. 2.9. The location of these newly created defect states in the

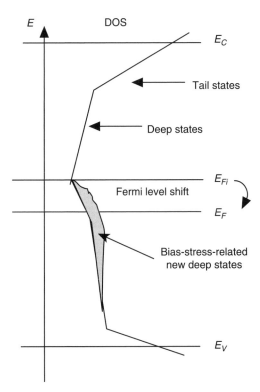

**Fig. 2.9.** Bias-induced metastable defect states in a-Si TFTs (adapted from reference 22). Note that most of the states are below $E_F$ near that valence band.

gap depends on the polarity of the applied stress bias and, in some cases, can cause the $V_T$ of the TFT to shift [22, 27].

The charge trapping and defect state creation mechanisms distinguish themselves from each other (for TFTs with an a-Si:N gate insulator) in that charge trapping in the nitride has been observed to occur at high gate bias voltages and long stress times. Here, the shift in the threshold voltage can be positive or negative, depending on the polarity of the trapped charge (electron or hole). In contrast, defect state creation dominates at lower stress voltages and at smaller stress times. At small stress voltages, Powell showed that while state creation in the lower part of the gap dominated at positive biases, state removal from the lower part of the gap occurred for negative bias voltages [22].

Powell distinguished between defect state creation and charge trapping in 1987 by experimenting with ambipolar TFTs [28]. If charge is trapped in the nitride under positive bias stress, the $V_T$ for both electron and hole conduction will shift to more positive values. In contrast, defect state creation in the a-Si layer manifests itself by the $V_T$ becoming more negative for hole conduction and more positive for electron conduction since the Fermi level has to move through the newly created states to establish the channel. Powell showed that for his TFTs, state creation was the dominant mechanism up until 50 V [22] and charge trapping became significant thereafter. The critical voltage at which charge trapping overtakes state creation depends on the bandgap of the a-Si:N gate insulator, which also supports the theory that the charge trapping process is nitride-related [29]. The defect creation process appeared to be independent of the nitride characteristics, implying that state creation takes place in the a-Si film [14].

Powell also determined that state creation is characterized by a power-law time dependence and is strongly affected by temperature. Charge trapping, on the other hand, has a logarithmic time dependence and is weakly dependent on temperature. A logarithmic time dependence and temperature independence is generally observed for charge trapping in an insulator where the injection current depends exponentially on the density of previously injected charge and the charge is trapped near the gate insulator interface [15, 30]. As noted before, the process of state creation observed in a-Si film was reported to have several features in common with observations of state creation through illumination or thermal equilibration. One proposed model is that Si dangling bonds are created by breaking weak Si–Si bonds, which are then stabilized by the dispersive diffusion of hydrogen [31]. The power-law time dependence is found to be consistent with a defect creation model [32].

The subthreshold slope in the transfer characteristics of $n$-channel transistors does not change after positive bias stress, confirming the location of the newly created states below the flat-band Fermi level (see Fig. 2.9). This is consistent with the idea that the created states are Si dangling bonds or deep state defects. The process of creating states below the flat-band Fermi level shifts the Fermi level nearer to midgap, which accounts for the increase in threshold voltage.

For negative biases, there are a few newly created states located above the flat-band Fermi level in addition to state removal in the lower part of the gap. This explains why there is some subthreshold slope degradation in $n$-channel TFTs during negative bias stress. For imaging applications using $n$-channel a-Si TFTs, TFT OFF voltages are chosen primarily by the requirement of low leakage current. The magnitude of typical TFT OFF voltages ranges from 0 to $-5$ V, implying that subthreshold slope degradation will be minimal. Therefore, although a-Si TFT metastability comprises both $\Delta V_T$ and subthreshold slope degradation, the focus in this chapter is on $\Delta V_T$ management.

***2.2.3.2. Positive Gate Bias Stress.*** Modeling the two $\Delta V_T$ mechanisms is summarized as follows: Defect state creation is characterized by a power-law dependence of $\Delta V_T$ over time, that is, $\Delta V_T \propto t^\beta$. Both Jackson and Powell [15, 28] claimed that their data mapped a relationship,

$$\Delta V_T(t) = A(V_{ST} - V_{Ti})^\alpha t^\beta \qquad (2.6)$$

for the defect state creation mechanism where $\alpha$ takes on a value of unity and $V_{Ti}$ is the $V_T$ before stress is applied. In contrast, charge trapping was associated with a logarithmic time dependence [32], which can be represented as

$$\Delta V_T(t) = r_d \log(1 + t/t_0) \qquad (2.7)$$

where $r_d$ is a constant and $t_0$ is some characteristic value for time.

On the other hand, Libsch and Kanicki [16] reported that an empirical stretched-exponential function could describe the stress–time and stress–voltage dependence of $V_T$ in a-Si TFTs irrespective of the $\Delta V_T$ mechanism,

$$V_T(t) = (V_{ST} - V_{Ti})[1 - \exp(-(t/\tau)^\beta)] \qquad (2.8)$$

where $\tau$ is some characteristic extractable time constant, and $\beta$ is the stretched-exponential exponent which is temperature-dependent. For a short effective stress time ($t \ll \tau$), the above function can be simplified to a form similar to Eq. (2.1),

$$\Delta V_T(t) = (V_{ST} - V_{Ti})\tau^{-\beta} t^\beta = A(V_{ST} - V_{Ti})^\alpha t^\beta \qquad (2.9)$$

The main difference is that Kanicki extracts nonunity values for $\alpha$ making the function empirical. Our results, as will be seen in the following sections, appear to be in concord with Powell's theory in that we noticed a unity value for $\alpha$ for the range of positive stress voltages applied as well as a power-law time dependence for the $\Delta V_T$.

***2.2.3.3. Negative Gate Bias Stress.*** Results for small negative biases ($-10 < V_{GS} < 0$ and $-10 < V_{GD} < 0$) are difficult to measure accurately since the positive $\Delta V_T$ that occurs during the TFT $I_{DS}-V_{GS}$ monitoring for $V_T$ extraction between bias stress applications corrupts the small (in mV) negative bias stress related $\Delta V_T$. For larger negative gate and drain voltages ($-25 < V_{GS} < -15$ and $-25 < V_{GD} < -15$), where both defect state creation and charge trapping mechanisms may be active, the following empirical relationship was observed with an error of 15%.

$$\Delta V_T^-(t) = -B|V_{ST}|^\alpha t^{\beta_2}, \qquad V_{GS} < 0, \qquad V_{GD} = V_{GS} \qquad (2.10)$$

where $B = 0.12 \times 10^{-3}$, $\alpha = 2.49$, $V_{ST} = V_{GS}$, and $\beta_2 = 0.28$. For imaging applications using a PPS architecture, the maximum negative bias on the a-Si TFT gate (around $-5$ V) is chosen primarily to reduce the leakage current. Here, the TFT functions as a simple switch, and the drain voltage rarely exceeds 1 V. Thus, $V_{GD} \sim V_{GS}$ and the above equation can be used to get reasonable accuracy in the $\Delta V_T$ prediction.

### 2.2.3.4. Effect of DC Bias Stress on Leakage Current.
It is unclear if other authors [14, 18, 33] have considered the impact of voltage stressing on TFT leakage current. However, commercial radiological X-ray detectors based on a-Si TFT PPS architectures have recently become available [6], implying that the effect of voltage stress on TFT leakage is manageable for imaging applications (where duty cycles are small). In general, however, the TFT leakage current for in-house fabricated TFTs stayed in the sub-pA range for all metastability measurements, which is adequate for imaging applications.

### 2.2.3.5. Pulse Bias Metastability.
A pulse can be characterized by its period (related to frequency) and pulse width (related to duty cycle) as shown in Fig. 2.10. Varying both of these parameters affects the magnitude of $\Delta V_T$ as will be discussed in the following subsections. The discussion and measurements in the previous sections are valid for constant stress bias voltages (DC). During circuit operation, however, the TFTs undergo pulse bias, which can change the metastability performance. The behavior of a-Si TFTs under pulse bias is examined with respect to frequency in this section.

Pulse-related $\Delta V_T$ has been widely reported to be relatively independent of frequency for positive pulse voltages [33–35]. For a unipolar (i.e., $+V/0$ or $-V/0$) pulse, a primary effect of frequency is to reduce the total stress time of the TFT; that is, for a 50% duty cycle, the pulse is ON only half of the time as compared to DC bias applied for the same time period. In addition, the OFF cycle provides a relaxation time for the TFT; this can further reduce the $\Delta V_T$, depending on the relaxation time required by the various trapped charge and/or defects. Note that a bipolar pulse can be viewed as an extension of the unipolar case where during the OFF cycle a negative voltage is applied to the TFT for the complement of the pulse duty cycle.

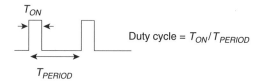

**Fig. 2.10.** Pulse definitions and duty cycle.

In contrast to positive gate bias pulses, the $\Delta V_T$ can be reduced significantly for a negative pulse bias in comparison to a constant negative bias for a similar effective application time. A model proposed in reference 33 attempts to explain this behavior by suggesting that electrons or holes accumulate in the a-Si channel, depending on the polarity of the applied pulse bias. This implies that as pulse frequency is increased past some threshold frequency, the $\Delta V_T$ should decrease significantly since carriers have not had sufficient time to accumulate. Originally, the model was proposed for only the charge trapping mechanism, which was in line with Kanicki's claim that charge trapping is the major phenomenon contributing to $\Delta V_T$ in a-Si TFTs. However, since accumulation of carriers in the TFT channel governs both deep defect creation and the charge trapping mechanisms, the negative pulse bias $\Delta V_T$ model does not contradict the defect pool model.

When a gate voltage is applied, carriers accumulate in the channel (some in the conduction band but mostly in tail states), allowing both charge trapping and deep defect creation to proceed. Since hole mobility is almost two orders lower than electron mobility in the a-Si material, the frequency response of the TFT is expected to be much higher for electron accumulation (via positive gate pulses) as opposed to hole accumulation (via negative gate pulses). Of course, carrier mobility is not the only cause of longer accumulation times for holes. For an $n$-channel transistor, the $n$-type source and drain regions are a primary source of channel electrons during the ON cycle, while thermal generation is the most likely source of holes during the OFF cycle.

Thus the $\Delta V_T$ decreases if the TFT is pulsed at a frequency greater than its 3-dB frequency since carriers do not have sufficient time to accumulate in the channel. For positive bias pulses, pulsing the TFT beyond its 3-dB frequency is a moot point since the TFT is generally operated at a frequency lower than the 3-dB point for proper performance. However, for negative bias pulses, the $\Delta V_T$ may decrease at pulse frequencies that are much less than the TFT electron carrier frequency response because the hole accumulation rate is much slower than electron accumulation.

To examine this phenomenon further, a simple RC model [33] can describe a-Si TFT operation under negatively pulsed bias stress as shown in Fig. 2.11. Here, the gate insulator capacitance, $C_i$, is in series with the a-Si semiconductor capacitance, $C_s$. An effective resistance, $R_s$, is included in parallel with $C_s$ to account for hole conductance in the a-Si layer. When a negative pulse is applied to the gate ($V_{ST}$), the total gate voltage is initially distributed across the gate insulator a-SiN and the a-Si layer because the a-Si layer acts as a semi-insulator under negative bias at higher frequencies. Eventually, hole carriers (via thermal generation) begin to accumulate near the a-Si/a-SiN interface and the voltage drop across the insulator, $V_i$, approaches $V_{ST}$. Here, $V_i(t)$ can be written as

$$V_i(t) = V_{ST}\left[1 - \frac{C_i}{C_i + C_s}\exp\left(\frac{-t}{\tau_h}\right)\right] \tag{2.11}$$

where $\tau_h = R_s C_s$ is the effective hole accumulation time constant. The most likely source of holes is thermal generation near the a-Si/a-SiN interface where the generated electrons are repelled into the bulk semiconductor by the applied negative gate bias.

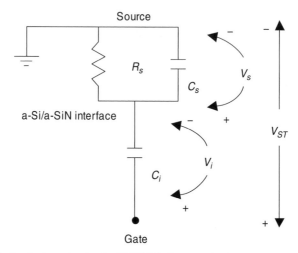

**Fig. 2.11.** Equivalent circuit of a-Si TFT during negatively pulsed gate bias stress.

Using Eq. (2.11), the initial and final values of $V_i$ are given as

$$V_i(0) = V_{ST} \frac{C_s}{C_i + C_s}, \qquad V_i(\infty) = V_{ST} \tag{2.12}$$

In order to relate the $\Delta V_T$ to the effective carrier concentration at the insulator–semiconductor interface, the accumulated hole concentration can be written as

$$N_{accum} = -N_i + N_s \tag{2.13}$$

where $N_i = C_i V_i/q$, $N_s = C_s V_s/q$, and $q$ is the usual electron charge. Noting that $V_{ST} = V_i(t) + V_s(t)$, $V_s(t)$ can be obtained in terms of Eq. (2.11) and $V_{ST}$. Using $V_i(t)$ and $V_s(t)$ in Eq. (2.13), the accumulated hole concentration can be simplified as

$$N_{accum}(t) = \frac{C_i V_{ST}}{q} \left[ \exp\left(\frac{-t}{\tau_h}\right) - 1 \right] \tag{2.14}$$

Then, the average accumulated carrier concentration under pulse bias stress, $N_{AC}$, with a pulse ON time, $T_{ON}$, can be expressed as

$$N_{AC} = \frac{1}{T_{ON}} \int_0^{T_{ON}} N_{accum}(t)\, dt \tag{2.15}$$

$$N_{AC}(T_{ON}) = \frac{-C_i V_{ST}}{T_{ON}} \left[ 1 - \frac{\tau_h}{T_{ON}} + \frac{\tau_h}{T_{ON}} \exp\left(-\frac{T_{ON}}{\tau_h}\right) \right] \tag{2.16}$$

$$N_{AC}(\infty) = \frac{-C_i V_{ST}}{T_{ON}} = N_{DC} \tag{2.17}$$

The above equation illustrates that as the negative pulse ON time, $T_{ON}$, increases and becomes comparable to $\tau_h$, $N_{AC}$ decreases. $N_{AC}$ is at its maximum when $T_{ON}$ is infinity (i.e., at DC).

Following the approach in reference 33, if the $\Delta V_T$ is assumed to be directly proportional to the effective carrier concentration at the a-Si/a-SiN interface ($N_{accum}$), then the pulse bias $\Delta V_T$ reduction factor can be computed as

$$\frac{N_{AC}(T_{ON})}{N_{DC}} = \frac{\Delta V_T(T_{ON})}{\Delta V_{T\_DC}} = 1 - \frac{\tau_h}{T_{ON}} + \frac{\tau_h}{T_{ON}}\exp\left(-\frac{T_{ON}}{\tau_h}\right) \quad (2.18)$$

The above equation was fit to negative pulse bias metastability data, and a value of 32 ms for $\tau_h$ was extracted in reference 33. In other words, for a negative pulse bias with a $T_{ON}$ of 32 ms, the $\Delta V_T$ corresponds to only 36% of the DC bias $\Delta V_T$ at the same negative voltage level [where the above Eq. (2.18) was used].

In general, smaller duty cycles induce less $\Delta V_T$ for the same pulse frequency because the effective stress time is further reduced. The reduced $\Delta V_T$ result holds true for both negative and positive pulse voltages. Due to lower hole mobility and accumulation times, the $\Delta V_T$ is also reduced if, in the process of reducing the duty cycle, the negative ON pulse does not provide sufficient time for the holes to accumulate in the channel; that is, the pulse OFF frequency drops below the hole carrier accumulation time constant, $\tau_h$, as noted in the previous section. In addition, detrapping and relaxation of defect states during the longer OFF cycle of small duty cycle ON pulses further reduce the overall TFT $\Delta V_T$.

### 2.2.4. Electronic Noise

***2.2.4.1. Thermal Noise.*** Van der Ziel [36] showed that the spectral intensity of the drain current fluctuation, $S_I(f)$, in a long channel MOSFET due to thermal noise is

$$S_I(f) = \frac{1}{L^2}\int_0^L 4kTg(y)\,dy \quad (2.19)$$

where $L$ is the channel length, $k$ is the Boltzmann constant, $T$ is the carrier temperature, $g(y)$ is the conductivity of the channel a distance $y$ from the source (in siemens-cm) and is directly proportional to the total charge in the channel, and $f$ is the frequency. At no drain bias, the conductivity of the channel is uniform and thus $g(y) = L \cdot g_{ds0}$, where $g_{ds0}$ is the zero drain bias conductivity ($(W/L)C_G\mu(V_{GS} - V_T)$ for $V_{DS} = 0$). Then, at zero drain source voltage, Eq. (2.19) becomes

$$S_I(f) = \frac{1}{L^2}\int_0^L 4kT(Lg_{ds0})\,dy = 4kTg_{ds0} \quad (2.20)$$

Since $g_{ds}$ is proportional to the total charge in the channel, $S_I(f)$ at larger drain biases can be determined by using the channel conductance $g(y)$ for a drain bias with Eq. (2.19) as shown below.

$$S_I(f) = \frac{2}{3} 4kT\mu \frac{W}{L} C_G \left[ \frac{(V_{GS} - V_T)^3 - (V_{GS} - V_T - V(y))^3}{(V_{GS} - V_T)^2 - (V_{GS} - V_T - V(y))^2} \right] \qquad (2.21)$$

Equation (2.3) can be solved at saturation ($V_{DS} = V(y) = V_{GS} - V_T$) to give

$$S_I(f) = \frac{2}{3} 4kT \left[ \mu \frac{W}{L} C_G (V_{GS} - V_T) \right] = \frac{2}{3} 4kT g_{ds0} \qquad (2.22)$$

which is the standard equation for MOSFET thermal noise quoted in the literature [36]. As the limit $V(y) = V_{DS} \rightarrow 0$, Eq. (2.21) reduces to the value given by Eq. (2.20) for zero drain bias. Therefore, Eq. (2.22) may be represented as

$$S_I(f) = (\gamma)4kT \cdot g_{ds0} \qquad (2.23)$$

where $\gamma$ is a factor that decreases for a transistor from a value of 1 in the linear mode to $2/3$ at saturation.

Measurements performed by Boudry and Antonuk [37] and verified by Karim et al. [38] confirmed that TFT thermal noise has some similarities to MOSFET thermal noise and can be modeled using Eq. (2.23),

$$g_{ds0} = \frac{dI_{DS}}{dV_{DS}} = M_0(V_{GS} - V_T)^n \qquad (2.24)$$

where $n = (\alpha - 1) = 1.38$ for the in-house TFTs. Substituting Eq. (2.24) into Eq. (2.23) yields

$$S_I(f) = (\gamma)4kTM_0(V_{GS} - V_T)^n \qquad (2.25)$$

where $M_0$ can be represented by the usual product of parameters, $\mu_{EFF} C_G (W/L)$.

**2.2.4.2. Flicker Noise.** When a constant voltage is applied across the terminals of a resistor, a fluctuation in the DC current is observed. This fluctuating current is in addition to the thermal noise, and exhibits the following general power spectral density (PSD):

$$S_I(f) = K \cdot I^2 / f^\Gamma \qquad (2.26)$$

where $I$ is the DC current, $\Gamma$ lies between 0.8 and 1.4 in general, and $K$ is a proportionality constant inversely proportional to the device area. The $1/f$ slope of the characteristic function has been verified over the range of frequencies from $10^{-6}$ to $10^6$ Hz [34]. No experimental evidence as yet has suggested that the spectrum will flatten out in the extreme low frequency range. The first recorded observations of $1/f$ noise were made by Johnson [39] in 1925. Since then, two somewhat different theories of $1/f$ noise have been developed. The first model, commonly known as the numbers fluctuation model, was proposed by McWhorter [40] in mid-1950s. The model attributed the generation of $1/f$ noise to fluctuations in the majority carrier density (number of carriers) near the semiconductor surface, due to a fluctuation in the occupancy of surface interface traps. It has been successfully used to model $1/f$ noise in surface channel devices such as the MOSFET. The numbers fluctuation model does not account for the $1/f$ noise observed in media such as ionic solutions, nor does it account for wire-wound resistors, which have no interface traps.

An alternative theory in the 1960s was proposed by Hooge and Hoppenbrouwers [41] to explain $1/f$ noise. The theory is based upon the mobility fluctuations within a homogeneous, conducting medium and is commonly known as the mobility fluctuation model. Hooge proposed the following empirical relation:

$$S_I(f) = \alpha_H \cdot I^2 / (N_{tot} f^{\Gamma}) \tag{2.27}$$

where $I$ is the drain current as before, $N_{tot}$ is the total number of charge carriers in the medium, and $\alpha_H$ is the Hooge dimensionless coefficient. Hooge's coefficient has been experimentally verified to have a value of $2 \times 10^{-3}$ in several homogeneous media. This value does not hold for some media; and Hooge and Vandamme proposed that $\alpha_H$ should be modified, depending on the impurity scattering in a material [42].

For a TFT operating in the linear mode, the Hooge flicker noise spectral density can be approximated as

$$S_{ID}(f) = \frac{\alpha_H}{f} \frac{q\mu}{L^2} I_D V_{DS} \tag{2.28}$$

where $N_{tot} = W \cdot L \cdot C_G \cdot (V_{GS} - V_T)$ in the linear mode. Equation (2.28) can now be expressed in terms of the terminal voltages only as

$$S_{ID}(f) = \frac{\alpha_H}{f} \frac{q\mu}{L^2} (M_0(V_{GS} - V_T)^n V_{DS}) V_{DS} = \frac{\alpha_H}{f} \frac{q\mu}{L^2} (M_0(V_{GS} - V_T)^n V_{DS}^2) \tag{2.29}$$

In contrast to the Hooge $1/f$ noise theory of mobility fluctuation, the McWhorter theory of number fluctuation is based on trapping and detrapping of carriers. For the trapping theory of carriers [43], the current noise spectral density for a MOSFET is expressed as

$$S_{ID}(f) = \frac{k^*}{f} \frac{\mu}{C_{OX} L^2} \frac{I_D V_{DS}}{(V_{GS} - V_T)} \tag{2.30}$$

**TABLE 2.2. $1/f$ Noise Spectral Densities for McWhorter and Hooge Theories in TFTs**

|  | $\Delta N$ Model | $\Delta \mu$ Model |
|---|---|---|
| $S_{ID(lin)}(f)$ | $\dfrac{k^*}{f}\dfrac{\mu}{C_{ox}L^2}(M_0(V_{GS}-V_T)^{n-1}V_{DS}^2)$ | $\dfrac{\alpha_H}{f}\dfrac{q\mu}{L^2}(M_0(V_{GS}-V_T)^n V_{DS}^2)$ |
| $S_{ID(sat)}(f)$ | $\dfrac{k^*}{f}\dfrac{\mu}{C_{ox}L^2}(M_0(V_{GS}-V_T)^{n+1})$ | $\dfrac{\alpha_H}{f}\dfrac{q\mu}{L^2}(M_0(V_{gs}-V_T)^{n+2})$ |

where $k^*$ takes into account the electron tunneling from insulator traps near the interface to the conducting channel. Equation (2.30) can be rewritten to obtain the noise current spectral density for a TFT operating in the linear region,

$$
\begin{aligned}
S_{ID}(f) &= \frac{k^*}{f}\frac{\mu}{C_{ox}L^2}\frac{(M_0(V_{GS}-V_T)^n V_{DS})V_{DS}}{(V_{GS}-V_T)} \\
&= \frac{k^*}{f}\frac{\mu}{C_{ox}L^2}(M_0(V_{GS}-V_T)^{n-1}V_{DS}^2) \qquad (2.31)
\end{aligned}
$$

Using Eq. (2.31), the noise voltage and noise resistance spectral densities for both mobility and carrier number fluctuation models can be determined [43, 44] and are summarized in the Table 2.2. Here, the $S_{ID(sat)}(f)$ for saturation mode TFT operation is calculated by substituting $V_{DS} = (V_{GS} - V_T)$ into $S_{ID(lin)}(f)$ as shown in Eq. (2.31).

The importance of the quantities shown in Table 2.2 appears when the dependence of noise spectral densities on the $(V_{GS} - V_T)$ term is examined. For both the carrier trapping and carrier mobility models, there is a difference in the dependence of $S_{ID}$ upon the $(V_{GS} - V_T)$ term, which provides the means to discriminate between the two theories.

Valenza and co-workers [45] carried out the various low-frequency noise experiments on a-Si TFTs and determined that the Hooge model of fluctuating mobility was responsible for $1/f$ channel noise in TFTs. Rhayem determined an experimental value of $1 \times 10^{-2}$ for the Hooge parameter $\alpha_H$ for a-Si TFTs, which corroborated earlier studies [46]. This number is about two to three orders of magnitude larger than the classical values observed in crystalline MOSFETS. In contrast, some results have been reported that contradict the Hooge theory of $1/f$ noise in TFTs proposed by Valenza. Specifically, Chen et al. [47] performed experiments similar to Valenza and concluded that $1/f$ noise followed a blend of number fluctuation and Hooge theories. Our experimental results, presented in the next section, appear to corroborate the Hooge theory of $1/f$ noise for our in-house fabricated a-Si TFTs.

*2.2.4.3. Noise in PPS Pixels.* The main sources of electronic noise are due to the readout TFT (thermal and flicker) and the charge amplifier. During readout, one TFT is turned ON and its associated sensor (selenium or photodiode) is connected to the data line with a TFT ON resistance of about 1 MΩ. This resistive component in the circuit generates thermal noise. The $1/f$ noise in a-Si:H TFTs and the shot noise from a-Si:H photodiodes increase with current under the conditions of high drain current and high sensor leakage current, respectively. However, even the photocurrent level is very low in these diagnostic X-ray imaging systems ($\sim$1 fA to 10 pA), and so $1/f$ and shot noise sources may be neglected as a first-order approximation. From Fujieda et al. [8], the final expression for equivalent input TFT ON resistance thermal noise in electrons is as given below ($C_{pixel}$ and $C_{amp}$ are the pixel and external charge amplifier input capacitances, respectively):

$$N_{th} = \sqrt{\frac{C_t}{q^2} kT \left[ 1 - \frac{2}{\pi} \tan^{-1} \left( \frac{2\pi R C_t}{\tau_{int}} \right) \right]}, \qquad C_t = \frac{C_{pixel} C_{amp}}{C_{pixel} + C_{amp}} \qquad (2.32)$$

For $C_{pixel} = 6$ pF, $C_{amp} = 100$ pF, $R_{ON\text{-}TFT} = 1.4$ M$\Omega$, and an integration time $\tau_{int} = 33$ $\mu$s, $N_{th}$ is estimated to be 620 electrons using the above equation and is typical for imaging applications.

The amplifier noise can be modeled as having a fixed noise component $N_{amp0}$ in addition to an input-capacitance-dependent component:

$$N_{amp} = N_{amp0} + \delta C_d \qquad (2.33)$$

Here $\delta$ is a constant determined by the design properties of the charge amplifier (e.g., input FET noise) and $C_d$ is the external capacitance loading the amplifier input node. This includes the parasitic capacitances on the data line such as $C_{GS}$ of readout TFT as well as the overlap capacitance of data and gate lines. A typical value for amplifier noise is $\sim$1000 electrons [5].

From the noise point of view, the readout TFT and amplifier noise are two independent noise sources that add in quadrature,

$$N_{tot}^2 = N_{th}^2 + N_{amp}^2 \qquad (2.34)$$

The total noise, $N_{tot}$, at the TFT drain is about 1200 electrons. Compared to the 620 noise electrons from the TFT, it can be concluded that the amplifier is the dominant noise source. The signal charge accumulating on $C_{pixel}$ for an a-Se photoconductor ranges from 1000 electrons (i.e., minimum 1 X-ray) to 49,100 electrons. This range is due to the low dose exposure range for fluoroscopy (0.1 $\mu$R to 10 $\mu$R). The lower end of range implies that part of the lower signal (1000 electrons) will be affected by electronic noise.

## 2.3. RECENT DEVELOPMENTS

Commercially available state-of-the-art flat panel imagers based on PPS designs are fabricated using a-Si TFT and photodiode array technology coupled to a scintillator layer [usually cesium iodide (CsI)]. Alternately, a smaller proportion use the direct detection approach where the photodiode and scintillator are replaced by an X-ray-sensitive photoconductor material (such as amorphous selenium) (a-Se) that can directly convert X-rays into electronic charge. While the PPS has the advantage of being compact and amenable toward high-resolution imaging, the small PPS output signals are swamped by external column charge amplifier and data line thermal noise, which reduce the minimum readable sensor input signal, thus limiting its use to higher-dose, general radiography modalities. Although the imager performance can, in principle, be improved by using a better, high-gain detector in place of a-Se or CsI, this section focuses on recent developments in pixel amplifier circuits to overcome the shortcomings of the flat panel imager in high-performance imaging applications such as real-time low-dose fluoroscopy or 3D mammography. The primary benefit in using a circuit approach as opposed to developing a better material is to leverage the capabilities of existing established fabrication infrastructure and processes.

### 2.3.1. Current Mode Active Pixel Sensor

Central to the current mode active pixel sensor (APS) illustrated in Fig. 2.12 is a source follower circuit, which produces a current output (C-APS) to drive an external charge amplifier. The C-APS operates in three modes:

- *Reset Mode*: The RESET TFT switch is pulsed ON and $C_{PIX}$ charges up to $Q_P$ through the TFT's on resistance. $C_{PIX}$ is usually dominated by the detector (e.g., a-Se photoconductor or a-Si photodiode detection layer [5]) capacitance.

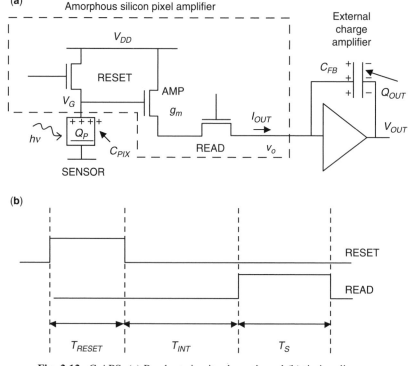

**Fig. 2.12.** C-APS: (a) Readout circuit schematic and (b) timing diagram.

- *Integration Mode*: After reset, the RESET and READ TFT switches are turned OFF. During the integration period ($T_{INT}$) the input signal, $hv$, generates photocarriers discharging $C_{pixel}$ by $\Delta Q_P$ and decreases the potential on $C_{pixel}$ by a small signal voltage, $\Delta V_G$.
- *Readout Mode*: After integration, the READ TFT switch is turned ON for a sampling time $T_S$, which connects the APS pixel to the charge amplifier and an output voltage, $V_{OUT}$, is developed across $C_{FB}$ proportional to $V_G + \Delta V_G$ and $T_S$.

Operating the READ and RESET TFTs in the linear region reduces the effect of inter-pixel threshold voltage ($V_T$) nonuniformities. Although the saturated AMP TFT causes the C-APS to suffer from FPN, using CMOS-like off-chip double sampling [48] and offset and gain correction techniques can alleviate the problem.

***2.3.1.1. Linearity.*** The linearity of the C-APS readout circuit is obtained from a sensitivity analysis of the change in output current, $I_{OUT}$, with respect to the input illumination, $hv$, in Fig. 2.12,

$$\gamma = \frac{dI_{OUT}}{dhv} = \frac{dQ_P}{dhv}\frac{dV_G}{dQ_P}\frac{dI_{OUT}}{dV_G} \tag{2.35}$$

where $\gamma$ is a constant for a linear sensor. The first term is linear if the detector gives a linear change in the charge on $C_{PIX}$, $\Delta Q_P$, with changing $hv$. The second term is linear if the voltage change $\Delta V_G$ across $C_{PIX}$ is linearly dependent on $\Delta Q_P$ where

$$\Delta Q_P = \Delta V_G \cdot C_{PIX} \tag{2.36}$$

Equation (2.36) is linear, provided that $C_{PIX}$ stays constant under the changing bias conditions where the TFT parasitic capacitances have been neglected.

However, the last term in Eq. (2.35) imposes a linear small signal condition on the allowable change in the AMP TFT gate bias, $V_G$. For a MOS device, the $I$–$V$ relationship can be written as

$$I_{OUT} + \Delta I_{OUT} = (K/2)(V_G + \Delta V_G - V_T)^2 \tag{2.37}$$

This can be expanded to

$$I_{OUT} + \Delta I_{OUT} = (K/2)(V_G - V_T)^2 + K(V_G - V_T)\Delta V_G + (K/2)(\Delta V_G)^2 \tag{2.38}$$

Separating the DC bias and focusing only on the small signal AC components, we obtain

$$\Delta I_{OUT} = K(V_G - V_T)\Delta V_G + (K/2)(\Delta V_G)^2 \tag{2.39}$$

Here, for linear circuit operation, the $(\Delta V_G)^2$ nonlinearity must be minimized, giving

$$\Delta V_G \ll 2(V_G - V_T) \tag{2.40}$$

For the APS, $V_G$ is the DC bias voltage at the AMP TFT gate and $V_T$ is its threshold voltage. Thus, for linear C-APS operation, the change in voltage at the integration node due to the X-ray input ($\Delta V_G$) must be kept small. The equations above also serve to illustrate that both biasing and signal components exist in the output current. The bias current component ($I_{OUT}$) comes from the steady-state value of the gate voltage ($V_G$) while the signal current ($\Delta I_{OUT}$) comes from the small signal change at the AMP TFT gate ($\Delta V_G$).

***2.3.1.2. Gain.*** When photons are incident on the detector, electron–hole pairs are created, leading to a change in the charge on the integration node. In small signal operation, the change in the amplifier's output current with respect to a small change in gate voltage, $\Delta V_G$, is

$$\Delta I_{OUT} = g_m \cdot \Delta V_G \tag{2.41}$$

where $g_m$ is the transconductance of the AMP and READ TFT composite circuit and $\Delta V_G$ represents the small signal voltage at the gate of the AMP TFT. Again, simple MOS Level 1 models are used here to obtain insight although more exact results are obtained via simulation using a previously developed a-Si TFT model [49, 50]. The composite circuit $g_m$ is obtained by equating the current through the AMP TFT operating in saturation and READ TFT operating in linear as shown in Eq. (2.42). This is done to derive $V_{DS} = V_B$ across the READ TFT, which then enables the determination of $R_{ON}$, $I_{OUT}$, and $g_m$.

$$(K_{AMP}/2)(V_{GA} - V_B - V_{TA})^2 = (K_{RD}/2)(2(V_{GRD} - V_{TRD})V_B - V_B^2) \tag{2.42}$$

Here, $V_{GA}$ is the AMP TFT gate voltage, $V_{TA}$ is the AMP TFT threshold voltage, $V_{GRD}$ is the READ TFT gate voltage, $V_{TRD}$ is the read TFT threshold voltage, and $R_{ON}$ is the ON resistance of the READ TFT. Solving and extracting the correct root for $V_B$, we obtain

$$V_B = \frac{\left[ K_{AMP}(V_{GA} - V_{TA}) + K_{RD}(V_{GRD} - V_{TRD}) - \left[ K_{RD} \left( \begin{array}{c} -2K_{AMP}V_{GA}(V_{GRD} - V_{TRD} + V_{TA}) \\ +2K_{RD}V_{GRD}V_{TRD} + K_{AMP}(V_{GA}^2 - V_{TA}^2) \\ -K_{RD}(V_{GRD}^2 + V_{TRD}^2) \\ +2K_{AMP}V_{TA}(V_{GRD} - V_{TRD}) \end{array} \right)^{1/2} \right] \right]}{(K_{RD} + K_{AMP})} \tag{2.43}$$

Given $V_B$, substituting back into the $I_{DS}$ equation for the AMP or READ TFT gives $I_{OUT}$ in the APS readout circuit branch. Then, dividing $V_B$ by $I_{OUT}$ gives the switch

**TABLE 2.3. Sample Values of APS Circuit Parameters for Different Supply Voltages $V_{GRD} = 15$ V, $W/L = 150/25$, $\mu_{EFF} = 0.17$ cm²/V·s**

|  | $V_{DD} = 6$ V | $V_{DD} = 8$ V | $V_{DD} = 12$ V |
|---|---|---|---|
| $V_B$ (V) | 0.48 | 1.0 | 2.43 |
| $I_{OUT}$ (µA) | 0.23 | 0.46 | 0.98 |
| $R_{ON}$ (MΩ) | 2.08 | 2.17 | 2.47 |
| $g_m$ (µs) | 0.067 | 0.096 | 0.145 |

resistance, $R_{ON}$, for the READ TFT. To quantify some results, Table 2.3 consists of sample values of APS circuit parameters at different bias values. Note the inherent feedback in the circuit that causes $V_B$ to increase for increasing supply voltage, $V_{DD}$, hence limiting the $V_{GS}$ across the AMP TFT, thereby reducing $I_{OUT}$ and $g_m$.

Particular care must be taken in designing the C-APS to ensure that bias component, $I_{OUT}$, does not saturate the column charge amplifier. For example, if the maximum output voltage that a charge amplifier, with a feedback capacitance of 5 pF, can hold is 15 V, then an $I_{OUT}$ of 2.5 µA for 30 µs will saturate the output. Since $g_m$ is related to $V_{DD}$, charge amplifier saturation for high $V_{DD}$ values limits the maximum achievable $g_m$. Using Eq. (2.36) and Eq. (2.41), the charge gain, $G_i$, stemming from the drain current modulation is

$$G_i = |\Delta Q_{OUT}/\Delta Q_P| = (\Delta I_{OUT} \cdot T_S)/\Delta Q_P$$
$$= (\Delta I_{OUT} \cdot T_S)/(\Delta V_G C_{PIX}) = (g_m T_S)/C_{PIX} \qquad (2.44)$$

The charge gain amplifies the input signal, making it resilient to external noise sources. The corresponding voltage gain can be calculated using Eq. (2.41) and assuming a constant $\Delta I_{OUT}$. Then, $\Delta V_{OUT}$ for the charge integrating circuit in Fig. 2.12 becomes

$$\Delta V_{OUT} = -\frac{1}{C_{FB}} \int_0^{T_S} \Delta I_{OUT}\, dt = \frac{\Delta I_{OUT} T_S}{C_{FB}} = \frac{(g_m \Delta V_G)T_S}{C_{FB}} \qquad (2.45)$$

Thus, the voltage gain, $A_V$, can be written as

$$A_V = \Delta V_{OUT}/\Delta V_G = (g_m T_S)/C_{FB} \qquad (2.46)$$

So for high charge and voltage gain, it is necessary to maximize $g_m T_S$. In contrast, decreasing $C_{PIX}$ is beneficial from both a signal gain and noise reduction point of view. However, a tradeoff exists for the traditional coplanar layout of TFTs and sensor where decreasing $C_{PIX}$ requires decreasing sensor area, which degrades pixel fill factor. For medical imaging applications using C-APS readout circuits,

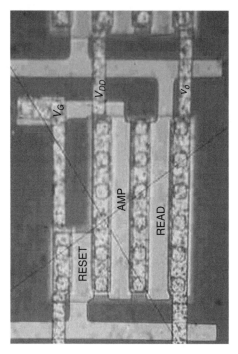

**Fig. 2.13.** Die micrograph of current mode APS pixel.

a small $C_{PIX}$ in addition to a high fill factor is required. Thus, continuous layer, thick a-Se sensors are preferable to achieve low-pixel-capacitance, high-fill-factor pixels.

The APS test pixel, consisting of an integrated a-Si amplifier circuit in a 250-$\mu$m $\times$ 250-$\mu$m pixel area, was fabricated in-house and is shown in Fig. 2.13. The small signal linearity is within 5% of the theoretical value and is shown in Fig. 2.14. In addition, small signal voltage gain ($A_V$) measurements were performed on the APS test circuit using the charge amplifier of Fig. 2.12 (a Burr–Brown IVC102 model), $V_{RD} = 20$ V, $V_G = V_{DD}$, and a $C_{FB} = 10$ pF for various READ pulse widths, $T_S$, and supply voltages, $V_{DD}$.

Theoretical voltage gain, $A_V$ ($A_V = \Delta V_{OUT}/\Delta V_G$) and experimental results agree reasonably well with a maximum discrepancy of about 10%. The verified theoretical model was extended to predict charge gain in Fig. 2.15 using Eq. (2.44) for different values of $C_{PIX}$, $C_{FB} = 10$ pF, and $A_V = 1.33$ (for $T_S = 30$ $\mu$s in Fig. 2.14). Theoretically, using a low sensor capacitance (i.e., small $C_{PIX}$) provides a higher charge gain, which minimizes the effect of external noise. In addition, minimizing $C_{PIX}$ also reduces the reset time constant (which comprises mainly of the RESET TFT on-resistance and $C_{PIX}$), leading to faster pixel resets and, hence, reducing image lag [51]. Like other current mode circuits, the C-APS, operating at a maximum of 30 kHz (for diagnostic fluoroscopy), can be susceptible to sampling clock jitter. However, the operation frequency is quite low and off-chip low-jitter clocks (e.g., using crystal oscillators) alleviate any concerns.

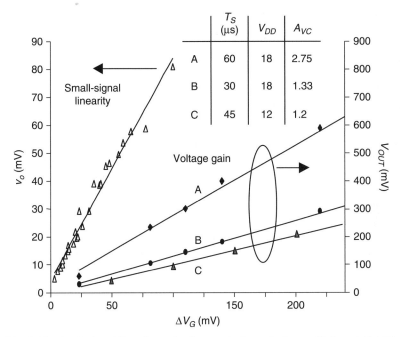

**Fig. 2.14.** Small-signal linearity and voltage gain for an in-house fabricated C-APS.

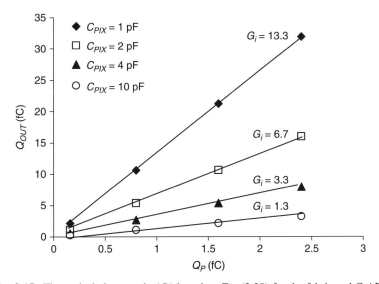

**Fig. 2.15.** Theoretical charge gain ($G_i$) based on Eq. (2.38) for the fabricated C-APS.

### 2.3.2. Application to Emerging Diagnostic Medical X-Ray Imaging Modalities

Recently reported sophisticated pixel amplifier readout circuits are summarized and illustrated in Fig. 2.16. In this section, we discuss the use of these multimode amplified pixel readout circuits based on the C-APS circuit described earlier to enable the emerging modalities of dual-mode R/F and 3D mammography tomosynthesis.

***2.3.2.1. Dual-Mode Radiography/Fluoroscopy (R/F).*** Increasing the signal-to-noise ratio (SNR) of AMFPIs can extend their application to low-dose, real-time fluoroscopy which is currently performed using bulky, small field-of-view (FOV), edge-distortion prone, CCD-based image intensifier units (10-cm × 10-cm FOV with 0.25-mm pixel pitch). More importantly, an imager capable of both chest radiography and real-time fluoroscopy would satisfy a radiologist's need to simultaneously obtain high-quality chest radiographs at a region of interest during a fluoroscopy examination. In order to apply flat-panel technology to real-time low-dose fluoroscopy applications, we start with the C-APS shown in Fig. 2.16c. The C-APS circuit has the potential to reduce the readout noise to levels that can meet the stringent requirements of X-ray fluoroscopy (i.e., <1000 input referred noise electrons).

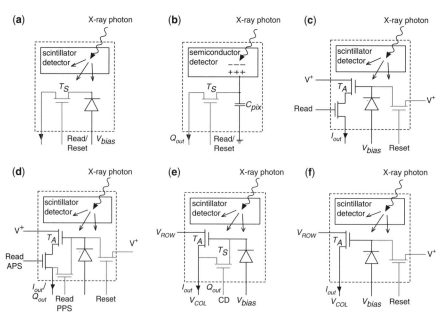

**Fig. 2.16.** Schematic diagram of AMFPI pixel architectures: (a) PPS architecture consisting of a TFT switch and pin photodiode with scintillator detector; (b) PPS architecture using photoconductor detector and direct detection method; (c) current-mediated, multimode active pixel sensor (C-APS) architecture; (d) multimode 4-TFT C-APS architecture; (e) multimode 2-TFT architecture with shared reset line; (f) alternate 2-TFT C-APS architecture.

A challenge with the C-APS circuit, however, is the presence of a small-signal input linearity constraint. While using such a pixel amplifier for real-time fluoroscopy (where the exposure level is small) is feasible, the voltage change at the amplifier input is much higher in chest radiography due to the larger X-ray exposure levels, which causes the C-APS output to be nonlinear. To combat this issue, we developed a hybrid amplified pixel architecture based on a combination of PPS and amplified pixel designs that, in addition to low-noise performance, also resulted in large-signal linearity and consequently higher dynamic range [52]. The additional benefit in large-signal linearity, however, came at the cost of an extra pixel transistor, RDP, as shown in Fig. 2.16d. Later, we optimized the four-transistor (4T) design to into a 3T multimode pixel design that can fit within a 0.15-mm pixel pitch [53]. The readout circuit is similar to the circuit shown in Fig. 2.16c and achieves the goals of low-noise performance and large-signal linearity without the need for an additional pixel transistor.

***2.3.2.2. 3D Mammography Tomosynthesis.*** AMFPIs have also gained traction in the mammography imager market. Mammography is a 2D projection imaging modality that depicts normal and pathological tissue structures within the breast. Mammography is the accepted gold standard for breast cancer detection; nevertheless, it suffers from several drawbacks that limit its sensitivity and specificity. In the dense breast for example, mammographic image quality is often limited by overlapping shadows of tissue structures which can obscure lesions (false negatives) or mimic a suspicious mass (false positive). This is referred to as anatomic noise—that is, signal due to structures that are not diagnostically important but which create a cluttered background. Realization that a significant percentage of breast cancers are not detected by screening mammography has prompted recent interest in 3D breast imaging or tomosynthesis. Initial studies show that tomosynthesis has the ability to reveal 16% more cancers than conventional mammography and reduce false positives by 85% [54] even at a radiation dose to the patient comparable to a conventional two-view mammogram. In order not to increase patient dose, the imaging system has to be able to read out multiple (ideally $\sim$100) images with each image using only 1% of the normal dose. However, this puts an extreme requirement on the detector with regard to the parasitic amplifier noise level.

Almost all mammographic detectors currently available are amplifier noise limited in some part of their operating range (i.e., dark part of image) even at normal doses, resulting in additional radiation necessary to make up for the deficiency. Increasing the number of readouts makes this problem worse. Overcoming amplifier noise in the subdivided images is therefore the key problem to implementing a practical tomosynthesis system. Tomosynthesis for cancer detection is a high-contrast imaging task necessitating (a) rapid image acquisition to prevent image blurring due to gantry angle change, (b) large-area (18 × 24 cm) and high-resolution (0.075 mm in-plane) imaging for accurate measurement of lesion/calcification size, and (c) patient positioning flexibility to facilitate detection.

The high-resolution 2T multimode amplified pixels (shown in Fig. 2.16e,f) are promising candidates for mammography tomosynthesis. Of keen interest is the use

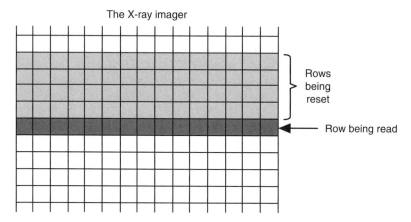

**Fig. 2.17.** Multiple rows reset simultaneously.

of only two TFTs to provide similar functionality to the 3T and 4T designs. The 2T multimode pixel architecture achieves amplification along with high resolution by either trading off higher complexity in gate driver circuitry for lower pixel complexity or making use of a new charge gated TFT device architecture. Full operating details of the various 2T pixel amplifier designs are reported elsewhere [55, 56].

An additional challenge posed by mammography tomosynthesis is that of high frame rate (ideally up to 100 frames per second). The readout speed of AMFPI arrays based on a PPS architecture is $RC$ time constant limited, with R representing the "ON" resistance of the a-Si TFT switch (on the order of mega ohms) and $C$ representing the detector capacitance. The $RC$ time constant is usually minimized by reducing $R$ and $C$, although reducing $C$ also reduces the maximum charge holding capacity of the imaging pixel. An alternate approach to higher frame rates is via pixel binning— that is, grouping neighboring pixels together to reduce the total number of pixels to read. Although pixel binning increases the frame rate by decreasing the time it takes to read out imager data, it also degrades image resolution.

To illustrate how C-APS-type architectures offer higher-speed readout to enable mammography tomosynthesis, consider the pixel architecture shown in Fig. 2.16c. The amount of $RC$ delay required to operate a PPS arises from the $RC$ time constant of the readout switch ON resistance and the detector capacitance. For a C-APS pixel, a similar $RC$ time constant exists for the reset switch ON resistance and the detector capacitance. However, unlike a PPS, the reset and read cycles are decoupled in a C-APS circuit. Then, instead of performing a typical C-APS reset cycle after each read cycle, the reset cycle is delayed and can be grouped together for multiple row reset as shown in Fig. 2.17.

Grouping the row reset cycles reduces the reset time proportionally and makes the frame time dependent on the pixel row readout cycle instead, where the readout time is directly related to the charge gain for C-APS circuits. Here, decreasing the pixel readout time proportionally reduces charge gain. Thus, charge gain can be traded off for faster frame rates where multimode pixel amplifier circuits provide utility by providing high resolution, charge gain, and faster frame rates to enable dose reduction.

In summary, the current mode on-pixel a-Si amplifier pixel readout circuits reported to date exhibit fast readout and higher immunity to external noise sources associated with the charge amplifier and data line in an X-ray imaging array. The evolution of the pixel amplifier from a four-, then three-, and now to a two-transistor design is presented where the two-transistor design shows promise for the emerging diagnostic X-ray imaging modalities.

## REFERENCES

1. S. Tao, K. S. Karim, P. Servati, and A. Nathan, Large area digital X-ray imaging, in *Sensors, Update*, 10: Sensors Technology–Applications–Markets. H. Baltes, G. K. Fedder, and G. Korvink, eds., Wiley-VCH Verlag GmbH, 2002.

2. P. G. LeComber, W. E. Spear, and A. Ghaith, Amorphous silicon field effect device and possible application, *Electron. Lett.* **15**, 179–181, 1979.

3. R. L. Weisfield, Amorphous silicon linear-array device technology: Applications in electronic copying, *IEEE Trans. Electron Dev.* **36**, 2935–2939, 1989.

4. R. A. Street, Large area electronics, applications and requirements, *Phys. Status Solidi A* **166**(2), 695–705, 1998.

5. *Handbook of Medical Imaging*, **1**, *Physics and Psychophysics*, J. Beutel, H. L. Kundel, and R. L. Van Metter, eds., SPIE, 2000, www.spie.org/bookstore/.

6. R. L. Weisfield, Amorphous silicon TFT X-ray image sensors, *IEEE Int. Electron Dev. Meeting (IEDM 1998) Technical Dig.*, pp. 21–24, 1998.

7. W. Zhao and J. A. Rowlands, Digital radiology using active matrix readout of amorphous selenium: Theoretical analysis of detective quantum efficiency, *Med. Phys.* **24**(12), 1819–1833, 1997.

8. I. Fujieda, R. A. Street, R. L. Weisfield, S. Nelson, P. Nylen, V. Perez-Mendez, and G. Cho, High sensitivity readout of 2D a-Si image sensors, *Japan J. Appl. Phys.* **32**, 198–204, 1993.

9. A. Miri, Development of a Novel Wet Etch Fabrication Technology for Amorphous Silicon Thin-Film Transistors, Ph.D. thesis, University of Waterloo, 1996.

10. R. V. R. Murthy, P. Servati, A. Nathan, and S. G. Chamberlain, Optimization of $n^+\mu$c-Si: H contact layer for low leakage current in a-Si thin film transistors, *J. Vac. Sci. Technol. B* **18**(2), 685, 2000.

11. R. A. Street, X. D. Wu, R. Weisfield, S. Ready, R. Apte, M. Nguyen, and P. Nylen, Two dimensional amorphous silicon image sensor arrays, *MRS Symp. Proc.* **377**, 757–766, 1995.

12. B. Park, K. S. Karim, and A. Nathan, Intrinsic thin film stresses in multi-layered imaging pixels, *J. Vac. Sci. Technol. A* **18**(2), 688–692, Mar 2000.

13. P. Servati, K. S. Karim, and A. Nathan, Static characteristics of a-Si:H dual-gate TFTs, *IEEE Trans. Electron Dev.* **50**(4), 926–932, 2003.

14. M. J. Powell, Charge trapping instabilities in amorphous silicon–silicon nitride thin-film transistors, *Appl. Phys. Lett.* **43**(6), 15, 1983.

15. W. B. Jackson and M. D. Moyer, Creation of near interface defects in hydrogenated amorphous silicon–silicon nitride heterojunctions: The role of hydrogen, *Phys. Rev. B* **36**, 6217, 1987.

16. F. R. Libsch and J. Kanicki, Bias-stress-induced stretched-exponential time dependence of charge injection and trapping in amorphous silicon thin-film transistors, *Appl. Phys. Lett.* **62**, 1286–1288, 1993.

17. F. R. Libsch and J. Kanicki, Bias-stress-induced stretched-exponential time dependence of charge injection and trapping in amorphous silicon thin-film transistors, *Extended Abstracts of the 1992 International Conference on Solid State Devices and Materials*, pp. 155–157, 1992.

18. F. R. Libsch, Steady state and pulsed bias stress induced degradation in amorphous silicon thin-film transistors for active-matrix liquid-crystal displays, *IEDM 1992 Technical Dig.*, pp. 681–684, 1992.

19. N. Lustig and J. Kanicki, Gate dielectric and contact effects in hydrogenated amorphous silicon-silicon nitride thin-film transistors, *J. Appl. Phys.* **65**, 3951–3957, 1988.

20. R. A. Street and C. C. Tsai, Fast and slow states at the interface of amorphous silicon and silicon nitride, *Appl. Phys. Lett.* **48**, 1672–1674, 1986.

21. N. Nickel, R. Saleh, W. Fuhs, and H. Mell, Study of instabilities in amorphous silicon thin-film transistors by transient current spectroscopy, *IEEE Trans. Electron Dev.* **37**, 280–283, 1990.

22. M. J. Powell, C. van Berkel, A. R. Franklin, S. C. Deane, and W. I. Milne, Defect pool model in amorphous silicon thin-film transistors, *Phys. Rev. B* **45**, 4160–4170, 1992.

23. M. Stutzmann, Weak bond-dangling bond conversion in defects in amorphous silicon, *Philos. Mag.* **56**, 63–70, 1987.

24. Z. E. Smith and S. Wagner, Band tails, entropy, and equilibrium defects in hydrogenated amorphous silicon, *Phys. Rev. Lett.* **59**, 688–691, 1987.

25. D. L. Staebler and C. R. Wronski, Reversible conductivity changes in discharge-produced amorphous Si, *Appl. Phys. Lett.* **31**, 292–294, 1977.

26. M. Stutzmann, W. B. Jackson, and C. C. Tsai, Light-induced metastable defects in hydrogenated amorphous silicon, *Phys. Rev. B* **32**, 23–47, 1985.

27. M. J. Powell et al., A defect-pool for near-interface states in amorphous silicon thin-film transistors, *Philos. Mag.* **63**, 325–336, 1991.

28. C. van Berkel and M. J. Powell, The resolution of amorphous silicon thin film transistor instability mechanisms using ambipolar transistors, *Appl. Phys. Lett.* **51**, 1094, 1987.

29. M. J. Powell, C. van Berkel, I. D. French, and D. H. Nicholls, Bias dependence of instability mechanisms in amorphous silicon thin film transistors, *Appl. Phys. Lett.* **51**, 1242, 1987.

30. R. H. Walden, A method for the determination of high field conduction laws in insulating films in the presence of charge trapping, *J. Appl. Phys.* **43**, 1178, 1972.

31. M. J. Powell, The physics of amorphous silicon thin film transistors, *IEEE Trans. Electron Devices* **36**, 2753, 1989.

32. M. J. Powell, C. van Berkel, and J. R. Hughes, Time and temperature dependence of instability mechanisms in amorphous silicon thin-film transistors, *Appl. Phys. Lett.* **54**(14), 1323, 1989.

33. C. Chiang, J. Kanicki, and K. Takechi, Electrical instability of hydrogenated amorphous silicon thin-film transistors for active-matrix liquid-crystal displays, *Japan J. Appl. Phys.* **37**(1), Part A, 4704–4710, 1998.

34. R. Oritsuki, T. Horii, A. Sasano, K. Tsutsui, T. Koizumi, Y. Kaneko, and T. Tsukada, Threshold voltage shift of amorphous silicon thin-film transistors during pulse operation, *Jpn. J. Appl. Phys.* **30**(12B), Part 1, 3719–3723, 1991.

35. C. Huang, T. Teng, J. Tsai, and H. Cheng, The instability mechanisms of hydrogenated amorphous silicon thin film transistors under AC bias stress, *Jpn. J. Appl. Phys.* **39**, Part 1, No. 7A, 3867–3871, July 2000.

36. A. van der Ziel, *Noise: Sources, Characterization, Measurement*, Prentice Hall, Englewood Cliffs, NJ, 1970.

37. J. M. Boudry and L. E. Antonuk, Current-noise-power spectra of amorphous silicon thin-film transistors, *J. Appl. Phys.* **76**, 2529–2534, 1994.

38. K. S. Karim, A. Nathan, and J. A. Rowlands, Feasibility of current mediated amorphous silicon active pixel sensor readout circuits for large area diagnostic medical imaging, in *Opto-Canada: SPIE Regional Meeting on Optoelectronics, Photonics and Imaging*, SPIE **TD01**, May 2002, pp. 358–360.

39. J. B. Johnson, The Schottky effect in low frequency circuits, *Phys. Rev.* **26**, 71–85, 1925.

40. A. L. McWhorter, $1/f$ noise and germanium surface properties, *Semiconductor Surface Physics*, University of Pennsylvania Press, Philadelphia, pp. 207–228, 1956.

41. F. N. Hooge and A. Hoppenbrouwers, Amplitude distribution of $1/f$ noise, *Physica* **42**, 331–339, 1969.

42. F. N. Hooge and L. K. J. Vandamme, Lattice scattering causes $1/f$ noise, *Phys. Lett.* **66A**, 316–326, 1978.

43. J. Rhayem, D. Rigaud, M. Valenza, N. Szydlo, and H. Lebrun, $1/f$ noise in amorphous silicon thin-film transistors: Effect of scaling down, *Solid State Elec.* **43**, 713–721, 1999.

44. J. Rhayem, M. Valenza, D. Rigaud, N. Szydlo, and H. Lebrun, $1/f$ noise investigations in small channel length amorphous silicon thin-film transistors, *J. Appl. Phys.* **83**(7), 3660–3667, 1998.

45. J. Rhayem, D. Rigaud, M. Valenza, N. Szydlo, and H. Lebrun, $1/f$ noise investigations in long channel length amorphous silicon thin-film transistors, *J. Appl. Phys.* **87**(4), 1983–1989, 2000.

46. G. Cho, J. S. Drewery, I. Fujieda, T. Jing, S. N. Kaplan, V. Perez-Mendez, S. Qureshi, D. Wildermuth, and R. A. Street, Measurements of $1/f$ noise in a-Si pin diodes and thin-film transistors, *MRS Symp. Proc.* **192**, 393–398, 1990.

47. X. Y. Chen, M. J. Deen, A. D. van Rheenen, C. X. Peng, and A. Nathan, Low-frequency noise in thin active layer a-Si thin-film transistors, *J. Appl. Phys.* **85**(11), 7952–7957, 1999.

48. S. K. Mendis, S. E. Kemeny, and E. R. Fossum, CMOS active pixel image sensor, *IEEE Trans. Electron Dev.* **41**(3), 452–453, 1994.

49. P. Servati, Modeling of Static and Dynamic Characteristics of a-Si TFTs, M.A.Sc. Thesis, University of Waterloo, 2000.

50. K. S. Karim, P. Servati, N. Mohan, A. Nathan, and J. A. Rowlands, VHDL-AMS modeling and simulation of a passive pixel sensor in a-Si technology for medical imaging, in

*Proceedings of the IEEE, International Symposium on Circuits and Systems 2001*, Sydney, Australia, **5**, May 6–9, 2001, pp. 479–482.

51. H. Tian, B. Fowler, and A. El Gamal, Analysis of temporal noise in CMOS photodiode active pixel sensor, *IEEE J. Solid-State Circuits* **36**(1), 92–101, 2001.

52. K. S. Karim, G. Sanaie, T. Ottaviani, M. H. Izadi, and F. Taghibakhsh, Amplified pixel architectures in amorphous silicon technology for large area digital imaging applications, *J. Korean Phys. Soc.* **48**(1), 85–91, 2006.

53. M. H. Izadi and K. S. Karim, High dynamic range pixel architecture for advanced diagnostic medical X-ray imaging applications, *J. Vac. Sci. Technol. A* **24**(3), 846–849, 2006.

54. E. Rafferty, Tomosynthesis: New weapon in breast cancer fight, Guest editorial in *Decisions in Imaging Economics, The Journal of Imaging Technology Management*, April 2004.

55. F. Taghibakhsh and K. S. Karim, Two transistor active pixel sensor for high resolution large area digital X-ray imaging, *IEEE International Electron Devices Meeting (IEDM) Technical Digest*, Dec 2007.

56. F. Taghibakhsh and K. S. Karim, Two-transistor active pixel sensor readout circuits in amorphous silicon technology for high resolution digital imaging applications, *IEEE Trans. Electron Dev.* **55**(8), 2121–2128, 2008.

# 3 Circuits for Digital X-Ray Imaging: Counting and Integration

EDGAR KRAFT and IVAN PERIC

## 3.1. INTRODUCTION

This chapter discusses the results of a project focusing on the exploration and realization of a new signal processing concept for medical X-ray imaging with direct conversion semiconductor sensors.

### 3.1.1. Image Formation

Medical X-ray imaging records the transmission image of a photon beam passing through the patient. The X-ray tube spectrum is generally polychromatic with photon energies typically ranging from a few kiloelectron volts up to a maximum value determined by the acceleration voltage, usually less than 150 keV. In this energy range, the interaction of the photons with matter is governed by two processes, Compton scattering and photo effect. Compton scatting is the inelastic scattering process between an X-ray photon and a free electron. Electrons bound in an atom can be considered free for photon energies much larger than the binding energy. The amount of energy transferred from the photon to the electron depends on the deflection angle. The photo effect is the absorption of a photon by an electron in the atomic shell. Since the energy of the photon is fully transferred to the electron, the electron is emitted from its atom with a remaining kinetic energy. This energy is equal to the difference between the photon energy and the binding energy of the electron.

It is interesting to note that the spectrum and intensity of the transmitted X-ray beam carry information about both the thickness and the composition of the material in its path. This information arises from the different contributions of the two absorption mechanisms to the total attenuation. The amount of absorption due to photo effect depends strongly on the atomic number, thus providing an indication of the composition of the investigated tissue. The intensity of Compton scattering, on the other

*Medical Imaging: Principles, Detectors, and Electronics*, edited by Krzysztof Iniewski
Copyright © 2009 John Wiley & Sons, Inc.

hand, depends on the electron density, which is proportional to the mass density for most materials [1]. One of the key benefits of the signal procession scheme discussed in this chapter is that it simultaneously acquires information about the beam intensity and its spectral composition, namely the mean photon energy.

### 3.1.2. X-Ray Detectors

The transmitted beam is detected through its absorption in the X-ray sensor. Modern digital X-ray (flat-panel) detectors can be divided into two groups: direct and indirect detectors [2].

***3.1.2.1. Indirect Detectors.*** Indirect detectors consist of a layer of scintillator material (such as gadolinium oxysulfide or cesium iodide) which is coupled to a large-scale array of active picture elements. Each pixel comprises a switching transistor and a photodiode that produces a charge proportional to the intensity of the light generated in the scintillator.

***3.1.2.2. Direct Detectors.*** Direct conversion detectors use a photoconductor, commonly amorphous selenium (a-Se), instead of the scintillator layer [3]. The photoconductor converts incoming X-ray photons directly to electron–hole pairs. In each pixel, the sensing instrument is often a simple charge storage capacitor and a collection electrode, which replace the photodiode of the indirect approach. The backside of the selenium layer is coated with a continuous electrode, which is used to apply the bias voltage (typically several kilovolts) necessary for the collection of the produced electron–hole pairs. Readout of the pixel array is performed row-wise by connecting the respective picture elements to the readout lines and measuring the signal with charge-sensitive amplifiers.

The main appeal of direct converting systems compared to the indirect approach lies in the high intrinsic spatial resolution and in the larger number of produced electron–hole pairs per X-ray photon. Both advantages are achieved through the avoidance of an additional conversion step. However, these benefits are only present to a limited extent for amorphous selenium as a conversion material: Compared to other direct conversion materials like cadmium telluride (CdTe), the number of electrons produced by an X-ray photon is smaller by a factor of $5-10$, depending on the applied electric field. Unfortunately, such materials offer only a very limited set of processing options, which prevents the implementation of elaborate pixel electronics on the sensor.

***3.1.2.3. Hybrid Pixel Detectors.*** Hybrid pixel detectors offer a solution to this problem by separating sensor and readout chip. In every pixel, the sensor electrode is connected to the input of its corresponding channel on the readout chip through a bump-bond connection. In contrast to the flat-panel technology discussed above, the readout chip is usually fabricated from crystalline silicon. This permits us to use industry standard processing technologies that provide a high level of electronics integration and allow more advanced signal processing circuits inside the pixels.

The separation of both parts removes the necessity to choose a sensor material in which active elements such as switches and amplifiers can be implemented. Instead, the sensor can be made from a wide range of materials and the same readout chip can be used with different sensor types. Sensor types used successfully in hybrid pixel detectors include Si, GaAs, CdTe, CdZnTe, diamond, and 3D-silicon detectors [4–6].

***3.1.2.4. Readout Concepts for Hybrid Pixel Detectors.*** Readout chips for hybrid pixel detectors used in medical imaging can usually be categorized as either integrating or photon counting systems.

Integrating systems accumulate the X-ray signal over a certain time interval and provide a measurement that is usually proportional to the total amount of energy deposited by the incident radiation. The flat-panel detector concepts discussed above are examples of integrating systems. The designs are well-suited for large rates and signal currents, but measuring small signals can be difficult due to the intrinsic electronic noise of the readout channel. Photon counting systems, on the other hand, measure the number of absorbed photons whose deposited energy exceeds a certain threshold. The lowest measurable flux is therefore a single photon per measurement interval. With rising photon flux, subsequent pulses pile up and it becomes increasingly difficult to distinguish individual charge pulses. Consequently, the number of unregistered events increases up to the point where no counts are registered at all. Unless multiple energy thresholds are used, measurements of the photon count rate do not yield any spectral information besides the minimal energy determined by the threshold.

The concept of simultaneous counting and integrating overcomes the limitations of the individual schemes by measuring both the absorbed photon flux and the deposited energy. This combination not only extends the dynamic range beyond the limits of the respective concepts, but also yields additional spectral information in terms of mean photon energy in the region where the operating ranges of counter and integrator overlap. Medical X-ray imaging applications can benefit from this concept through improved contrast and the ability to determine the hardening of the tube spectrum due to attenuation in the imaged object.

## 3.2. CIRCUIT IMPLEMENTATION

The results discussed later in this chapter were obtained with a prototype chip with an $8 \times 8$ pixel matrix, fabricated in AMS 0.35-$\mu$m technology. The various test circuits on the prototype required pixel dimensions of 500 $\mu$m $\times$ 250 $\mu$m. On a full-scale imaging chip, final dimensions are expected to be 300 $\mu$m $\times$ 300 $\mu$m. Results with a previous prototype generation are presented in reference [7].

A pixel providing counting and integrating X-ray imaging (CIX) contains three basic elements: a photon counter, an integrator, and a special feedback circuit that provides both signal shaping for the photon counter and signal replication for the integrator.

### 3.2.1. The Photon Counter

The signal processing chain of the photon counting channel (Fig. 3.1) consists of a charge-sensitive amplifier (preamplifier) with a 10-fF feedback capacitor, a two-stage comparator with differential output, and a 16-bit ripple counter. Incoming charge accumulates on the feedback capacitor until it is removed by the feedback circuit, which is basically a differential pair acting as a voltage-controlled current source. This circuit provides a continuous reset of the amplifier. In the absence of an input signal, the output voltage of the preamplifier settles to $V_{CountBaseline}$. A negative signal of charge $Q$ on the input node raises the output voltage by $\Delta V$ (ignoring effects of finite charge collection time and ballistic deficit for now):

$$\Delta V = Q/C_{Fb}$$

The increased output voltage activates the feedback current source, which delivers a current $I_{Fb}$ to the input node. This current compensates the signal charge, thereby decreasing the output voltage. Full compensation of the signal charge is achieved when the output voltage reaches $V_{CountBaseline}$, thereby also turning off the feedback current source. Assuming that the delivered current is approximately constant, the time needed for the return to baseline $t_{rtb}$ is given by

$$t_{rtb} = Q/I_{Fb}$$

The resulting pulse shape on the amplifier output is approximately triangular, with a falling slope determined by $I_{Fb}$. Contrary to the behavior of a CR–RC shaper, the return-to-baseline time of this feedback circuit scheme is not independent of the pulse size, but approximately proportional to it. Since the differential pair in the feedback circuit delivers smaller currents when the output voltage approaches the baseline, the actual time interval is somewhat longer. The comparator stage switches to a logical

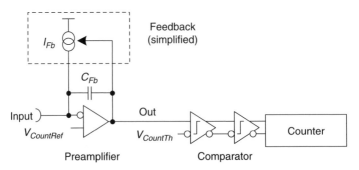

**Fig. 3.1.** Schematic of the photon counter. Absorption of a photon in the conversion layer causes a negative charge pulse at the input node. The electrons accumulate on the feedback capacitor $C_{Fb}$ until they are removed by the feedback circuit (continuous reset). A two-stage comparator triggers if the rise in the preamplifier output potential exceeds the threshold $V_{CountTh}$, thereby incrementing the 16-bit ripple counter.

high state while the preamplifier output exceeds the threshold voltage $V_{CountTh}$. Each positive transition triggers a single count event in the connected photon counter. Provided that the interval between subsequent photons is long enough for the preamplifier output to fully return to its baseline ($t > t_{rtb}$), a photon is counted if its deposited energy exceeds the threshold $E_{Th}$:

$$E_{Th} = w_i \cdot C_{Fb} \cdot (V_{CountTh} - V_{CountBaseline})/q_e$$

Here, $q_e$ is the elementary charge of an electron and $w_i$ is the *intrinsic-pair creation energy*, a material property that describes the average energy necessary to produce one electron–hole pair in the conversion layer. In CdTe, a typical value [8] is $w_i =$ 4.64 eV. A 60-keV photon will hence deposit about 2 fC. This value is thus often used as an input pulse size for the characterization.

### 3.2.2. The Integrator

The integrator implementation shown in Fig. 3.2 is similar to the *sigma–delta converter* concept, which is often used in high-precision, low-frequency measurement applications [9]. Recent results with a chip based on a similar implementation are discussed in reference 10. The first stage of the integrator signal processing chain is an amplifier-comparator stage similar to the one found in the single photon counter. One difference of the photon counter is the clock-synchronized operation of the feedback circuit. It uses a charge pump to remove a charge packet of defined size $Q_{pkt}$ from the integrator input each time the accumulated charge on the feedback capacitor exceeds a threshold given by $V_{IntTh}$. In a somewhat simplified sense, this type of feedback converts a continuous input current to a frequency of pump actions (see Fig. 3.3).

Two counters record the number of charge packets $N_{pkt}$ and the elapsed time $\Delta t$. The time interval is computed from the number of clock cycles $N_t$ between the first

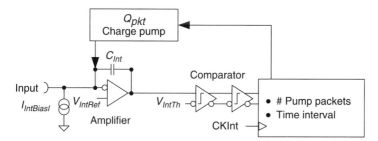

**Fig. 3.2.** Simplified schematic of the integrator circuit. The input signal is integrated on the $C_{Int}$ capacitor until the amplifier output voltage exceeds a certain threshold $V_{IntTh}$. This triggers the clock synchronous operation of the charge pump, which removes a charge packet of defined size $Q_{pkt}$ from the integrator input. Counters record the number of charge packets and the time interval between first and last pump event. The injection of an additional bias current $I_{IntBiasl}$ allows measuring smaller input signals.

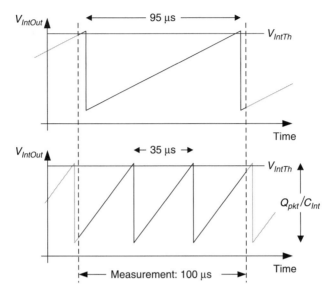

**Fig. 3.3.** Illustration of the integrator operation for two different input currents: The signal charge is accumulated on the integration capacitor $C_{Int}$ until the output voltage $V_{IntOut}$ exceeds a certain threshold $V_{IntTh}$. This triggers a charge pump that removes a charge packet of defined size $Q_{pkt}$ on the next clock cycle. Even though both input currents happen to produce only a single packet within the measurement duration (here: 100 μs), they can be distinguished by the difference in the recorded time interval.

and the last pump action in the measurement cycle. The measurement of the current $I_{Signal}$ is then given by

$$I_{Signal} = \frac{N_{pkt} \cdot Q_{pkt}}{\Delta t} = \frac{N_{pkt} \cdot Q_{pkt} \cdot f_{CK}}{N_t}$$

Here, $f_{CK}$ is the integrator clock frequency. As will be explained below, a convenient property of this method of current measurement is that the *absolute* discretization error decreases as the input signal gets smaller, giving rise to a nearly constant relative resolution throughout the full dynamic range. Common analog-to-digital converters with a constant bin size do not possess this property due to the inherently large *relative* discretization errors at small values.

Since the charge on the feedback capacitor is generally not fixed, the same number of charge packets can occur at different currents (and vice versa). Large currents, however, will trigger the charge pump in shorter time intervals. This is illustrated in Fig. 3.3.

The dynamic range of the integrator is determined by the charge packet size, the clock frequency, and the measurement duration $t_{meas}$. Small signals can be measured if they produce at least one charge packet (i.e., two pump events) within the measurement interval.

$$I_{min} = \frac{Q_{pkt}}{t_{meas}}$$

Smaller currents are certain to fail with regard to producing a valid measurement. However, *twice* this minimal current is needed to ensure a proper result under all circumstances. The reason for this is that the initial charge on the integration capacitor at the commencement of the measurement is only defined to within one pump packet. For illustration, consider the case of a pump event that is triggered shortly before the start of the measurement. In this case the current $I_{min}$ will cause only one pump event near the end of the measurement interval and cause will thus fail to produce a valid measurement. A twice as large current, on the other hand, will cause the first event before half the measurement duration and cause the second event before the end. It will thus meet the minimal requirements independent of the initial charge on the integration capacitor.

The largest measurable signal is the current that causes a pump action on every clock cycle. With $N_{max}$ being the number of clock cycles within the measurement interval, the maximum current is therefore approximately an $N_{max}$-fold multiple of the minimal current:

$$I_{max} = \frac{N_{max} \cdot Q_{pkt}}{t_{meas}} = N_{max} \cdot I_{min}$$

The discretization precision is determined by $N_{max}$ as well. A detailed explanation is given in reference 11. For example, at a clock rate of 20 MHz and a measurement duration of 100 μs, the discretization precision computes to

$$\frac{1}{N_{max}} = \frac{1}{2000} = 0.0005 \,\hat{=}\, 10.97 \text{ bits}$$

This means that a current is measured with a relative precision of about 0.05%. This does not, however, imply a statement about the *accuracy* of the measurement. In this example, the dynamic range extends over approximately log $(2000) = 3.2$ decades. A 2-ms-long measurement, on the other hand, yields a discretization precision of about 15.3 bits and a theoretical dynamic range covering 4.6 orders of magnitude. The absolute values of $I_{min}$ and $I_{max}$ are determined by the choice of the packet size $Q_{pkt}$.

A common means to extend the lower limit to even smaller currents is the introduction of an additional bias current $I_{IntBiasI}$. This small current is fed directly into the integrator input and is ideally just large enough to ensure two pump actions even when no additional signal current is present. Offline calibration subtracts the bias current from the integrator measurement in order to yield the correct signal current. As a result of the usage of this bias source, the minimal current is in principle only limited by the discretization precision of the bias current measurement. The electronic noise inherent to real measurements, however, becomes increasingly dominant for such small input currents. A meaningful number for the minimal measurable current can thus only be stated for a given signal-to-noise-ratio.

### 3.2.3. The Feedback Circuit

A simplified diagram of the feedback circuit is shown in Fig. 3.4. The circuit's main purposes are signal shaping for the photon counter, signal replication for the integrator, and leakage current compensation. The key elements of the feedback circuit are the two differential pairs. Pair 1 provides the feedback for the photon counting amplifier and signal replication for the integrator. Pair 2 is responsible for leakage current compensation. Both differential pairs share the same basic behavior: The two current drains at the bottom of each branch drain precisely half the current entering from the current source above. If the gate potentials of both PMOS transistors match, the pair is balanced and no current will flow into or out of the nodes between the transistors and the current drains. A mismatch between the gate voltages, however, shifts the current from one branch into the other, causing some additional current in one branch and missing current in the other. This additional/missing current must leave/enter the branch through the node above the respective current drain.

***3.2.3.1. Feedback and Signal Duplication.*** The operation of the feedback circuit can be understood by once again following the chain of events caused by a negative charge pulse $Q$ arriving at the input node from the sensor. Let us assume that both differential pairs are balanced in the beginning. This implies that the preamplifier output voltage (Out) equals $V_{CountBaseline}$ and there is no current entering or leaving the differential pair through the nodes above the current drains. Since the input node is connected to the inverting input of the amplifier, the arrival of the signal charge $Q$ causes an increase in the amplifier output voltage by $Q/C_F$, so that the input voltage remains constant at $V_{CountRef}$. This voltage increase is seen on the left branch of the first differential pair and shifts some feedback current from

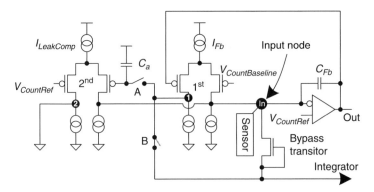

**Fig. 3.4.** Simplified schematic of the feedback circuit in static leakage current compensation mode (switch A open, switch B closed). The first differential pair provides feedback for the photon counting amplifier and signal duplication for the integrator, while the second pair delivers a static current, thereby compensating the sensor leakage current. The leakage current is sampled in absence of a real signal by opening switch B and closing switch A. This adjusts the voltage on $C_a$ until the compensation current (from the second differential pair) matches the leakage current to the input node.

the left branch into the right. The additional current in the right branch flows into the input node, thereby canceling the original charge pulse. During the cancelation process, the output voltage of the amplifier decreases until it reaches $V_{CountBaseline}$. At that point, the first differential pair is balanced again and no further current flows into the input node. This is how the first differential pair provides the feedback (*continuous reset*) for the amplifier of the photon counter.

Note that the imbalance in the differential pair causes not only a current flowing out of the right branch into the amplifier input node, but also an identical current *entering* the left branch from node 1. The integral over both current pulses is identical in size (both matching the original input pulse), but of opposite sign. Hence, if the integrator is connected to node 1 (switch B closed, switch A open), it receives a charge pulse of equal size and sign as the input pulse. This is how the first differential pair provides signal duplication.

Between the input nodes of the preamplifier and the integrator, there is also a diode connected NMOS transistor (*bypass*), seen in the bottom right of Fig. 3.4. This transistor becomes conductive if the input node potential drops more than a threshold voltage below the integrator input voltage (about 1.2 V), a situation that can occur only if the sensor signal becomes too large to be compensated by the feedback current. Hence any current exceeding the limits of the feedback bypasses the feedback via the diode connected transistor, allowing the integrator a proper measurement of the input signal.

***3.2.3.2. Static Leakage Current Compensation.*** Biased semiconductor sensors usually exhibit some degree of leakage current flowing into the readout electronics even when no real signal is present. This current can cause shifts in the output voltage baseline of the amplifier and decrease the dynamic range. It is therefore desirable to compensate the leakage current *prior* to the signal processing. In our feedback design, this is done by the second differential pair. Note that the right output node of this pair is connected to the preamplifier input node. The current delivered to the input node corresponds to an imbalance in the second pair. The magnitude of this current is determined by the voltage on the *sampling capacitor* $C_a$. Provided that switch A is open, the voltage will remain constant, thereby freezing the current to the input node to a constant value (which is the reason why this mode is called *static* leakage current compensation). During a separate sampling phase (explained below), this voltage is adjusted so that the current compensates the leakage current. The current in the left branch of the second differential pair is simply drained to ground. More details on alternative feedback configurations can be found in references 7 and 11.

***3.2.3.3. Sampling.*** The sampling of the feedback current is performed in the absence of real signals, when all current flowing to the sensor is due to leakage current. Closing the sampling switch A and disconnecting the integrator (switch B) leaves node 1 connected only to the right gate of the second differential pair and the sampling capacitor $C_a$. There can thus be no DC current flowing into or out of the first differential pair. This implies that the leakage current can only be compensated by the current from the right branch of the second differential pair (which is connected to the input node). Through this mechanism, sampling adjusts the voltage on $C_a$ until the leakage current is matched by the current leaving the right output node of the second differential pair. The current on the left branch of the second differential pair (node 2) is again ignored and simply drained to ground.

For illustration purposes, one can follow the chain of events triggered by a sudden increase in the sensor leakage current. At first, this increase is compensated by the first differential pair, shifting current from the left branch to the right. This current decreases the potential on the sampling capacitor $C_a$. As a result, the second pair delivers more current to the input node. The first pair reacts by delivering slightly less current. This slows down the voltage decrease on $C_a$ until it reaches an equilibrium. At that point, the first differential pair is balanced again and the second pair has adjusted to the new leakage current, following the adjustment of the voltage on $C_a$. Once stable conditions are reached, the sampling switch A can be opened, thus storing the voltage on the sampling capacitor and consequently setting the compensation current delivered to the input node to a constant value. For stability reasons, $I_{Fb}$ is much smaller than $I_{LeakComp}$ and the sampling capacitor $C_a$ is large compared to $C_{Fb}$. The first differential pair reacts quickly but with limited current, while the second pair reacts slower but can handle significantly larger currents. If the integrator is reconnected to node 1 after sampling, there will be no current flow to the integrator unless a signal arrives at the input node. Since the leakage current is already compensated at the input node, there is also no baseline shift at the amplifier output.

Note that the design of switch A requires special attention, because any current leaking through that switch and any charge injection during switching will corrupt the current delivered to the input node. Details on the design are given in reference 11.

## 3.3. EXPERIMENTAL RESULTS

### 3.3.1. Photon Counter Measurements

*3.3.1.1. Dynamic Range.* The dynamic range of a photon counter starts with a single photon during the measurement interval and reaches up to photon fluxes at which the pileup of subsequent pulses becomes dominant, leading to a corresponding decline in counting efficiency.

In the case of artificially generated, equidistant test pulses, this decline is very steep, because any remaining charge of the previous pulse quickly adds up and saturates the preamplifier. The maximum count rate is thus well-defined and depends only on the shaping duration, which is determined by the feedback strength and pulse height, but not by the threshold setting.

In the more realistic case of Poisson-distributed pulse spacings (as are present in real X-ray signals), the decline depends on the threshold settings. If large threshold voltages are chosen, the decline is more gradual than in the case of equidistant pulse spacings. The maximum number of recorded counts per frame matches roughly with the frequency of the steep decline measured with equidistant pulse spacings. Figure 3.5 shows a comparison of both behaviors with 2.1-fC input pulses, a threshold

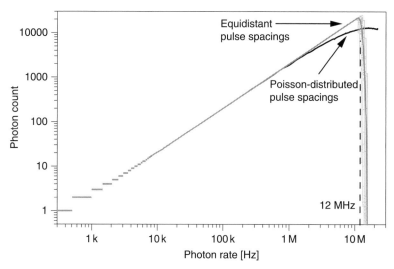

**Fig. 3.5.** Dynamic range of the photon counter, tested with 2.1-fC input pulses. *Gray lines*: Response to pulses of equidistant spacings (superposition of the all measurements in all pixels); a steep decline marks the maximum count rate at about 12 MHz. *Black*: Poisson-distributed pulse spacings cause a more gradual decline, with a maximum photon count at a similar frequency.

setting of about half the pulse size, and a feedback current of 91 nA. The maximum count rate with equidistant pulse spacings is about 12 MHz.

***3.3.1.2. Electronic Noise.*** Threshold scans with 2-fC charge pulses produced by a switched capacitor showed an electronic noise equivalent to approximately 78 $e^-$ at minimal feedback settings. Larger $I_{Fb}$ feedback bias currents cause an additional noise of about 0.65 electrons per nanoamperes. The $I_{LeakComp}$ bias current in the second differential pair increases this value with a slope between about 0.6 and 1.3 electrons per nA. Investigations on a previous prototype chip measured the dependence on the capacitive load at the preamplifier input node [7]. This was made possible by five 100-fF capacitors which could be connected to the input node. Starting from an ENC of 119 $e^-$ without additional load, the noise increases by approximately 0.375 electrons per femtofarad.

***3.3.1.3. Noise Count Rate.*** The noise count rate of the photon counter is measured by sweeping the comparator threshold voltage $V_{CountTh}$ around the baseline voltage at the preamplifier output ($V_{CountBaseline}$). If no additional input signal is present, all recorded counts can be attributed to noise. The result of such a measurement is shown in Fig. 3.6. The noise count rate reaches its maximum when the comparator threshold matches the baseline. Its absolute value is influenced by the speed of comparator and preamplifier and the magnitude of the feedback current. The noise

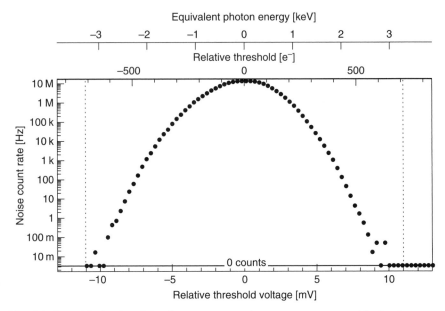

**Fig. 3.6.** Logarithmic plot of the photon counter noise count rate, measured in the absence of input signals by sweeping the comparator threshold voltage around the preamplifier output baseline voltage (compare Fig. 3.1). All counts are caused by the inevitable electronic noise. Low digital-to-analog crosstalk is a key requirement for this measurement. Noise count rates below one count per millisecond are obtained for thresholds above 6 mV. Less than one noise count per minute is achieved above 9 mV. The standard deviation of the noise count rate distribution is comparable to the ENC obtained from threshold scans.

count rate distribution is symmetrical in the voltage difference, which is expected because a count event requires both a positive and a negative transition in the comparator output.

It is, however, a sign of very small digital-to-analog crosstalk that the chip can measure this function at all. In designs with a larger crosstalk, the digital activity caused by a noise hit can trigger additional comparator transitions. This counting only stops if the threshold voltage is increased beyond the level of crosstalk (hysteresis). No such effect was encountered here. The shape of the noise count rate can be fitted by a Gaussian distribution. Its standard deviation is comparable to the ENC obtained from threshold scans. The noise count rate adheres to the Gaussian shape even at threshold voltages further away from the baseline, as can be seen from the inverse-parabolic shape in the logarithmic plot of the same data set in Fig. 3.6. This plot allows an accurate prediction of the noise count rate at a chosen threshold voltage. For example, a threshold voltage of 6 mV (400 $e^-$) above the baseline will yield an expectation value of less than one noise hit during a 1-ms measurement. If the threshold voltage is chosen larger than 9 mV, the noise count rate drops below a single hit per minute.

### 3.3.2. Integrator Measurements

The measurements presented in this section characterize the integrator on its own, without the influence of the feedback circuit. Two charge injection methods were used:

a. *Pulsed* current injection with a current chopper, connected to the integrator either directly or via the feedback circuit. If pulse width and pulse spacing are sufficiently large (larger than a few nanoseconds), the pulse size is determined by pulse duration and the strength of the injection current source.
b. *Continuous* current injection with current sources whose magnitude is controlled by external references connected to the input of current mirror circuits. Two circuits with designed mirror ratios of 4000 : 1 and 400 : 1 are available.

***3.3.2.1. Dynamic Range.*** The dynamic range of the integrator was found to be in good agreement with the theoretically expected value determined by measurement duration, integrator clock frequency, and pump packet size as explained above. A measurement of the dynamic range is provided in the next section (Fig. 3.8). The upper current limit of 200 nA results from the chosen integrator clock rate of 20 MHz and the pump size 10 fC of the charge pump. The frame duration of 2 ms sets the lower integrator current limit to 10 pA. This limit was, however, removed by inserting a 200-pA offset current so that the minimal current is only limited by the signal-to-noise ratio. Note that the wavy pattern visible in Fig. 3.8 in the region below 10 pA is an artifact caused by the quantized nature of the input current (pulse current injection). In this region, the current consisted of only 1 – 10 charge pulses throughout the frame duration.

***3.3.2.2. Noise Performance.*** The noise of the integrator current measurement was determined from the standard deviation of 100 subsequent measurements of the same input current. Figure 3.7 shows a measurement of the signal-to-noise ratio as a function of the input current, which was produced by an externally controlled current source injecting into the integrator directly (injection type b). An additional offset current (here: 770 nA) extended the dynamic range to lower currents and causes a constant noise floor of about 0.4 pA at input signals that are small compared to the offset current. Larger input current cause an noise increase up to about 10 pA, reached at input currents of 200 nA.

Note that this corresponds to a signal-to-noise ratio of 20,000 : 1. In this example, the precision of discretization was about 1 : 60,000, given by the 6-ms frame duration and the clock rate of 10 MHz. As can be seen in Fig. 3.7, the statistical contribution arising from the number fluctuations ($\sqrt{N}$, marked by the dashed line) exceeds the integrator noise by more than an order of magnitude. The integrator noise contribution can thus usually be neglected for X-ray-generated input signals.

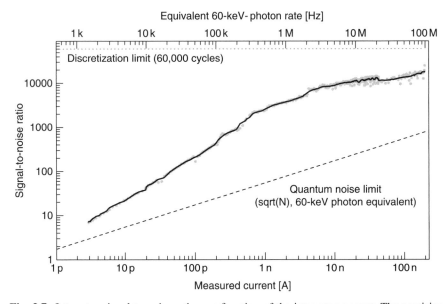

**Fig. 3.7.** Integrator signal-to-noise ratio as a function of the integrator current. The precision of discretization (*dotted line*, **top**) is about 1 : 60,000 due to the frame duration of 6 ms and the clock rate of 10 MHz. The signal-to-noise ratio is far better than the quantum noise limit for a signal produced by 60-keV photons (*dashed line*, **bottom**). An additional offset current of 770 pA extends the dynamic range to lower currents and is responsible for a noise floor at 0.4 pA.

### 3.3.3. Simultaneous Photon Counting and Integration

This section discusses the performance of the CIX chip in measurements with simultaneous photon counting and integration. The usage of pulsed input signals allows investigating the correlation between photon counter and integrator measurements. It also allows determining the energy resolution of the reconstructed average pulse size. The section will conclude with a discussion of measurements with polychromatic input signals whose energy distribution approximates actual X-ray spectra.

*3.3.3.1. Total Dynamic Range.* The combined dynamic range and the energy resolution of a CIX pixel in simultaneous counting and integration mode were determined from a measurement with an input signal consisting of equidistant 2.1-fC charge pulses whose frequency was varied from 500 Hz to 20 MHz. Signals beyond the count rate limit of the photon counter were generated with a constant current source. This allowed us to deliver currents to the integrator up to and beyond its maximum current. Table 3.1 summarizes the used operation parameters.

With these settings, the CIX pixel covers a total dynamic range from about 1 pA to 200 nA, as can be seen in Fig. 3.8. The photon counter handles photon rates up to about 12 MHz, equivalent to a current of 25.2 nA. At smaller frequencies, all pixels

**TABLE 3.1.  Operation Parameters for the Measurement of the Combined Dynamic Range and the Energy Resolution in Simultaneous Counting and Integration Mode**

| Property | Setting |
|---|---|
| Input signal | 2.1 fC (13,000 e$^-$, 60.4-keV photons) $>$30 nA: constant current source |
| Pulse spacing | equidistant |
| Photon counter threshold | 1.12 fC (7,000 e$^-$, 32.5 keV) |
| Integrator clock | 20 MHz |
| Frame duration | 2 ms (40,000 clock cycles) |
| Pump type | Switched current source |
| Pump packet size | 10 fC (62,400 e$^-$) |
| Feedback mode | Static leakage current compensation |
| Feedback bias current | $I_{Fb} = 91$ nA |
| Compensation bias current | $I_{LeakComp} = 29$ nA |

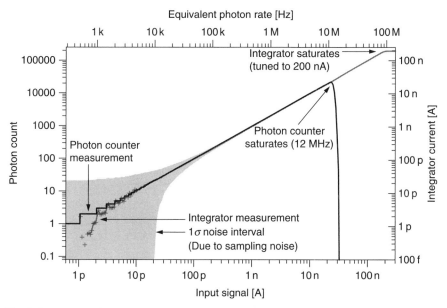

**Fig. 3.8.** The combined dynamic range of photon counter and integrator covers signals from about 1 pA up to 200 nA (equivalent to rates of 60 keV photons between 500 Hz and 95 MHz). In the high-signal regime, the photon counter saturates at about 12 MHz, while the integrator measures currents up to the chosen value of 200 nA. At small input signals, the integrator noise, marked by the interval around the measurement, becomes dominant. The integrator noise is mainly caused by the sampling step before every measurement. It retains a constant value of (23.6 $\pm$ 3.4) pA throughout the dynamic range.

achieve 100% count efficiency. The integrator can measure currents up to the chosen maximum of 200 nA (equivalent to a photon rate of about 95 MHz). In the low current regime, the current measurement becomes increasingly noise dominated, as can be seen from the one-standard-deviation noise interval. The interval around the measurement denotes the average noise of the current measurements. It is dominated by the contribution from the sampling step and retains a constant amplitude of (23.6 $\pm$ 3.4) pA throughout the dynamic range. As was already mentioned above, the wavy pattern visible in the integrator measurement at currents below 10 pA is an artifact caused by the quantized nature of the input current.

***3.3.3.2. Pulse Size Reconstruction.***   Figure 3.9 shows the average pulse size that was reconstructed from the count rate and current measurements in Fig. 3.8. As can be seen from the interval around the measurement, the noise in the current measurement (and thus the noise in the pulse size measurement) becomes dominant at signals below about 100 pA. Even though a single measurement in this range will not produce a reliable value, the average of multiple measurements does still yield sensible results down to currents of about 20 pA. At large input signals, the pulse size is overestimated due to the breakdown of the count efficiency caused by the saturation of the photon counter.

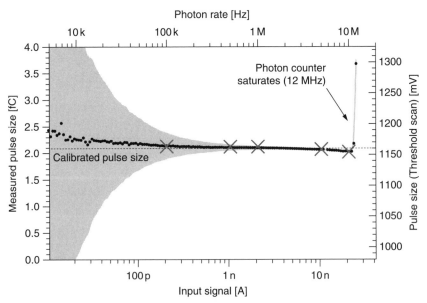

**Fig. 3.9.** Average pulse size, reconstructed from photon counter and integrator measurements. At low frequencies, the measurement becomes dominated by the noise in the current measurement (marked by the interval), even though sensible results can still be obtained by averaging multiple measurements. The visible deviations of the measured pulse size from the calibrated value (*dashed line*) are due to variations in the input pulse size. This was confirmed using independent threshold scans (*cross marks, right scale*).

The measured pulse size also shows a certain dependence on the input pulse rate, as can be seen by the slight deviations from the calibrated value (noted by the dashed line). Independent pulse size measurements with threshold scans (noted by the crosses), however, confirm that this deviation is caused by variations in the input pulse size rather than by inaccuracies in the measurement [11]. The presence of this deviation in the measured pulse size is thus a good sign, demonstrating the sensitivity of the pulse size measurement even to slight changes.

### 3.3.3.3. Spectral Resolution.

Two input signals with different average pulse sizes can be distinguished from each other if the difference in their pulse size measurements is large compared to the noise of the measurements. The width of the interval shown in Fig. 3.9 can thus be used as a measure for the spectral resolution of the CIX pixel.

The pulse size can be measured with a signal-to-noise ratio better than 10:1 in the range between 250 pA and 25 nA, corresponding to 60-keV photon rates between 120 kHz and 12 MHz. The form of the curve matches the expected shape for a measurement with signal independent noise in the current measurement, 100% count efficiency, and a photon counter saturation frequency of 12 MHz.

### 3.3.3.4. Spectral Hardening.

One of the most interesting questions for the evaluation of the signal processing concept is, To what extend it will be able to provide information about spectral hardening in an actual X-ray spectrum? This was investigated with an input signal that simulates the transmission spectra of a 90-keV X-ray tube after passing through different absorber materials. In this case, the first absorber was a copper sheet of 0.5-mm thickness, while the second absorber was 11.6-mm aluminum. The simulation also took into account the thickness and material of the conversion layer, which was 1 mm of CdZnTe. Both signals were

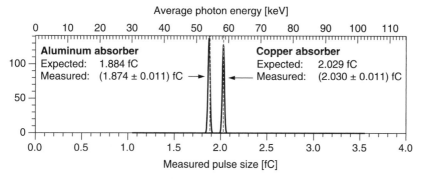

**Fig. 3.10.** Comparison of the spectral hardening observed behind two different absorbers. The absorber thicknesses were chosen so that both signals are indistinguishable in their beam intensity. Within the achieved precision of 0.011 fC, the measured average photon energies (*black curves*) agree with the expected value computed from the transmission spectra (*dashed lines*). The clear separation of both curves indicates that the spectral hardening is determined successfully and allows distinguishing both signals.

**TABLE 3.2. Comparison of the Measured and Expected
Mean Energies Behind the Cu and Al Absorbers**

| Absorber | Expected | Measured | Unit |
|----------|----------|----------|------|
| Aluminum | 54,215 | 53,927 ± 317 | eV |
|          | 1.884  | 1.874 ± 0.011 | fC |
|          | 11,760 | 11,698 ± 69 | $e^-$ |
| Copper   | 58,388 | 58,416 ± 317 | eV |
|          | 2.029  | 2.030 ± 0.011 | fC |
|          | 12,665 | 12,672 ± 69 | $e^-$ |

normalized to a photon rate of 250 kHz, which corresponds to an initial incident flux of 7.3 million photons/(mm$^2$ s) for Cu and to 6.8 million photons/(mm$^2$ s) for Al. The threshold was set to a photon energy equivalent of about 12 keV.

The input signal was generated by providing a 65-ms-long pseudorandom sequence of digital strobe signals to the current chopper. Different photon energies were produced by varying the duration of the strobe pulses. A calibration beforehand determined the corresponding photon energy for a set of the trigger durations between 1 and 30 ns. The input signal resembled a real X-ray signal in terms of the randomness of the arriving signals, the total photon flux, and the energy distribution.

The results of this measurement (Fig. 3.10 and Table 3.2) showed that the mean energy was measured with a precision of about 0.6%. Both spectra can clearly be distinguished, since their energy difference of 4.17 keV exceeds the achieved precision of 0.32 keV 13-fold. In order to obtain a separation of one standard deviation between both spectra, a measurement duration of 377 μs would have been sufficient (1.5 ms for a separation of two standard deviations). Shorter measurements do not allow a reliable separation of both spectra, since the average photon rate of 250 kHz yields an expectation value of only a 25 photons per 100 μs.

## 3.4. CONCLUSION

A readout scheme for direct conversion X-ray imaging using simultaneous photon counting and integration was proposed. The two channels are combined into a single pixel using a special feedback circuit that also provides leakage current compensation. The combination of the two channels extends the dynamic range by about one order of magnitude beyond the limits of the individual channels. Spectral information is obtained in the overlap region, which covers about two orders of magnitude. This allows to determine the hardening of the tube spectrum due to attenuation by the imaged object. The feedback circuit is able to handle the sensor leakage currents and delivers accurate reproductions of the input signal. Its additional noise contribution to the current measurement is usually small compared to the quantum noise due to photon number fluctuations in the input signal. The crosstalk between the integrator circuit and the photon counter is small enough to allow simultaneous operation of both channels.

# REFERENCES

1. R. E. Alvarez and A. Macovski, Energy-selective reconstructions in X-ray computerized tomography, *Phys. Med. Biol.* **21**, 733–744, 1976.

2. J. A. Rowlands and J. Yorkston, Flat panel detectors for digital radiography, in J. Beutel, H. L. Kundel, and R. L. Van Metter (eds.), *Handbook of Medical Imaging*, Vol. 1: *Physics and Psychophysics*, SPIE Society of Photo-Optical Instrumentation Engineers, Bellingham, WA, 2000.

3. S. Kasap and J. A. Rowlands, Direct-conversion flat-panel X-ray image sensors for digital radiography, in *Proceedings of the IEEE*, IEEE, New York, Vol. 90, pp. 591–604, April 2002.

4. M. Lindner et al., Comparison of hybrid pixel detectors with Si and GaAs sensors, *Nucl. Instr. Meth. A*, **466**, 63–73, 2001.

5. M. Löcker et al., Single photon counting X-ray imaging with Si and CdTe single chip pixel detectors and multichip pixel modules, *IEEE Trans. Nucl. Sci.* **51**, 1717–1723, 2004.

6. G. Pellegrini et al., Performance limits of a 55 μm pixel CdTe detector, *IEEE 2004 Nucl. Sci. Sympo. Confe. Rec.* **4**, 2104–2109, 2004.

7. E. Kraft et al., Counting and integrating readout for direct conversion X-ray imaging— Concept, realization and first prototype measurements, *IEEE Trans. Nucl. Sci.* **54**, 383–390, 2007.

8. eV Products, Material properties—Semiconductor detector materials, Saxonburg Blvd., Saxonburg PA.

9. P. Aziz, H. Sorensen, and J. van der Spiegel, An overview of sigma–delta converters, *IEEE Signal Processing Mag.* **13**, 61–84, 1996.

10. R. Lutha et al., A new 2D-tiled detector for multislice CT, *Proc. SPIE Medi. Imag.* **6142**, 275–286, 2006.

11. E. Kraft, Counting and integrating microelectronics development for direct conversion X-ray imaging, PhD thesis, Bonn University, Germany, 2008.

# 4 Noise Coupling in Digital X-Ray Imaging

JAN THIM and BÖRJE NORLIN

In modern mixed-signal system design, there are increasing problems associated with noise coupling caused by switching digital parts to sensitive analog parts. The problems associated with on-chip noise coupling have been discovered in digital X-ray photon counting pixel detector readout systems, where the level of integration of analog and digital circuits is very high on a very small area, and it would appear that these problems will continue to increase for future system designs in this field. This chapter handles the problems associated with noise coupling in digital X-ray systems, where the focus lies in the prediction of noise coupling in a design chain.

## 4.1. CHARACTERIZATION OF NOISE PROBLEMS IN DETECTOR SYSTEMS

The present development of digital X-ray medical imaging systems strives toward goals such as higher spatial resolution, lower noise, and shorter acquisition times. The primary goal of this development is to provide a lower dose to the patients while retaining equal or improved image quality. It has been demonstrated that with energy-resolved X-ray imaging, it is possible to achieve a dose reduction of about 50% [1]. Energy-resolved images can be obtained if the detection system is used for processing each individual photon, and thus it is possible to determine its energy (single-photon processing readout) instead of accumulating the charge generated by the absorbed photons (charge integrating readout). In the first case the response is proportional to the number of detected photons, possibly sorted in different energy bins that could be given different weights in the image or even subtracted in order to enhance the contrast at certain energies. In the second case the response is merely proportional to the total amount of absorbed X-ray energy. Also there is a fundamental difference in the noise properties between the two detection methods. In a

*Medical Imaging: Principles, Detectors, and Electronics*, edited by Krzysztof Iniewski
Copyright © 2009 John Wiley & Sons, Inc.

charge-integrating detector the electronic noise is added to the signal. In a single-photon processing detector, the electronic noise only affects the threshold, causing an uncertainty in the energy of the photon. The hybrid pixelated semiconductor detector was introduced in the early 1990s as an alternative to the strip detectors in energy-resolved imaging applications with high X-ray flux [2]. The photon processing electronics for each pixel should be contained within the area of that pixel, which places severe constraints on the design of the readout. An early example of such a pixelated system is the MEDIPIX1 [3]. This system is based on the photon-counting principle with a low threshold where all photons with energies above the threshold are counted. However, due to the reduced weight of the high-energy photons, the image is improved when compared to the charge integrating readout. The MEDIPIX project, coordinated by the CERN microelectronics group, can be taken as an example to show the evolution of single photon processing detectors.

With reference to spatial resolution, MEDIPIX1 was unable to compete with a charge-integrating system, since the complexity of the electronics in each pixel limits the pixel size to 170 $\mu$m $\times$ 170 $\mu$m. However, as Moore's law predicts a constant decrease in component size, it was only a matter of time before the MEDIPIX2 with 55-$\mu$m $\times$ 55-$\mu$m pixels and dual thresholds which allowed for the implementation of an energy window could be designed and manufactured [4]. However, the reduced pixel size introduces a complication in the single-photon processing detectors. When the pixel size becomes comparable to the size of the charge cloud generated by the absorbed photon, problems occur. The size of the charge cloud is a function of both the absorption process and the diffusion of charges during their drift through the detector. The charge absorbed in the border region will leak to neighboring pixels, a process referred to as charge sharing, which will limit the spatial resolution in both the integrating and counting detector systems. The most important effect is that it will destroy the spectral information in single-photon processing detectors. In systems with a large pixel size ($>$100 $\mu$m), charge sharing is mainly a border problem that only, to some extent, affects the spectral information. For detector systems with small pixels, charge sharing leads to a significant loss of the spectral information (see Fig. 4.1). One hit will be counted as two hits with lower energies resulting in a shift of the actual spectrum toward lower energies [5].

One way to preserve the spectral information is to introduce charge summing over neighboring pixels in the readout [6]. This particular approach will solve the charge-sharing problem but at the cost of higher spectral noise and a lower count rate. However, charge summing requires higher component density in the readout than was possible when MEDIPIX2 was designed. However, at a later stage and with Moore's law still being active, it was possible to introduce a prototype of MEDIPIX3 including charge-sharing compensating pixel-to-pixel communication [7].

The purpose behind this discussion of the MEDIPIX project is to illustrate how electronic systems for X-ray imaging systems evolve, which is illustrated by Table 4.1. The first idea (high count rate and photon counting readout) results in a complication (low resolution) that, when solved, gives rise to new complications (charge sharing) resulting in a final much more complex solution than that originally intended. As soon as it is possible to increase the component density, creative ideas for

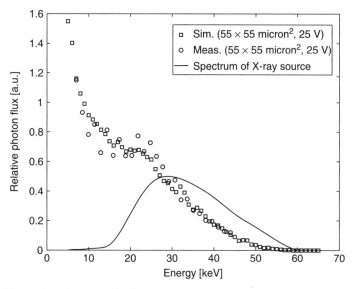

**Fig. 4.1.** Comparison between simulated and measured spectra for the Medipix2 system. The bias voltage of 25 V corresponds to the depletion voltage of the detector. The simulation confirms that the distortion of the spectrum is an effect of the charge sharing. The solid line shows the real spectrum of the source.

**TABLE 4.1. The Evolution of the MEDIPIX Project, an Example of How Substrate Noise Models Gain Importance in Design of X-Ray Imaging Readout Electronics**

| | Medipix 1 | Medipix 2 | Medipix 3 |
|---|---|---|---|
| Year available | 1998 | 2002 | Prototype 2007 |
| Pixel size | 170 $\mu$m $\times$ 170 $\mu$m | 55 $\mu$m $\times$ 55 $\mu$m | 55 $\mu$m $\times$ 55 $\mu$m |
| Process | 1-$\mu$m SACMOS with 2 metal layers | 0.25-$\mu$m 6-metal layers CMOS | 0.13-$\mu$m 8-metal layers CMOS |
| Transistor density (transistors/$\mu$m$^2$) | 0.014 | 0.182 | 0.364 |
| Feature | Photon counting with one low threshold | Photon counting with low and high thresholds | Charge summing to overcome charge sharing |
| Consideration | Resolution too low to compete with film-based systems | Charge sharing due to small resolution degenerates energy information | Substrate noise had to be considered in the chip design |

the inclusion of new features will appear. It is, for example, desirable for physicists to have access to reasonably good spectroscopic information in each pixel, since it can be used in several applications concerning material resolving imaging [8]. Although a physical limit for the resolution due to the size of the charge cloud after photon absorption does exist, pixel sizes smaller than that limit can provide usable information that enables interpolation of the point of photon impact.

We believe that this striving toward higher and higher complexity in the readout circuit for each pixel will continue. It is likely that the next drawback in X-ray imaging system development will involve limitations due to substrate noise in the readout electronics. Therefore it is of great importance to be aware of the physics behind this limitation.

## 4.2. NOISE MECHANISMS IN READOUT ELECTRONICS

The increasing level of integration between analog and digital parts in mixed-signal systems is beginning to cause problems with on-chip noise coupling. Noise currents caused by switching digital circuits are spread to the sensitive analog components via the power distribution network and through the substrate by current injection. With reference to this aspect, readout systems for radiation pixel detectors represent one of the worst-case scenarios, because of the extreme level of integration required to read out pixel data [9]. As an example, the European Medipix3 readout system comprises approximately 1100 transistors per pixel in a 0.13-$\mu$m technology where, on a pixel surface of 55 $\mu$m $\times$ 55 $\mu$m, about half of the pixel readout surface consists of analog circuits and half consists of digital circuits [7]. For these types of highly integrated System-on-Chip (SoC) designs, on-chip noise coupling is a possible show-stopper, because the performance of the system is limited by unwanted signal interactions.

The key to counteracting or avoiding these problems is by means of prediction. How do we seek out the problems associated with noise coupling before they cause design iterations and the remanufacturing of hardware? Today, the solution is often to insert countermeasures in all those places where noise coupling problems are suspected. This often causes the manufacture of the system to become overly expensive, since there is a large risk of an excessive number of processing steps. Another solution would be to simulate noise in the design process, starting at an early design level in order to grasp the potential problems from the outset [10]. This enables the designer to adapt the system at various design levels in order to avoid large design iterations caused by noise coupling problems.

The specific mechanisms that cause on-chip noise coupling in SoC designs such as digital X-ray imaging readout systems can be described using models that fall into two different categories, noise models and physical properties (see Fig. 4.2 [11]). The noise models describe the different injection mechanisms in and out of components, while the physical properties describe the actual media over which the injected currents spread. The next two subsections will define these models, which are necessary in order to fully describe noise coupling in a SoC design (Fig. 4.2).

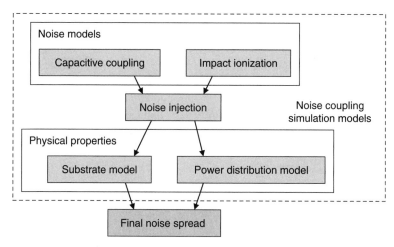

**Fig. 4.2.** Schematic image of the different models needed in an on-chip noise coupling simulation method.

### 4.2.1. Noise Models

There has been a great deal of previous work, performed over many years, involving the extraction of viable substrate noise injection mechanisms at both the layout and circuit level. Several injection mechanisms in substrate noise coupling that have been discovered, and these must be considered:

First there is *impact ionization*, which basically yields a current flowing out of the bulk node of the transistor and into the substrate. This current is generated by fractions of carriers in the depleted region of a saturated transistor which gain sufficient energy to become "hot" [12]. They then scatter and create additional electron–hole pairs. The holes generated in the NMOS transistor are then swept to the substrate, forming the current into the substrate.

*Capacitive coupling* to the substrate is caused by voltage fluctuations in the source and drain of the MOS transistor which are coupled through the junction capacitances. There is also a capacitive coupling over the gate oxide via channel capacitances. Together these form a current injected into the substrate [13].

*Gate-induced drain leakage* (GIDL) occurs when there are high fields across the gate-drain overlap region. These form a deep-depletion layer in the drain; and when the voltage drop across this layer is sufficient, the valence electrons start to tunnel between bands, resulting in the creation of holes, which are swept into the substrate [14].

The hot electrons that are not subjected to impact ionization are likely to release their excess energy by emitting a photon [15]. Electron–hole pairs can then be created when the photons are reabsorbed, causing a *photon-induced current* (PIC) [16].

In addition to the capacitive coupling, there will also be a *diode leakage current* at the junctions of the MOSFETs source and drain, which are in actuality reverse-biased diodes [13]. This current is then injected into the substrate.

From these five substrate injection mechanisms, it has been shown that the substrate currents caused by capacitive coupling and impact ionization are dominant, in that order, and that capacitive coupling becomes relatively more important at high frequencies [17]. The next two subsections will discuss these two important substrate injection mechanisms.

Finally, from the power distribution network, there can also be power supply noise (i.e., power bounce) and ground noise (i.e., ground bounce) directly injected into the substrate via substrate contacts. Such a noise injection mechanism has the same amplitude or greater than that of the capacitive-coupled injection. The problem is usually tackled with well-designed power and ground networks in addition to appropriately allocated substrate contacts [18].

***4.2.1.1. Capacitive Coupling.*** In *capacitive coupling*, an injection of current occurs over the capacitances of junctions and gate channels. In a *p*-substrate—specifically, in an *n*MOS transistor—coupling occurs through the drain-bulk and source-bulk junctions. In a *p*MOS transistor, the coupling also occurs through the *n*-well and *p*-substrate junction capacitance [19]. Additionally coupling also occurs through the gate oxide channel capacitance [17]. The coupling mechanisms of capacitive coupling are illustrated in Fig. 4.3.

For modeling and simulation, a numerical representation of the capacitive noise coupling can be derived by making the assumption that in a noise coupling simulation, working on higher design levels, an estimation of the power supply current $i$ is required. Using this information as a starting point, the targeted model for estimating the substrate injection current for a technology $\mathbf{T}$ and a block area $A$ can be stated as

$$i_{inj,cap}(t) = f(i(t), \mathbf{T}, A) \qquad (4.1)$$

In reference 20, it is shown that from Eq. (4.1), a more concrete estimation of the injecting current from a number of identical transistors on a single chip can be derived. This estimating equation can be stated as follows:

$$i_{inj,cap}(t) = \frac{kC_j}{\dfrac{1}{V_{dd}} \displaystyle\int_{t_0}^{t_d+t_0} i_{ds}(t)\, dt} i_{ds}(t) \qquad (4.2)$$

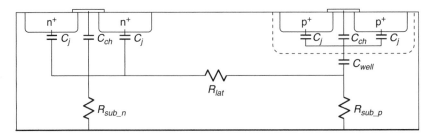

**Fig. 4.3.** Model of capacitive noise coupling between an *n*MOS source and a *p*MOS receiver.

where $k$ is the number of noise transmitting transistors, $C_j$ is the junction capacitance for the specific transistors, $i_{ds}$ is the drain-source current, $V_{dd}$ is the voltage supply, and $t_d$ and $t_0$ are the initial and end times for the transistor voltage switch.

***4.2.1.2. Impact Ionization.*** In the case of *impact ionization*, this injection mechanism is depicted in Fig. 4.4. In Fig. 4.4a, fractions of carriers in the depletion region of the *n*-channel MOSFET gain sufficient energy to become "hot." When this occurs, they scatter and create additional electron–hole pairs. The positively charged holes generated in the NMOS transistor are then swept to the substrate, thus forming a current into the substrate. Impact ionization is measured, in general, on the bulk node of the *n*MOS transistor (see Fig. 4.4b), since the current out of the *p*MOS bulk node is at least one order of magnitude lower than that of the NMOS [17]. There are many representations, and models for impact ionization and a small selection of these are described in this section.

In references 21 and 22 a macroscopic transport model for what is called the hot electron subpopulation (HES) is presented along with a "nonlocal" impact ionization model [23]. These are derived from the Boltzmann transport equation and an empirically determined equation, and they are based on the premise that the population of electrons with energies larger than the gap energy plays a dominant role in the impact ionization mechanism.

Looking at reference 24, a model for impact ionization using the tail electron density is found. A set of tail electron hydrodynamic (TEHD) equations is introduced, in which numerical aspects and model parameters are described. Based on Monte Carlo simulations, a model for impact ionization is introduced, expressed as a function of the tail electron density.

There is a simplified model based on the two previously described. Tang and Nam [25] propose a simplification to the HES hydrodynamic model, where the three transport equations from references 21, 22, and 24 are reduced to one. A neat expression of

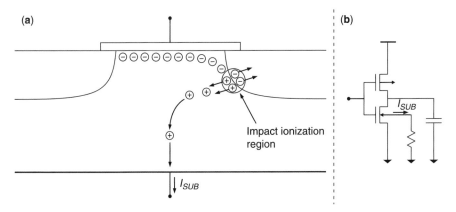

**Fig. 4.4.** Impact ionization in an *n*-channel MOSFET illustrated by a substrate layout (a) and a circuit schematic (b).

the impact ionization coefficient $\alpha$ is presented:

$$\alpha(\omega_2) = \begin{cases} 9.59 \times 10^5 \, e^{-1.34 \times 10^2/E_{\mathit{eff}}}, & \text{for } \Delta\omega_2 > 0.164 \\ 1.156 \times 10^4 \, e^{-0.9 \times 10^2/E_{\mathit{eff}}}, & \text{otherwise} \end{cases} \quad (4.3)$$

where $\omega_2$ is the average energy of the hot-electron subpopulation whose energy exceeds the impact ionization threshold energy. The only problem involves the determination of $\omega_2$, which should, most probably, be performed empirically for each simulation case.

Perhaps the most promising impact ionization model for use at higher design levels is that proposed by Sakurai et al. [26], which is a combination of work done by Sing and Sudlow [27] and Hu [28]. The injecting substrate current for this model is described as

$$I_{SUB} = I_{ds} * a(V_{ds} - V_{dsat})^b \quad (4.4)$$

where the fitting parameters $a$ and $b$ are dependent on the processing technology. $I_{SUB}$ stands for the substrate current out of the $n$-channel MOSFET bulk node, $I_{ds}$ for drain-source current, $V_{ds}$ for drain-source voltage, and $V_{dsat}$ for drain saturation voltage. The parameters of Eq. (4.4) are further described in reference 26.

### 4.2.2. Physical Properties

The two major media involved in noise transport have been found to be the power distribution network and the substrate [29]. This subsection is dedicated to providing an explanation as to how these can be seen in a simulation environment.

*4.2.2.1. Power Distribution Networks.* The basic idea of noise coupling in a power distribution network is described in Fig. 4.5. A noise-generating source module is connected to the power distribution network model, which describes the physical connection and how the current is transferred between noisy on-chip modules and the power distribution lines. The noise current is fed to the network by the source module, transferred, and thereby shaped by the network model, which describes the electrical properties of power distribution lines. The noise current is then fed into

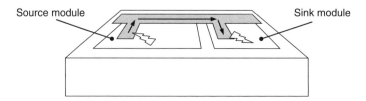

**Fig. 4.5.** Basic idea of the power distribution network modeling technique. The noise current is coupled over the network model from a source module to a sink module.

the receiving sink module, which has a connection model equal or similar to the one in the source module. The sink module is then affected by the coupled current, which is added to the existing functional currents in the components of the module.

The models of power distribution networks have changed significantly over the years. When the gate delay was the dominant factor, the interconnects were modeled as short-circuits. However, when the interconnect capacitances became comparable to those of the gate capacitances, the interconnects were modeled as capacitances to ground, where these capacitances were directly related to the length of the interconnect power distribution lines [30].

The two more recent and currently dominant models are the *RC* and the *RLC* models [31], which will be discussed in greater detail in this subsection. There are also considerations to be made concerning capacitances and inductances between different power lines in a layout design [32, 33]. This, however, requires information about the metal layer layout of the chip to be fabricated and thus belongs to the final stages of the layout design level.

Over time, the resistance of interconnects became comparable to those of the open resistance for a transistor, and thus a new interconnect model had to be considered— the *RC* model. Since the resistances in the interconnects became significant, a resistance was added in series with the capacitance from the earlier purely capacitive model, forming the *RC* model shown in Fig. 4.6a, where *R* and *C* are approximations of the distributed values. As can be seen in Fig. 4.6a, the power distribution line model forms a series of low-pass filters, which of course shape the signals over the power distribution line accordingly. This model is sufficient for clocked signal transition times $t_r > 2T_0$, where $T_0$ is the time of flight [34].

As an era of shorter rise times is now being entered into, the *RC* model is beginning to become obsolete for certain applications. There are factors that support the expansion of the *RC* model into an *RLC* model having an inductance in series with the *RC* link, as shown in Fig. 4.6b [34]. These factors are that transition times becomes shorter and shorter, wider lines at higher metal layers reduce the interconnect resistances, copper with a lower resistivity than aluminum has been introduced, and the devices are becoming ever faster.

*R*, *L*, and *C* are as for those in the earlier models approximations of the distributed values. Now, which model should be used in different simulations? For modern

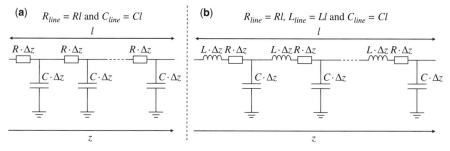

**Fig. 4.6.** (a) *RC* and (b) *RLC* interconnect model.

applications, it is the *RC* model and the *RLC* model that are in use. According to Ismail et al. [34], the *RC* model is sufficiently accurate if the following conditions are satisfied:

$$\frac{t_r}{2\sqrt{L_{line}C_{line}}} < l < \frac{2}{R_{line}}\sqrt{\frac{L_{line}}{C_{line}}} \tag{4.5}$$

where $l$ is the length of the interconnect, $t_r$ is the transition time, and $R_{line}$, $C_{line}$, and $L_{line}$ are the resistance, capacitance, and inductance of the interconnect line. The conditions are fulfilled if the attenuation is sufficiently large to make reflections negligible and if the waveform transition time is slower than twice the time of flight.

***4.2.2.2. Substrates.*** A chip substrate is a complex media to simulate at low levels, since the accuracy of the simulations is scaled against the 3D nature of the substrate, resulting in a cubic dependency. This increases the number of cells to be included in the matrix solution, resulting in longer computational simulation times for a given size of the simulated cells. This is why a great deal of research has been involved in extracting higher-level simulation models for substrates. Today, different models exist ranging from resistive network models with one common ground node, all the way down to device simulation models. We will investigate different physical issues causing noise coupling and look at a few different macromodels developed for substrate modeling on the circuit and layout levels.

The first and most obvious physical property is the resistivity of the substrate which can be modeled as resistors in a substrate mesh cell (see Fig. 4.7a) [35]. This is a simple and rapid substrate model to simulate since DC analysis can be used. A more complex and accurate model is that shown in Fig. 4.7b, which includes capacitors in the substrate mesh cell [36]. There are several places where the capacitors become more vital, for example between the gate and substrate, and between the wells and

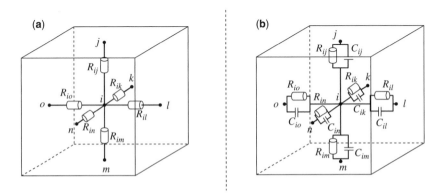

**Fig. 4.7.** Comparison between (a) resistive substrate mesh cell model and (b) resistive/capacitive substrate mesh cell model.

surrounding substrate. This makes it possible to reduce the number of capacitors by only including those that are more vital [37, 38].

Let us now study some of the existing substrate macromodels in use during previous years to describe the physical properties of the substrate.

The *Asymptotic Waveform Evaluation (AWE) macromodel* is based on a nodal analysis approach, where the center of a mesh cube is named $i$ and $j$ stands for one of the sides of the cube. The following equation describes the parallel connection of a resistance $R$ and a capacitance $C$ between the center node and one of the sides:

$$\sum_j \left[ R_{ij}(\Psi_i - \Psi_j) + C_{ij} \left( \frac{\partial \Psi_i}{\partial t} - \frac{\partial \Psi_j}{\partial t} \right) \right] = 0 \tag{4.6}$$

where $\Psi_i$ and $\Psi_j$ are the potentials in the nodes $i$ and $j$ from Fig. 4.7b [39]. The substrate macromodel is computed as an $n \times n$ admittance matrix:

$$\begin{bmatrix} y_{11}(s) & \cdots & y_{1n}(s) \\ \cdots & \cdots & \cdots \\ y_{n1}(s) & \cdots & y_{nn}(s) \end{bmatrix} \begin{bmatrix} v_1(s) \\ \cdots \\ v_n(s) \end{bmatrix} = \begin{bmatrix} i_1(s) \\ \cdots \\ i_n(s) \end{bmatrix} \tag{4.7}$$

where $n$ represents the number of ports [40]. However, in order to simulate substrate coupling in a given circuit, a combination of the AWE macromodel and the nonlinear circuit is required; that is, a new matrix must to be formed for every time point in the simulation (transient simulation). As a consequence, the simulation becomes a complex task, but by using modern matrix solvers, the simulation time takes a fraction of that taken in device simulation.

A *DC macromodel* is presented by Verghese et al. [18], where it is shown that, in heavily doped bulk processes, the relaxation time of the substrate, $\tau = \rho' \varepsilon$, where $\rho'$ is a constant resistivity used to describe carrier flow in the substrate and $\varepsilon$ is the permittivity of the substrate, is of the order of $10^{-11}$ s. This means that the intrinsic capacitances can be ignored for operating frequencies up to a few gigahertz and switching times on the order of 0.1 ns. This then enables the substrate to be modeled as a purely resistive mesh, provided that the well capacitances, field oxide capacitances, and die attached capacitances are modeled as lumped circuit elements. What is then obtained is a purely resistive macromodel that is able to simulate in DC mode, cutting simulation times by an order of 10 as compared to the AWE macromodel. The DC macromodel uses mesh cells such as that shown in Fig. 4.7a. The simulation results, presented by Verghese et al., from the AWE and DC macromodels are indistinguishable and have a maximum error of approximately 20% from device simulations, with only approximate values of bonding pad and chip-to-package capacitances [41]. This type of resistive mesh has also been used by Panda et al. [42].

In heavily doped bulk and lightly doped epitaxial layer processes the substrate model can be further simplified. A single node can be used for the bulk substrate connected to surface nodes as for example in Fig. 4.8, resulting in a *modified single node substrate model* [43].

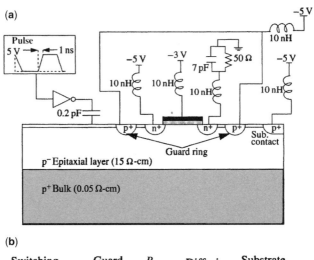

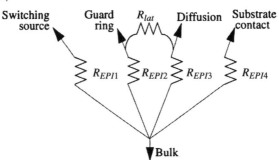

**Fig. 4.8.** (b) Single-node model of the substrate in (a).

The epitaxial resistances can be calculated as a parallel combination of component area resistance $R_{AREA}$ and the perimeter resistance $R_{PER}$ due to current flow:

$$R_{AREA} = \frac{\rho T}{A} \tag{4.8}$$

$$R_{PER} = \frac{\rho}{P} \tag{4.9}$$

where $\rho$ is the resistivity of the epilayer, $T$ is the effective thickness of the epilayer, $A$ is the active surface area, and $P$ is the perimeter of the active area. The resulting formula for the epitaxial resistances is

$$R_{EPI} = \left(\frac{k_1 \rho T}{(L+\delta)\cdot(W+\delta)}\right) \left\|\left(\frac{k_2 \rho}{2(W+L+2\delta)}\right)\right. \tag{4.10}$$

where $W$ and $L$ are the width and length of the active surface area. $k_1$, $k_2$, and $\delta$ are fitting parameters extracted from empirically measured resistances. According to

Su et al., the values $k_1 = 0.96$, $k_2 = 0.71$, and $\delta = 5.0\,\mu$m would yield simulation results within 15% of measured results [38].

The model in Fig. 4.8b involves a modification to the original single-node model. If the distance between two nodes is smaller than approximately four times that of the epilayer thickness, the lateral resistance between transistors/contacts/wells becomes significant, so the lateral resistance $R_{lat}$ is added to increase the accuracy of the single-node model and is given by

$$R_{lat} = \frac{1}{y_{g-d}} \tag{4.11}$$

where $y_{g-d}$ is the conductance between the guard ring and the diffusion area in Fig. 4.8a [18].

This substrate model has also been used on IMEC by van Heijningen et al. [44].

A compromise of the DC macromodel and the modified single node model becomes, what could be called, the *circuit-level single-node substrate model*. This model has a resistance from every surface node to the single substrate node and lateral resistances between surface nodes [45]. This would result in long simulation times if lateral resistors were placed between all nodes; and in order to reduce the number of resistors between surface nodes, it has been documented that triangulation techniques should be used to draw the resistive mesh in the lateral plane [46]. One suitable technique is a modified version of the Delaunay triangulation technique, which basically starts by inserting an edge (resistor) on the shortest distance between two vertices (nodes) and continues by inserting the second shortest edge, with the premise that it does not cross any of the previous edges, and so on [47]. A simple example of this technique is shown in Fig. 4.9, where the resistor $R_{a-c}$ is not inserted because it would cross $R_{b-d}$. In this case, only one resistor is saved for the simulation, but in the case of 900 nodes in a $30 \times 30$ mesh the total number of resistors (lateral + vertical) would reduce from $404{,}550 + 900$ to $2700 + 900$.

Trials have also been performed where capacitances have been incorporated into this model and, in addition, have been placed when well crossings occur [37].

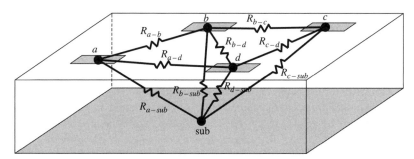

**Fig. 4.9.** Circuit-level single-node model example where the resistance $R_{a-c}$ has been eliminated by the modified Delaunay triangulation technique.

The substrate resistance extraction model for the circuit-level single-node model is based on an interpolating model for known resistance values, with extraction functions as

$$R_{sub} = \frac{1}{G_{sub}} = \frac{1}{L \cdot P^m \cdot A^n} \tag{4.12}$$

$$R_{dir} = \frac{Kd^p}{\sqrt{A_a} + \sqrt{A_b}} \tag{4.13}$$

where $L$, $m$, $n$, $K$, and $p$ are empirical fitting parameters, and $R_{sub}$ is the substrate resistance to the substrate node for a contact with perimeter $P$ and area $A$. $R_{dir}$ is the direct resistance between two contacts with distance $d$ and areas $A_a$ and $A_b$ [48]. This model has been used in the application Subspace, originating at Delft University of Technology. The accuracy for this model has been reported to have an approximate 9% maximum error compared to layout-level simulations for heavily doped processes with epilayer. For lightly doped processes the figure increases to an approximate 25% maximum error [47]. However, trials have been made to increase the accuracy of this model. In reference 49, it is reported that the accuracy can be increased to an approximate 4% maximum error for heavily doped processes by setting the parameter $p = 1$ and a re-fit of the parameter $K$, thus making the dependence of $R_{dir}$ linear to the node distance $d$.

## 4.3. SIMULATION MODELS IN VARIOUS DESIGN LEVELS

In order to simulate noise coupling in digital X-ray readout systems, and for that matter in all SoC designs, a selection to find a suitable model for each design level must be made from the available models. The question one must ask is how accurate the model must be for a particular design level. It is obvious that lower levels of accuracy are required in behavioral-level explorations than in transistor-level circuit simulations, but where the line is drawn may vary from case to case. Nevertheless, some conclusions can be drawn by the accuracy and simulation speed of the various models described in Section 4.2 to define suitable design levels for these models.

For capacitive coupling injection simulation the expression in Eq. (4.2) is well-suited for use from a behavioral design level down to at least the circuit level. Any level lower than that requires full-out layout-level simulations.

Regarding impact ionization, the hot electron subpopulation (HES) and tail electron hydrodynamic (TEHD) expressions presented in Section 4.2.1.2 are too complex computationally for the behavioral level, but would be suitable for circuit-level simulations. The model proposed by Sakurai et al. is to be preferred on higher design levels up to the behavioral level, and this has been attempted in reference 49.

The simulation of power distribution networks can be carried out with simple $RC$ and $RLC$ models according to the conditions specified by Ismail et al. for all design levels except the layout level, where a more accurate model is required in order to take field distortions into account.

For substrate modeling the circuit-level single-node model is sufficiently rapid for simulations on the behavioral level down to circuit level, where the DC macromodel should be used. Lower design levels would require device simulators.

## 4.4. READOUT ELECTRONICS NOISE COUPLING IN DIGITAL X-RAY SYSTEMS

In order to illustrate noise coupling in digital X-ray readout systems, a photon-counting pixel detector readout system is taken as an example. This section handles this design example and the manner in which it is affected by readout noise coupling problems.

Submicron technology has enabled X-ray imaging with image sensors composed of photon-counting pixels to evolve [3]. In these image sensors, each pixel is implemented as a single-channel radiation detector, which means that each detected X-ray photon in the pixel is counted. The number of counts will represent the pixel value. A count is processed such that when a photon hits the detector, the photon energy is converted into a charge pulse, $i_{in}$, which is integrated by a charge sensitive preamplifier, forming $V_{charge}$ in (1) in Fig. 4.10 [50]. This is then pulse-shaped in (2)—that is, first high-passed filtered and then low-passed filtered, thus reducing the system bandwidth and reducing the noise in the detected signal before discrimination, forming $V_{RC}$ [51]. In this design example, a window-discriminator (3) with two thresholds in order to

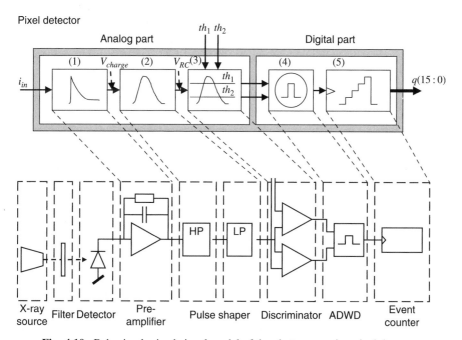

**Fig. 4.10.** Behavioral mixed-signal model of the photon-counting pixel detector.

define the discrimination window is used. The signal is fed to two analog comparators that output binary signals ($th_1$ and $th_2$), indicating whether or not the signal exceeds the threshold values. All the analog circuits have been implemented using current-mode circuits.

A conditionally generated clock signal to the event-counter (5) is generated by the All-Digital Window Discriminator (ADWD) [52] in (4) on the basis of the sequences of $th_1$ and $th_2$. The ADWD is a simple asynchronous controller generating the clock signal without reference to either internal or external timings. The event-counter is a Linear Feedback Shift Register (LFSR) consisting of 12–16 bits. It is built on dynamic Flip-Flops with a single-phase clock, each consisting of six transistors.

A pixel with 13 bits consists of a total of 215 transistors of which 112 are for the digital part. The idle power dissipation (static) is 1.6 μW and the analog part is its main contributor. The active power, when an event is detected, is 3.6 μW. The pixel has been implemented in a 120-nm CMOS technology and operates at 1.2 V.

Parts (1), (2), and (3) in this design example are analog components, in which the most sensitive node is in the preamplifier of the charge integrator. Parts (4) and (5) are digital components that generate the switching noise that is then spread to the analog nodes.

### 4.4.1. Noise Coupling Effects on the Design Example System

In a photon-counting pixel detector readout system, the discrete events from each photon are registered in a counter. The counter is the major contributor to digital noise in the system. These mixed-signal noise effects are either truncated by the discriminator or, if the noise contribution is too large, an extra event is counted. To demonstrate the effect that mixed-signal noise coupling would have on an ordinary grayscale X-ray image, a dental X-ray photo is used, as depicted in Fig. 4.11a. The noise caused by the registration in a pixel's counter can cause an erroneous event either in the same pixel or in those adjacent to it. For self-generated events, noise in

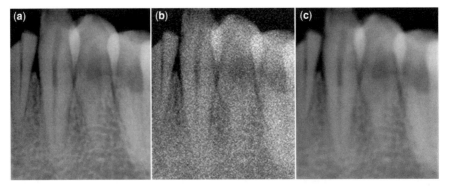

**Fig. 4.11.** (a) Example of dental image. Images distorted by erroneous events caused by mixed-signal noise effects, where (b) depicts self-generated noise within a pixel and (c) depicts noise coupling to adjacent pixels.

the image would appear as Gaussian as illustrated in Fig. 4.11b; and for noise coupling between pixels, it would appear as blurring, which is illustrated in Fig. 4.11c.

In a color X-ray system, the effects of the noise coupling would be delays in the discriminator phase and would result in color value errors from the photon energy value counter. This would appear as blurring in the image or even as color bleeding.

Simulations with models from Section 4.2 have been made of the design example photon-counting pixel detector readout system [9]. Injection current noise caused by

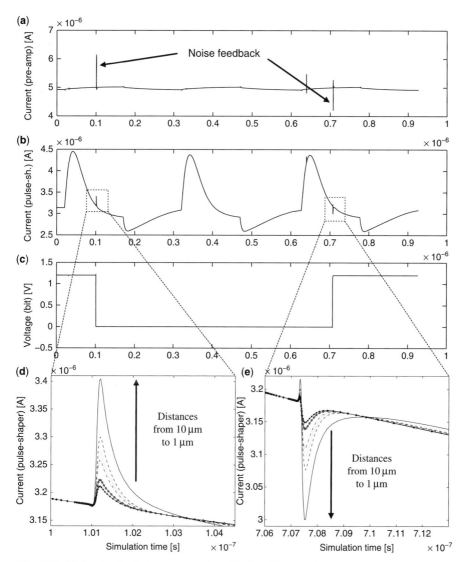

**Fig. 4.12.** Noise feedback from one register bit for different distances between register and preamplifier. The response from preamplifier and pulse-shaper is shown.

the voltage transitions in the counter shift registers have been fed back through the substrate into the preamplifier. The total output from the preamplifier and from the pulse shaper including the substrate noise have then been measured. Two different aspects have been simulated. The noise from one register bit was, firstly, fed back into the preamplifier, and the response was simulated for various distances between the register and preamplifier. Then, secondly, the noise from all the register bits was fed back into the preamplifier for a 12-bit and a 16-bit system, with a distance of 5 μm, and the responses from these were simulated.

Figure 4.12 illustrates the case where the distances have been varied for a one-bit noise feedback simulation. The output response from the preamplifier is shown in Fig. 4.12a, where the noise levels are very high compared to the amplifier transition current. In Fig. 4.12b the output from the pulse-shaping filter is plotted and, as expected, the noise levels are greatly reduced by the filter. The transition voltage of the register bit is shown in Fig. 4.12c. In Figs. 4.12d and 4.12e, enlargements of the noise peaks from Fig. 4.12b are plotted for a number of distances from 10 μm to 1 μm between the register bit and the preamplifier. As could be anticipated, the

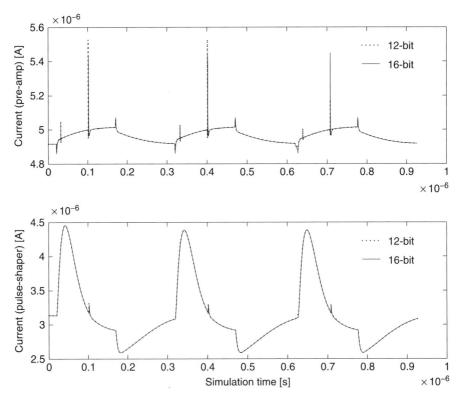

**Fig. 4.13.** Noise feedback from 12-bit and 16-bit register bit for a distance of 5 μm between register and preamplifier. The response from preamplifier and pulse-shaper is shown.

amount of noise fed back from the register bits is highly dependent on the distance to the sensitive analog nodes. Shorter distances cause higher noise levels.

The simulation results from the 12-bit and 16-bit feedback cases are shown in Fig. 4.13. The plots show that the noise levels in the 16-bit case have increased in number. For those cases in which the 12-bit and 16-bit registers both have transitions, the amplitude values are very similar.

This example illustrates that in a design process of a digital X-ray readout system, careful consideration is required regarding on-chip noise coupling. The noise coupling simulations throughout the entire design process have the ability to save funds and resources due to costly design iterations and remanufacturing.

## REFERENCES

1. D. Niederlöhner, J. Karg, J. Giersch, and G. Anton, The energy weighting technique: Measurements and simulations, *Nucl. Instrum. Methods Phys. Res. A* **546**(1), 37–41, 2005.

2. F. Anghinolfi et al., A 1006 elements hybrid silicon pixel detector with strobed binary output, *IEEE Trans. Nucl. Sci.* **39**, 654–661, 1992.

3. M. Campbell, E. H. M. Heijne, G. Meddeler, E. Pernigotti, and W. Snoeys, Readout chip for a 64 × 64 pixel matrix with 15-bit single photon counting, *IEEE Trans. Nucl. Sci.* **45**(3), Part 1, 751–753, 1998.

4. X. Llopart, M. Campbell, R. Dinapoli, D. San Segundo, and E. Pernigotti, Medipix 2: A 64-k pixel readout chip with 55-/spl mu/m square elements working in single photon counting mode, *IEEE Trans. Nucl. Sci.* **49**(5), Part 1, 2279–2283, 2002.

5. H.-E. Nilsson, E. Dubaric, M. Hjelm, and K. Bertilsson, Simulation of photon and charge transport in X-ray imaging semiconductor sensors, *Nucl. Instrum. Methods Phys. Res. A* **487**, 151, 2002.

6. H.-E. Nilsson, B. Norlin, C. Fröjdh, and L. Tlustos, Charge sharing suppression using pixel to pixel communication in photon counting X-ray imaging systems, *Nucl. Instrum. Methods Phys. Res. A*, 6 February 2007.

7. R. Ballabriga, M. Campbell, E. H. M. Heijne, X. Llopart, and L. Tlustos, The Medipix3 prototype, a pixel readout chip working in single photon counting mode with improved spectrometric performance, *IEEE Trans. Nucl. Sci.* **54**(5), Part 2, 1824–1829, 2007.

8. B. Norlin, C. Fröjdh, and H.-E. Nilsson, Spectral performance of a pixellated X-ray imaging detector with suppressed charge sharing, *Nucl. Instrum. Methods Phys. Res. A*, 30 January 2007.

9. J. Lundgren, S. Abdalla, M. O'Nils, and B. Oelmann, Evaluation of mixed-signal noise effects in photon-counting X-ray image sensor readout circuits, *J. Nucl. Instrum. Methods Phys. Res. Sect. A* **563**(1), July 2006.

10. M. O'Nils, J. Lundgren, and B. Oelmann, A SystemC extension for behavioral level quantification of noise coupling in mixed-signal systems, *Invited to the 2003 International Symposium on Circuits and Systems (ISCAS'03)*, Vol. 3, Bangkok, Thailand, May 25–28, 2003, pp. III-898–III-901.

11. J. Lundgren, S. Abdalla, M. O'Nils, and B. Oelmann, Power distribution and substrate noise coupling investigations on the behavioral level for photon counting imaging readout circuits, *J. Nucl. Instrum. Methods Phys. Res. Sect. A* **576**(1), 113–117, 2007.

12. J. R. Brews, The submicron MOSFET, in *High-Speed Semiconductor Devices*, S. M. Sze, ed., John Wiley & Sons, New York, pp. 139–210, 1990.

13. J. Briaire and K. S. Krisch, Principles of substrate crosstalk generation in CMOS circuits, *IEEE Trans. Comput. Aided Des. Integr. Circuits Syst.* **19**(6), 645–653, 2000.

14. T. Y. Chan, J. Chen, P. K. Ko, and C. Hu, The impact of gate-induced drain leakage current on MOSFET scaling, *IDEM Technical Dig.* **33**, 718–721, 1997.

15. A. Toriumi, M. Yoshima, M. Iwase, Y. Akiyama, and K. Taniguchi, A study of photon emission from *n*-channel MOSFET's, *IEEE Trans. Electron. Devices* **ED-34**, 1501–1508, 1987.

16. S. Tam and C. Hu, Hot-electron-induced photon and photocarrier generation in silicon MOSFET's, *IEEE Trans. Electron. Devices* **ED-31**, 1264–1273, 1984.

17. J. Briaire and K. S. Krisch, Substrate injection and crosstalk in CMOS circuits, in *Proceedings of the IEEE Custom Integrated Circuits Conference*, May 1999, pp. 483–486.

18. N. K. Verghese, T. J. Schmerbeck, and D. J. Allstot, *Simulation Techniques and Solutions for Mixed-Signal Coupling in Integrated Circuits*, Kluwer Academic Publishers, Dordrecht, 1995.

19. A. Samavedam, A. Sadate, K. Mayaram, and T. Fiez, A scalable substrate noise coupling model for design of mixed-signal ICs, *IEEE J. Solid-State Circuits* **35**(6), 895–904, 2000.

20. J. Lundgren, T. Ytterdal, M. O'Nils, Simplified gate level noise injection models for behavioral noise coupling simulation, in *Proceedings of the European Conference on Circuit Theory and Design*, Cork, Ireland, August 29–September 1, 2005.

21. P. G. Scrobohaci and T.-W. Tang, Modeling of the hot electron subpopulation and its application to impact ionization in submicron silicon devices—part I: Transport equations, *IEEE Trans. Electron Devices* **41**(7), 1197–1205, 1994.

22. P. G. Scrobohaci and T.-W. Tang, Modeling of the hot electron subpopulation and its application to impact ionization in submicron silicon devices—part II: Numerical solutions, *IEEE Trans. Electron Devices* **41**(7), 1206–1212, 1994.

23. J. Slotboom, G. Streutker, M. van Dort, P. Woerlee, A. Pruijmboom, and D. Gravesteijn, Nonlocal impact ionization in silicon devices, *IEDM* **91**, 5.4.1–5.4.4, 1991.

24. J.-G. Ahn, C.-S. Yao, Y.-J. Park, H.-S. Min, and R. W. Dutton, Impact ionization modeling using simulation of high energy tail distributions, *IEEE Electron Device Lett.* **15**(9), 348–350, 1994.

25. T.-W. Tang and J. Nam, A simplified impact ionization model based on the average energy of hot-electron subpopulation, *IEEE Electron Device Lett.* **19**(6), 201–203, 1998.

26. T. Sakurai, K. Nogami, M. Kakumu, and T. Iizuka, Hot-carrier generation in submicrometer VLSI environment, *IEEE J. Solid-State Circuits* **SC-21**(1), 187–192, 1986.

27. Y. W. Sing and B. Sudlow, Modeling and VLSI design constraints of substrate current, *IDEM* **26**, 732–735, 1980.

28. C. Hu, Hot-electron effects in MOSFETs, *IEDM* **29**, 176–181, 1983.

29. X. Aragonés, J. L. González, and A. Rubio, *Analysis and Solutions for Switching Noise Coupling in Mixed-Signal ICs*, Kluwer Academic Publishers, Dordrecht, 1999.

30. Y. I. Ismail, E. G. Friedman, and J. L. Neves, Equivalent Elmore delay for *RLC* trees, *IEEE Trans. Comput. Aided Des. Integr. Circuits Syst.* **19**(1), 2000.

31. Y. I. Ismail, E. G. Friedman, and J. L. Neves, Performance Criteria for Evaluating the Importance of On-Chip Inductance, *Proc. IEEE Int. Symp. Circuits Syst.* **2**, 224–247, 1998.

32. H. B. Bakoglu, *Circuits, Interconnections and Packaging for VLSI*, Addison-Wesley, Reading, MA, 1990.

33. A. V. Mezhiba and E. G. Friedman, Inductive properties of high-performance power distribution grids, *IEEE Trans. VLSI Syst.* **10**(6), 762–776, 2002.

34. Y. I. Ismail, E. G. Friedman, and J. L. Neves, Figures of merit to characterize the impedance of on-chip inductance, *IEEE Trans. VLSI Syst.* **7**(4), 442–449, 1999.

35. T. A. Johnson, R. W. Knepper, V. Marcello, and W. Wang, Chip substrate resistance modeling technique for integrated circuit design, *IEEE Trans. Comput. Aided Des.* **CAD-3**(2), 126–134, 1984.

36. N. K. Verghese and D. J. Allstot, Rapid simulation of substrate coupling effects in mixed-mode ICs, *Proc. IEEE*, 18.3.1–18.3.4, 1993.

37. H. H. Y. Chan and Z. Zilic, A practical substrate modeling algorithm with active guardband macromodel for mixed-signal substrate coupling verification, *Int. Conf. Electron. Circuits Syst.* **3**, 1455–1460, 2001.

38. D. K. Su, M. J. Loinaz, S. Masui, and B. A. Wooley, Experimental results and modeling techniques for substrate noise in mixed-signal integrated circuits, *IEEE J. Solid State Circuits*, **28**(4), 420–430, 1993.

39. S. Kumashiro, *Transient simulation of passive and active VLSI devices using asymptotic waveform evaluation*, Ph.D. thesis, Carnegie Mellon University, 1992.

40. S.-Y. Kim, N. Gopal, and L. T. Pillage, AWE macromodels of VLSI interconnect for circuit simulation, in *IEEE/ACM International Conference on Computer-Aided Design, Digest of Technical Papers*, 8–12 November 1992.

41. B. R. Stanisic, N. K. Verghese, R. A. Rutenbar, L. R. Carley, and D. J. Allstot, Addressing substrate coupling in mixed-mode ICs: Simulation and power distribution synthesis, *IEEE J. Solid-State Circuits* **29**(3), 226–238, 1994.

42. R. Panda, S. Sundareswaran, and D. Blaauw, On the interaction of power distribution network with substrate, in *Proceedings of the International Symposium on Low Power Electronics and Design*, 6–7 August 2001, pp. 388–393.

43. N. K. Verghese and D. J. Allstot, Verification of RF and mixed-signal integrated circuits for substrate coupling effects, in *Proceedings of the IEEE Custom Integrated Circuits Conference*, 5–8 May 1997.

44. M. Van Heijningen, M. Badaroglu, S. Donnay, M. Engels, and I. Bolsens, High-level simulation of substrate noise coupling generation including power supply noise coupling, in *Proceedings of the 37th Design Automation Conference*, 5–9 June 2000, pp. 446–451.

45. E. Charbon, R. Gharpurey, P. Miliozzi, R. G. Meyer, and A. Sangiovanni-Vincentelli, *Substrate Noise, Analysis and Optimization for IC Design*, Kluwer Academic Publishers, Dordrecht, 2001.

46. M. de Berg, M. van Kreveld, M. Overmars, and O. Schwarzkopf, *Computational Geometry, Algorithms and Applications*, Springer, Berlin, 1997.

47. A. J. van Genderen, N. P. van der Meijs, and T. Smedes, Fast computation of substrate resistances in large circuits, in *Proceedings of the European Design Test Conference*, 1996, pp. 560–565.

48. A. J. van Genderen, N. P. van der Meijs, and T. Smedes, *SPACE Substrate Resistance Extraction User's Manual*, Report ET-NS 96-03, Circuits and Systems Group, Department of Electrical Engineering, Delft University of Technology, The Netherlands.

49. J. Lundgren, Simulating behavioral level on-chip noise coupling, Ph.D. thesis.

50. G. Gramegna, et al., CMOS preamplifier for low-capacitance detectors, *IEEE Trans. Nucl. Sci.* **44**(3), 318–325, 1997.

51. M. A. Abdalla, A new biasing method for CMOS preamplifier-shapers, in *Proceedings of the 7th IEEE International Conference on Electronics Circuits and Systems*, 2000, pp. 15–18.

52. B. Oelmann et al., Robust window discriminator for photon-counting pixel detectors, *IEE Proc. Optoelectron.* **149**(2), 65–69, 2002.

# PART II
# Nuclear Medicine (SPECT and PET)

# 5 Nuclear Medicine: SPECT and PET Imaging Principles

ANNA CELLER

## 5.1. INTRODUCTION

A physician's ability to diagnose and treat disease, as well as a medical researcher's capability to advance current frontiers of knowledge in many areas of medical and biological sciences, is defined by his or her ability to investigate and visualize the inner workings of living systems. Imaging technologies have become a central part of both everyday clinical practice and medical and biological investigations. Their applications range from everyday clinical diagnosis to advanced studies of biological systems at a molecular level within basic, translational, and clinical research. Their relevance lay in many diverse areas, such as, for example, diagnosis and staging of cancer, assessment of the cardiovascular system, and multiple applications in neurosciences and molecular genetics.

Medical imaging represents an increasingly important component of modern medical practice, because early and accurate diagnosis can substantially influence patient treatment strategies and improve this treatment outcome and, by doing so, hopefully decrease both mortality and morbidity in many diseases. Additionally, on a global scale, it can facilitate patient management issues and improve health-care delivery and the effectiveness of utilization of resources.

Depending on the physics phenomena that are being used to obtain necessary imaging information, different modalities measure and visualize different characteristics of the investigated object [1, 2]. Attenuation of electromagnetic radiation lies behind X-ray imaging and computed tomography (CT); sound wave transmission and reflection are used in ultrasound (US) imaging; and magnetic resonance imaging (MRI) shows mainly body water contents because hydrogen in water molecules is responsible for the majority of the MRI signal. In general, however, unless special contrast agents are used, the images created by all these modalities display anatomy of the body rather than its function. This situation is changing rapidly with increasing

*Medical Imaging: Principles, Detectors, and Electronics*, edited by Krzysztof Iniewski
Copyright © 2009 John Wiley & Sons, Inc.

popularity of functional MRI (fMRI), which is sensitive to blood flow in the body; Doppler ultrasound measures organ movements (blood flow, heart beat); and more recently, CT angiography is performed with ultra-fast CT systems.

In this context, nuclear medicine (NM) imaging techniques, thanks to their inherently molecular characteristics and excellent sensitivity, have an important role to play, because they provide three-dimensional functional and in vivo information about biodistribution of tracer molecules labeled with radioactive isotopes [1–4]. For this reason, NM methods are often referred to as molecular imaging because they are able to localize molecular receptor sites, study biological markers expressed by diseased cells, and image the presence and extent of specific disease processes. The sensitivity of NM is several orders of magnitude better than MRI and CT for detection of metabolic changes in vivo [5–8]. Additionally, NM offers an exceptional opportunity to combine diagnosis with treatment: When the same molecule is labeled with one isotope, it can be used to diagnose tumor; when it is used with another molecule, can be used to carry chemo- or radiotherapy agents to treat the tumor; and, in parallel, it can be used to monitor effectiveness of this treatment.

## 5.2. NUCLEAR MEDICINE IMAGING

The general term "nuclear medicine" encompasses several different imaging techniques, ranging from simple planar and whole-body studies, to positron emission tomography (PET) and single-photon emission computed tomography (SPECT) [3–4, 9]. All these diagnostic techniques create images by measuring electromagnetic radiation emitted by the tracer molecules which have been labeled with radioactive isotope and introduced into a patient's body. Additionally, nuclear medicine includes internal radiotherapy (IRT) procedures where radioactive high-dose drinks or injections are being used in cancer treatment.

In general, any nuclear medicine imaging study may be classified as (a) single-photon imaging, which measures photons emitted directly from a radioactive nucleus, and (b) coincidence imaging of pairs of 511-keV photons emitted in exactly opposite directions. These photons are created by positron annihilation following positron decay of the investigated nucleus. PET studies are acquired by cameras containing several small detectors arranged in multiple rings around the patient, while single-photon imaging studies use Anger cameras, with one, two, or even three large-area detectors. Since practically all currently available Anger cameras are able to perform rotations and reconstruct tomographic SPECT images, these systems are often referred to as "SPECT cameras."

In general, depending on a clinical situation, the nuclear medicine data will be acquired using one of many different acquisition modes, each requiring a different data processing and analysis method [9].

- In static (planar) single-photon imaging the camera is positioned next to the investigated organ, such as heart or lungs, and the data are collected for several minutes. A two-dimensional map of radiotracer distribution is created. Planar images, however, suffer from poor contrast because photons coming from different layers in the body overlap and contributions from different organs cannot be separated.

- Tomographic, three-dimensional (3D) images are created when series of two-dimensional (2D) planar images (projections) acquired at different angles around a patient are combined and reconstructed. These projections are acquired sequentially by rotating detectors of Anger camera (SPECT) or simultaneously by rings of PET detectors.

- Dynamic imaging allows the user to investigate temporal changes in activity distribution, such as uptake of the tracer by an organ and its washout (for example, studies of kidney function, lung ventilation) to study processes related to body physiology. A series of planar (single-photon) or tomographic (PET or SPECT) images are acquired over time.

- Whole-body studies involve acquisitions of sequences of planar or tomographic images over different regions of the patient and creation of 2D or 3D composite images of a radiotracer distribution in all organs of a body.

- Gated studies are done by synchronizing photon acquisition from, for example, cardiac perfusion study with an ECG signal and/or with output signal from a monitoring device which tracks patient breathing movements. This allows for creation of 2D or 3D series of images, each corresponding to a particular phase of the moving organ. They can be viewed separately or as a movie.

For each clinical study, a software protocol is used to specify acquisition parameters such as the size of detection matrix, the number of projection angles, and the temporal duration of each view, as well as other parameters related to the data acquisition and rotation of the camera. The computer also digitizes and stores the collected data, and it performs data filtering and image reconstruction. Finally, it provides image analysis and visualization of the projections or reconstructed data.

## 5.3. RADIOTRACERS

The key element to the successful nuclear medicine imaging is proper selection of a radioactive tracer (radiopharmaceutical) [9, 10]. Radiotracer is composed of atoms of radioactive isotope attached to molecules of a pharmaceutical; they both need to be specific to their task. The isotope emits gamma photons ($\gamma$ photons) with energy high enough for a large proportion of the radiation to exit the body but energy low enough to be stopped by the detector. The radioisotope must also be relatively easy to produce, must be simple to attach to the pharmaceutical, and must have a half-life long enough for easy handling but short enough to not last too long in a patients body. It is also important for its emissions to not create high dose burden to the patient and the imaging technologist.

The majority of radioactive nuclei decay through emission of (a) an electron, (b) a positron (positive electron), or (c) through an electron capture. The daughter nucleus is usually left in an excited state and decays by emitting cascades of photons (and

**TABLE 5.1. Examples of the Most Popular Radioisotopes Used in Clinical SPECT Studies**

| Isotope | Half-Life | Main γ-Emissions (keV) | Photon Abundances (%) | Examples of Clinical Applications |
|---|---|---|---|---|
| Tc-99m | 6 h | 140 | 89 | Brain, heart, liver, lungs, bones, cancer, kidneys, thyroid |
| Ga-67 | 3.26 days | 93 | 38 | Abdominal infection, lymphoma, cancer imaging |
|  |  | 185 | 21 |  |
|  |  | 300 | 17 |  |
| In-111 | 2.80 days | 171 | 91 | Infections, cancer imaging |
|  |  | 245 | 94 |  |
| I-123 | 13.2 h | 159 | 83 | Thyroid, brain, heart metabolism, kidney |
| I-131 | 8.02 days | 364 | 81 | Thyroid, cancer imaging, metastasis detection, brain |
| Tl-201 | 3.04 days | 70 | X-rays | Myocardial perfusion |
|  |  | 167 | 11 |  |
| Xe-133 | 5.24 days | 81 | 37 | Lung ventilation, brain imaging, cerebral blood flow |

conversion electrons) with energies and intensities that are uniquely determined by the nuclear characteristics of the decaying nuclei. Thus, by measuring the energies of the emitted photons, a unique identification of the radiotracer can be made, while determination of the origin of their emission provides information about this radiotracer distribution, and signal intensity (number of recorded photons) relates to the amount of the radiotracer at any given location [3, 4]. Tables 5.1 and 5.2 list some of the most popular single photon (planar and SPECT) and positron emitting (PET) imaging isotopes, respectively. Information about their physical characteristics (half-lives, energies of emissions) and main clinical applications is provided.

**TABLE 5.2. Examples of the Most Popular Radioisotopes Used in Clinical and Research PET Studies**

| Isotope | Half-Life (minutes) | Mean Positron Energy (MeV) | Mean Range in Water (mm) | Examples of Clinical and Research Applications |
|---|---|---|---|---|
| F-18 | 110 | 0.24 | 1.0 | Brain studies, cancer imaging, metastasis detection, |
| O-15 | 2 | 0.73 | 1.5 | Blood flow, brain, heart research |
| N-13 | 10 | 0.49 | 1.4 | Brain, heart research |
| C-11 | 20 | 0.39 | 1.1 | Brain research |
| Rb-82 | 1.27 | 1.4 | 1.7 | Myocardial perfusion |

## 5.4. DETECTION SYSTEMS

Theoretically, by measuring photons emitted by the radioactive atom attached to a molecule, one should be able to obtain an exact information about the location of this molecule; however, in practice, the accuracy and resolution of NM imaging are severely restricted by both physical effects and deficiencies of the equipment. An extensive research effort of both academic centers and industry continuously focuses on development of techniques that could improve this situation [4, 11, 12].

Currently, almost all commercially available scanners employ scintillation crystals with photomultipliers to measure this radiation [3, 4]. Sodium iodine NaI(Tl) is certainly the most popular detector for the Anger camera. Modern PET systems use lutetium oxyorthosilicate (LSO), gadolinium oxyorthosilicate (GSO), or bismuth germanate (BGO); however, in the past, sodium iodide (NaI), cesium fluoride (CsF), and barium fluoride (BaF$_2$) were also tested for PET applications [3, 4]. Although the use of semiconductor detectors for medical imaging was advocated a long time ago, the first commercial systems equipped with CZT detectors have become available only recently [13, 14].

Behind the scintillating crystal(s) an array of photomultiplier tubes must be placed, the associated electronics with analog-to-digital converters. Their role, from the practical "imaging" point of view, is threefold. First, the electronic identification of the signal and its acceptance as "good event" requires pulse high analysis to measure the photon energy. The objective is to eliminate all undesired events (background and scattered photons). Then, the number of detected photons and the coordinates of each interaction in the detector(s) are being recorded. These data provide information about the amount of radiation and the directions of photons' arrivals. Additionally, in PET, for each photon the timing information is used to assign coincidence pairs. Alternatively, the data can be collected in a series of time frames for dynamic studies.

## 5.5. CLINICAL SPECT CAMERA—PRINCIPLES OF OPERATION

As mentioned, a typical SPECT system consists of one or more NaI scintillation detectors, which rotate around a patient to acquire information about the two-dimensional distributions of the detected photons (projections) from a number of angles. A standard detector thickness is $3/8$ to $1$ in. and its size is about $40\,cm \times 50\,cm$. These detectors are attached to a gantry, which in older systems often involved so-called "closed gantry" configuration and gradually evolved to more "open" systems. These are much more versatile and easy to adapt to different patient needs. Similarly, although in the past Anger cameras used smaller scintillators with round shapes, all current systems have large rectangular detectors. Lastly, several of the modern cameras combine functional (using radiotracers) and anatomical (with X-rays) imaging capabilities into a single hybrid SPECT-CT system. Figure 5.1 shows examples of such older as well as some of the newest systems.

Even though SPECT uses a wide range of radioisotopes (see Table 5.1), they all emit photons with energies within the 70- to 360-keV range. Additionally, the most

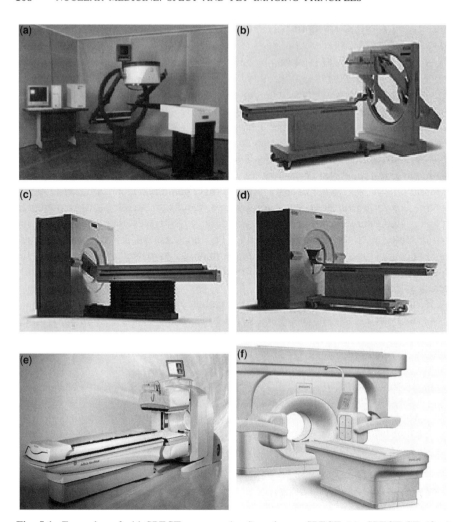

**Fig. 5.1.** Examples of old SPECT cameras (a–d) and new SPECT (e), SPECT-CT (f, g) and dedicated cardiac (h) systems (all figures were copied from the manufacturers websites). a, GE 400 Single Head Nuclear Camera (from http://www.tridentimaging.com/cgi-bin/ tridentimaging/ge400.html). b, Siemens Diacam (from Siemens old webpages). c, Siemens MS2. d, Siemens MS3. e, GE Infina-Hawkeye. f, Philips Precedence. g, Siemens Symbia. h, Philips Skylight. i, Siemens C.Cam specialized cardiac camera.

popular imaging isotope $^{99m}$Tc emits photons with energy 140 keV; therefore, understandably, the majority of SPECT systems are optimized for measuring this energy.

In order to determine the exact direction from which the detected photons arrive, a thick lead collimator is used [3, 4]. It consists of a sheet of lead containing thousands of holes and lying parallel to the crystal surface. A typical Anger camera configuration with a collimator, a detector, a large number of photomultipliers (usually 50–90), and

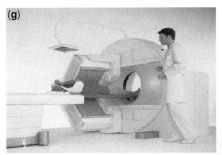
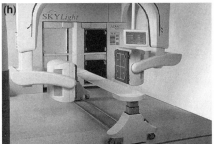

**Fig. 5.1.** *Continued.*

associated electronics is presented in Fig. 5.2. For most clinical applications, parallel hole collimators are used, with holes that are perpendicular to the surface of the crystal; however, several other collimator types are being used (fan-beam, cone-beam, pinhole). Each camera usually has more than one set of exchangeable collimators because their design is optimized for different applications. Collimators used for $^{99m}$Tc imaging have thinner septa than those which are used for studies involving isotopes with high-energy emissions such as $^{67}$Ga and $^{131}$I. Although collimators with longer septa have lower sensitivity, their better resolution is particularly useful in studies where small image details are important, such as bone scans. The use of converging collimators, such as fan-bean and cone-beam, improves sensitivity and resolution of the study, but distorts imaging geometry therefore special algorithms are required for image reconstruction. For studies of small organs (for example, thyroid gland), pinhole collimators are used because they create a magnified image but only for a small field of view (FOV). Table 5.3 compares the characteristics of some parallel hole collimators.

Use of a parallel hole collimator essentially means that only photons normal to the detector surface are transmitted through the collimator holes (photon a in Fig. 5.2). At the same time, photons arriving at any other angle are absorbed by the collimator septa (photon b). As a result, only photons traveling normal to the detector can have any influence on the produced image, providing a representation which is directly related to the spatial distribution of the radiopharmaceutical that emitted the radiation.

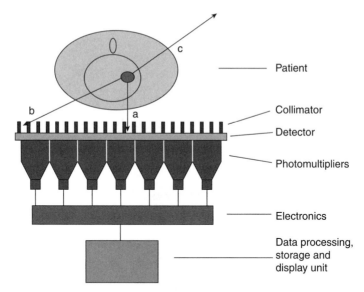

**Fig. 5.2.** The simplified schematic of a gamma camera. Three photons originating from a tumor in a patient head are shown. Photon a is parallel to the collimator septa is therefore transmitted through it and recorded by the detector. Photon b is absorbed by the collimator and photon c completely misses the camera.

Although theoretically only photons parallel to the collimator septa should pass through the collimator holes and reach the detector, in practice the collimator accepts all photons within a small cone corresponding to the acceptance angle of each collimator hole. This means that the resolution of the camera is distance-dependent and deteriorates as the distance between the source and the collimator increases [3, 4]. Additionally, since only a very few photons are emitted within the collimator acceptance

**TABLE 5.3. Physical Characteristics of Some Selected Parallel Hole Collimators**

| Collimator | LEGP | UHRES | MEGP | UHE |
|---|---|---|---|---|
| Hole diameter (mm) | 1.4 | 1.4 | 2.95 | 2.7 |
| Septal length (mm) | 24.7 | 32.8 | 48.0 | 60.0 |
| Septal thickness (mm) | 0.18 | 0.15 | 1.14 | 2.3 |
| Number of holes | 86,400 | 89,600 | 12,900 | 10,500 |
| Geometrical resolution at 10 cm (mm) | 8.0 | 6.3 | 10.7 | 9.2 |
| System resolution at 10 cm (mm) | 8.8 | 7.4 | 11.3 | 10.0 |
| Relative sensitivity | 1 | 0.6 | 0.8 | 0.4 |

*Abbreviations*: LEGP, low-energy general purpose; UHRES, ultra-high resolution; MEGP, medium-energy general purpose; UHE, ultra-high energy.

angle, the efficiency of the detection process is very low, usually on the order of $10^{-4}$. As a result of these two effects, nuclear medicine images notoriously suffer from poor resolution and low counts (or high noise levels).

The sensitivity of SPECT imaging, besides the effect related to the collimator acceptance angle, is also related to the efficiency of the detector material for detecting photons with a given energy and to the solid angle which is subtended by the detecting system. In this respect, 3/8-in. NaI crystals used in SPECT cameras provide high (close to 80–90%) sensitivity, but only for the detection of relatively low energy 140-keV photons; this sensitivity decreases to only 25–28% when 511-keV photons are being measured.

Although NaI(Tl) crystals are highly efficient at converting radionuclide emission photons into light photons, the amount of light produced is still quite small. The role of photomultiplier tubes (PMT) is to convert this light signal to an electric signal and amplify it. The tubes, which form a hexagonal array behind the crystal, are also used to localize detected photon and position each event on a discretized two-dimensional grid of camera bins. The amount of electrical signal produced in each PMT is proportional to the amount of light it sees, which in turn depends on its distance from the place where the photon interacted with the crystal. Therefore, the spatial location of the photon detection can be determined by combining the electrical signal from each photomultiplier tube in a logic circuit that weights the signals appropriately. This method of event localization is called Anger logic [3, 4].

Other factors that affect image resolution in SPECT include (a) the intrinsic resolution of the detector, (b) electronic noise, (c) issues related to the data representation, (d) reconstruction, and (e) processing (acquisition matrix, image voxel size, filtering, partial volume effects).

## 5.6. CLINICAL PET—PRINCIPLES OF OPERATION

Isotopes used in PET emit positrons that, after traveling for a small distance, annihilate with atomic electrons. Each annihilation event creates two gamma rays emitted in almost exactly opposite directions (Fig. 5.3). Positron emission tomography uses the coincidence detection of these nearly collinear annihilation photons to provide projection data from which tomographic images are reconstructed. No collimator is necessary as the source is assumed to be positioned along the line that connects the two detectors that have recorded both annihilation photons (Fig. 5.4).

A PET camera (Fig. 5.5) is composed of an array of scintillation detectors surrounding an imaging volume [1–4]. When a positron annihilates and one detector is hit by a photon, all other detectors are checked for the detection of a second photon. If a second photon is detected within a coincidence time window ($\sim$6–15 ns), one coincidence event is counted. If no second photon is detected, no coincidence event is registered. Then, by connecting detectors that recorded two coincident photons, a line called the "line of response" (LOR) is created. When multiple detectors are used, the coincidence counts from many parallel LOR are combined to produce a single projection and several projections from a ring PET system are

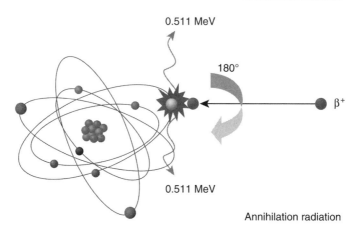

**Fig. 5.3.** Schematic representation of positron annihilation process. Position (denoted $\beta^+$) interaction with an atomic electron results in creation of two 511-keV photons that are emitted at 180° to each other.

needed to tomographically reconstruct a 3D image corresponding to the radiotracer distribution in the body.

Although currently the resolution of PET imaging is better than that of SPECT, the physics of positron decay and the detection of the resulting annihilation photons restrict the accuracy with which the position of the radiation source can be determined.

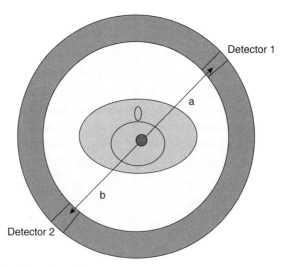

**Fig. 5.4.** Principle of operation of a PET system. The line linking two detectors which recorded the coincidence arrival of two annihilation photons creates a line-of-response (LOR) and determines the location of the positron-emitting atom.

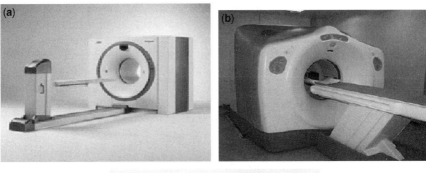

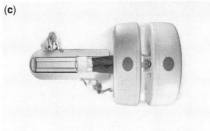

**Fig. 5.5.** Examples of PET and PET-CT systems (all figures were copied from the manufacturers websites). (a) Siemens Biograph 16. (b) GE Discovery CT. (c) Philips Gemini GXL.

Because positrons are emitted with a certain kinetic energy, the distance between the emission site and the annihilation point may be not negligible, resulting in uncertainty in source origin determination. Additionally, although in the center-of-mass system the two annihilation photons are emitted exactly in opposite directions, in the laboratory system there is an approximately 0.3° angle between them [15].

Further resolution loss may be related to the detector size and the depth of interaction error. For this reason, the size and spacing of the individual detectors is one of the most important factors that determines PET spatial resolution. In early PET systems, a single scintillator crystal was coupled to a single PMT, so that the minimum size of a detector was limited by the size of the PMT. Later, cut-block detectors were developed. In this design, each scintillator block is cut into an array of elements and several elements are read by a few PMTs [3, 4, 16]. This solution increases the number of detectors and decreases their sizes without significantly increasing the number of corresponding PMTs. The cuts are reaching various depths and are filled with a reflecting material. A unique pattern created by this design allows for identification of the crystal element where interaction took place. For example, the block detector in a typical whole body tomograph (CTI-Siemens ACCEL) uses an 8 × 8 matrix of 6.45-mm × 6.45-mm × 25-mm crystals (64 detector elements total) read out by only four PMTs. Additionally, neighboring detector blocks may share their PMTs. In this configuration, the identification of the event location is done using a technique similar to Anger logic used in SPECT data acquisition.

In modern PET cameras, with a large number of detectors, checking each event for coincidence with every other detector would be impractical. Instead, each detected event is given a time-stamp and sent to the digital coincidence processor. After a fixed time, each event in the buffer is tested for coincidence with every other event by comparing the differences in detection time with a predetermined coincidence window, and only then will the coordinates of LOR be calculated for each identified coincidence event. The event may be counted by incrementing the value in a matrix stored in computer that corresponds to a given LOR. Alternately, a file is created where event information, including its position, energy, and detection time, is stored in a list-mode format.

The detectors in PET cameras are arranged in a series of adjacent rings, and coincidences between detectors in different rings may be processed in two different ways. In 2D acquisition mode, thin annular lead shielding rings called "septa" are placed axially between the rings of crystals. Their role is to absorb photons arriving at large oblique angles. In this configuration, the shielding rings physically prevent true coincidences between all but the two neighboring detector rings, so coincidences are allowed only between detectors within the same ring or between the two neighboring rings. The 2D acquisition mode, which is often used for brain imaging, significantly reduces contributions from scattered photons and photons from radioactivity outside the field of view of the camera.

Most of the modern PET systems acquire data in "3D mode," where no interslice septa are used and coincidences are allowed between all the detectors in any rings. The 3D acquisition mode increases the sensitivity of the tomograph by a factor of 6–7 as it allows events along all the oblique LORs to be included. This increases the total number of counts collected, thus improving the statistical quality of the acquired data. However, the price is high. The 3D mode increase in sensitivity of the system is restricted by the electronic dead-time and a large number of random and scattered events that overwhelm the detector and the electronics. Additionally, the amount of data is huge, much larger than in 2D mode. Processing of these data and image reconstruction becomes much more complex because of the increased number of LORs and the difficulties caused by the oblique LORs intersecting multiple transverse slice planes. Therefore, some of the reconstruction algorithms use data "binning" to simplify the process [17]. This simplification, however, may result in compromised image resolution.

## 5.7. COMPARISON OF SMALL ANIMAL SCANNERS WITH CLINICAL SYSTEMS

In many situations, such as, for example, in research studies in genomic, molecular imaging, investigations of physiological processes, research involving development of new pharmaceuticals and new therapies animal models are used [18]. Imaging of small animals, such as mice and rats, is especially attractive as they are easy to handle, studies are inexpensive and several excellent small animal models of disease have been developed. These studies, however, require cameras with significantly better spatial resolution and sensitivity.

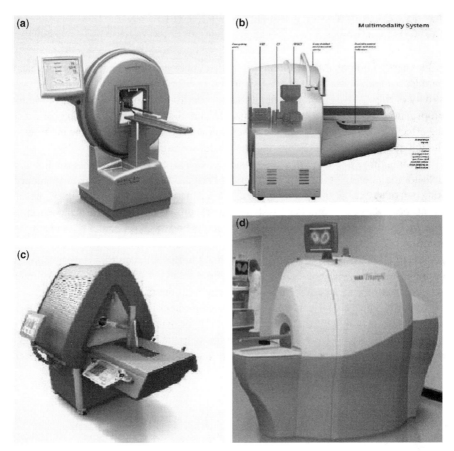

**Fig. 5.6.** Small animal hybrid systems: (a) Bioscan Nano-SPECT. (b) Siemens Inveon SPECT/ PET/CT. (c) MI-Labs U-SPECT/CT. (d) Gamma Medica Flex SPECT/PET/CT.

In order to improve resolution, preclinical (small animal) SPECT systems (Fig. 5.6) use pinhole collimators [18, 19]. Since imaging with single pinhole results in dramatically low efficiency, usually multiple pinholes in nonoverlapping or overlapping configuration are used. These systems have much better resolution (0.5–2 mm) and sensitivity (~0.3%) than clinical SPECT cameras (10–20 mm and 0.01–0.03%, respectively); the price is, however, a small field of view (5–8 cm).

Small-animal PET systems have very good sensitivity (1–3%), because the animal can be almost completely surrounded by multiple rings of detectors; however, due to discussed above physics limitations the best resolution of preclinical PET systems (1–2 mm) is slightly worse than that achieved in SPECT systems. Standard clinical PET cameras have resolution and sensitivity equal to 5–8 mm and 1–3%, respectively.

## 5.8. ELECTRONIC COLLIMATION PRINCIPLE AND COMPTON CAMERA

High-energy photons that are emitted in nuclear decays cannot be collimated by standard lenses and/or lead collimators without seriously limiting the resolution and sensitivity of the imaging system. Therefore, in order to identify the direction of photon propagation, an electronic collimation has been proposed [20]. Its principle is based on the effect of Compton scatter and the well-known formula that relates the energies of the initial and secondary photon and the scattering angle.

The imaging system in this case consists of two layers of detectors: In the first thin detector the photon emitted from the patient body undergoes Compton scatter, depositing part of its energy and is completely absorbed in the second detector (Fig. 5.7). Both events occur basically simultaneously and are recorded using coincidence electronics. The energy deposited in each of the two detectors and the positions of both events are measured. This information allows the user to define a conical surface which corresponds to all possible directions of the original photon. By recording several photons coming from any given source, location multiple cones are created. The point of intersection of these cones corresponds to the location of the radioactive source.

Although the general concept of using electronic collimation has been proposed already a long time ago, the detectors available at that time precluded construction of a Compton camera which could compete with traditional systems. Recently, a number of new detectors became available, but the problem of a Compton camera

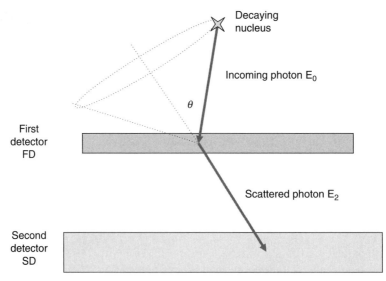

**Fig. 5.7.** The principle of electronic collimation. Photon originating from the decaying nucleus undergoes Compton scatter in the first detector (FD) and is absorbed by photoeffect in the second detector (SD). Information about the positions of each detection and deposited energies allow us to define a conical surface on which the original nucleus is located.

design that would be practical for medical imaging still remains unsolved. The common difficulties that have been identified are related primarily to a very complex reconstruction problem. This is mainly because Compton scatter allows the user to assign events to conic surfaces only; thus possible location of the radioactive source can be anywhere on the surface of a cone, which creates serious ambiguity in the reconstruction. Other problems are caused by the Doppler effect, which for low-energy photons seriously affects the accuracy of the measured energies and positions [21]. Further difficulties are related to the design and construction of the camera. They include the selection of the most appropriate material for both detectors, the design of these detectors' geometry, the choice of system electronics (both coincidence and readout), and the final assembly of the camera system. It is expected that the recent advancements in CZT detectors will lead to significant progress in Compton camera development [22, 23].

## 5.9. HYBRID SPECT–CT AND PET–CT SYSTEMS

The possibility to perform combined nuclear medicine and X-ray scans always seemed very attractive because such studies could provide simultaneous and perfectly registered information about the anatomy and function of a patient. Therefore, the first attempts to build such combined systems [24] met with a lot of interest and support from the imaging community.

Although the progress has been slow, currently many such systems are available and fast gaining popularity and recognition (Fig. 5.1e–g). In particular, with the introduction of hybrid imaging into clinical practice, accurate localization of previously obscure radiotracer uptake sites has became possible leading to anato-functional diagnostic imaging of cancer. Furthermore, such dual-modality studies allow for accurate assessment of radiotracer distribution in tumors which can potentially lead to full data quantitation and much more accurate calculations of the absorbed dose that it is currently possible. Presently, almost all newly installed PET and a large majority of SPECT cameras are combined hybrid NM-CT systems. More importantly, encouraged by this very successful example, Pichler et al. [25] have reported the first attempts to combine PET-MRI into a single unit.

## 5.10. PHYSICS EFFECTS LIMITING QUANTITATIVE MEASUREMENT

The two most important effects that degrade the accuracy and quality of nuclear medicine images are photon attenuation and scatter [4, 12]. Both effects are a result of photon interactions with the tissues in a patient body. Attenuation refers to the fraction of photons that are completely removed or diverted from their original path to the detector. Attenuated photons may be removed through photoelectric absorption or be diverted through scattering in which the photons energy and its direction of propagation change. If the scattering angle is large, also the photon energy decrease is large and the scattered photon may be eliminated by the energy discrimination of the camera. However, some subset of the scattered photons will be still accepted by the camera electronics. They are usually recorded in an incorrect part of the camera, providing false information about the source location and intensity. While the number of scattered events in SPECT studies is relatively low (~20–30%) and depends only weakly on the acceptance angle of the collimator, in PET their relative contribution increases dramatically when septa are removed and 3D acquisition mode is used (~20–30% in 2D PET and 40–60% in 3D PET). Attenuation and scatter affect both modalities SPECT and PET, but their relative importance may be different in different studies because they depend strongly on the energy of the imaged photons, on distribution of the attenuating medium in the body, and on the characteristics of the imaging system. In general, the probability of attenuation and scatter in an object may be represented by its attenuation map. Such maps are best obtained through transmission scans that can be done using tomographic acquisition with an external radioactive source or X-ray system (using, for example, hybrid SPECT/CT and PET/CT).

Other factors that negatively influence the quantitative accuracy of PET images are related to random coincidences and detector and electronics dead-time. Random coincidence events occur when single photons from unrelated annihilations are detected within the coincidence timing window. They incorrectly identify the event as a single annihilation and produce a wrong LOR for this event, there by degrading image quality and measurement accuracy.

## 5.11. TOMOGRAPHIC RECONSTRUCTION METHODS

In nuclear medicine, tomographic image reconstruction techniques are used to obtain three-dimensional images representing distribution of radiotracer in a patient's body. These images are obtained from sets of projection data that are recorded at different angles around the patient. The quality of the image and the accuracy of the resulting data strongly depend on the reconstruction method. Although analytical reconstruction methods (such as, for example, filtered back projection) are fast, they do not allow the reconstruction technique to model the SPECT or PET imaging processes [12, 26]. This can be done when iterative methods are used, where an accurate reconstruction may include realistic modeling of the patient, the data acquisition process, and the statistical nature of radioactive decay and photon interactions.

### 5.11.1. Filtered Back-Projection Reconstruction

Filtered back projection (FBP) is an analytical technique that is still the most commonly used method for image reconstruction in the clinical environment [3, 4, 26, 27]. It employs a very simple model of the imaging process which assumes that the number of photons recorded in any given detector bin represents the sum of contributions from the activity located along a line perpendicular to the detector surface. Back-projection procedure redistributes all these photons (counts) back along a line drawn through the images space. The points of intersection for the back-projected

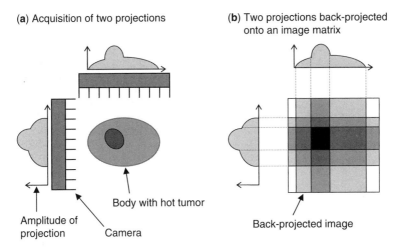

**(a)** Acquisition of two projections

**(b)** Two projections back-projected onto an image matrix

Body with hot tumor

Amplitude of projection

Camera

Back-projected image

**Fig. 5.8.** A simplified representation of the process of filtered back projection in 2D.

lines are assumed to correspond to potential source locations. The projection and back-projection processes are shown schematically in Fig. 5.8. The back-projection process results in a well-known star artifact around every source location unless the projections are convolved with a sinc filter (ramp filter in Fourier domain) prior to back projection.

As mentioned, the FBP method is fast and simple and is much less demanding than iterative algorithms. However, there are many problems associated with its use. Because the method does not account for the statistical noise in the data and the high-frequency noise is enhanced by the ramp filter, smoothing filter must be included in the reconstruction, which, in turn, degrades image spatial resolution. More importantly, however, FBP uses a simplistic model of the imaging process which does not allow for incorporation of any realistic data compensation techniques; therefore its images lack quantitative accuracy and often contain significant artifacts.

### 5.11.2. Iterative Reconstruction Algorithms

Iterative reconstruction algorithms iteratively search for the best estimate of the activity distribution which would correspond to the measured projection data and a particular model of the data acquisition process. Depending on the algorithm, this search is done by maximizing or minimizing a cost function that describes the goodness of fit between the estimated image and the measured projections. Reconstruction begins with an initial simplistic estimate of the image; for example, a uniform activity distribution is assumed. This estimate is then forward projected and the resulting projections are compared with the measured projections to produce image corrections that are subsequently back projected and applied to calculate an update of the estimated image. Figure 5.9 presents a block diagram illustrating the principle of iterative reconstruction. An additional smoothing filter may be applied to the final image to remove noise.

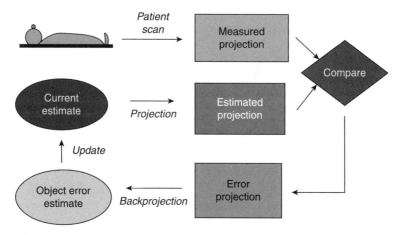

**Fig. 5.9.** A simplified representation of the principle of iterative reconstruction.

The iterative process is repeated until a convergence criterion is reached (if defined). Regularizers can also be included that enforce some *a priori* knowledge about the object and/or its activity distribution. For example, different assumptions about the smoothness of activity changes within the body can be implemented. The matrix which describes the probability that a photon emitted from any element (voxel) in the object is detected in any projection bin in the detector is called the system matrix. The system matrix can contain all geometrical details of the detection system (for example, collimator acceptance angle and detector efficiency), the details of the imaged object (distribution of the attenuating medium, object shape), and other data acquisition parameters (model of random events), and it can account for statistics of data collection (Poisson statistics). Therefore, through the use of a system matrix, the iterative reconstruction has the potential benefit that the physics of the acquisition process may be incorporated into the mathematics of the projector/back-projector calculations as the effects of counting statistics, attenuation, scatter, collimator blurring (in SPECT), or random events (in PET) will be included in the reconstruction process.

The iterative reconstruction method that includes the modeling of the Poisson counting statistics in the acquisition process is maximum-likelihood expectation maximization (MLEM) [26, 28, 29]. One of the disadvantages of MLEM algorithm is that it does not converge to a unique solution, but continues to iterate by attempting to fit the best distribution to the noise in the projection data, thereby degrading the noise characteristics of the reconstructed image. As a result, it is not obvious at what number of iterations the iterative process should be stopped. This optimal number of iterations cannot be known prior to reconstruction because it depends on the spatial frequencies and intensities present in the unknown activity distribution as well as the exact form of the system model used. Therefore, in practice the number of iterations is usually determined empirically. Another serious limitation of the MLEM algorithm is the time necessary to perform a reconstruction. The problem is exacerbated as the system matrix used becomes more complicated (less sparse) to better model the physics of data acquisition process. A commonly used technique to accelerate MLEM is the Ordered Subsets-Expectation Maximization (OSEM) algorithm [30]. As the MLEM algorithm uses all the projection angles in each iterative step of the reconstruction process, OSEM in each iteration employs only a small subset of projections and progressively uses different subsets in each iteration. Since in OSEM the image is updated more frequently than MLEM, OSEM reconstruction is generally much faster. Unfortunately, OSEM has the same problems associated with lack of convergence criterion as does MLEM.

## 5.12. DYNAMIC IMAGING

Dynamic studies usually generate sequences of images where each image is intended to represent the spatial distribution of the tracer at some particular moment. Mostly planar imaging is used in clinical practice where curve fitting or compartmental analysis can be applied directly to the images to obtain required dynamic information [31–33]. In planar imaging, however, all information about organ depth in the body is lost, contributions from different layers overlap, and, since proper attenuation correction is impossible, quantitative analysis is impossible.

Much better results are obtained using tomographic techniques such as PET [4] or ring camera SPECT [34]. There all projections for each image are acquired simultaneously, and the images are reconstructed using conventional methods and then analyzed by similar methods to those used with planar imaging. Dynamic PET studies are rarely performed for clinical diagnostic purposes, and the majority of investigations are motivated by the neurological research questions about brain metabolism and function. A dynamic PET study usually involves a series of short acquisitions where the data from each time frame is independently reconstructed. Then, the resulting set of dynamic images can be visualized and used to estimate physiological parameters. To each voxel or a larger region of interest (ROI) a time-varying tracer density is associated and modeled using functions that are based on assumptions about the physiological processes generating the data.

In the case of rotating camera SPECT, the projections required for a single image cannot all be collected simultaneously, so that the projections, taken at different times, represent different distributions of the tracer. This results in artifacts when the projections are used for image reconstruction and may lead to errors in diagnosis [35].

Multidetector SPECT systems, with up to three heads, have higher sensitivity and are able to collect complete data sets more quickly than the prevalent single head cameras. Exploitation of their capability to improve the performance of conventional dynamic imaging techniques has been investigated for myocardial perfusion studies [36] but due to very short acquisition times the images are dramatically noisy. Recently, a new camera design has been proposed that employs a set of columns composed of CZT detectors with tungsten collimators. The columns move back and forth independently, allowing for fast acquisition of good-quality data [37], which potentially may lead to the next generation of fast dynamic SPECT acquisitions. There is a number of different approaches aiming at obtaining dynamic functional information from SPECT data. They may estimate parameters of the dynamic models directly from the projection data without image reconstruction [38]. Alternatively, they may use mathematical optimization techniques to reconstruct series of dynamic images directly from projection data acquired using single rotation of the standard SPECT camera [39].

## 5.13. QUANTITATIVE IMAGING

Quality of nuclear medicine images can be greatly improved by compensating for photon attenuation. Unfortunately, some studies have shown that AC, although often very effective in solving diagnostic doubt, may sometimes worsen the situation by accentuating other artifact-producing effects such as scatter, motion, and spatial resolution loss [12]. This enhancement of artifacts related to secondary effects creates major difficulties in diagnosis. As a result, although available in many clinical centers, AC is often not being used because corrections for other degrading factors are lacking or are inadequate. Therefore, it is extremely important that the corrections methods that are implemented in clinics are free of errors and well understood by the users.

In general, attenuation correction for PET is much easier to perform than is the case for SPECT. This is because the probability of recording a coincidence event in PET depends only on the total probability of photon attenuation along the length of the LOR and is independent of the point of emission. Therefore, because the knowledge of source location is not required, the emission sinograms may be simply corrected for attenuation prior to the reconstruction. This, however, must be done using an accurate patient-specific attenuation map obtained through transmission scanning. Performing similarly accurate attenuation correction in SPECT is more difficult and may be done only by incorporating patient-specific attenuation maps into the system matrix in iterative reconstruction procedure.

Achieving accurate scatter correction is similarly difficult in PET and SPECT. Ideally scatter correction should use a realistic model of the camera and the exact physics description of photon interactions in the object (patient body). Since this approach is very computationally demanding, a number of simplified methods exist [4, 11]. Very efficient, although not very accurate, techniques try to estimate scatter contributions from the data acquired in additional energy windows set below the main photopeak. Other approaches try to model scatter using convolution, Monte Carlo estimates, or simplified models of the object.

Additionally, in PET the contribution from random events and count losses from detector dead-times, as well as detector sensitivity differences, must be taken into account. SPECT does not require any of these corrections. However, in order to provide absolute values of activity concentration, both PET and SPECT data must be properly calibrated. The activity quantitation calibration requires scanning a point source or a cylinder containing a uniform, known radioactivity concentration.

An example of clinical situation where absolute activity quantitation would be needed is the identification of balanced triple-vessel disease in heart perfusion studies. In this case all three of the major heart arteries are similarly restricted: The uptake of the tracer into the myocardium is roughly uniform throughout the heart which appears healthy, whereas in fact the entire heart is ischemic and tracer uptake is uniformly decreased in all areas. Similarly, in oncology, the accurate measurement of tumor physiology has important implications in assessing the malignant potential and staging of tumors. The knowledge of absolute tumor uptake would be helpful in evaluating delivery of radiopharmaceuticals to targeted tumors during cancer treatment.

## 5.14. CLINICAL APPLICATIONS

As mentioned in the introduction, with a large number of imaging modalities currently being available in clinics, it is extremely important to realize and correctly weight their respective abilities in diagnosing disease. Especially in an area of rising medical costs and shrinking medical budgets, strengths and limitations of every technique must be taken into account in order to optimize patient care and avoid unnecessary duplication of the procedures [40]. Obviously, in this analysis the physical bases of image creation must be considered.

In this context, since the advantage of nuclear medicine studies is in its ability to investigate body and/or particular organ's metabolism, function, and physiology, the focus of nuclear medicine studies must be on answering this type of questions. There is no place here to discuss, or even list, all nuclear medicine medical procedures, so only a brief review of few most important applications will be provided. A more detailed discussion of its clinical applications may be found in reference 9 and in numerous other books and publications.

One of the most important clinical uses of SPECT is in the diagnosis of coronary arterial disease (CAD). In general terms, these studies involve evaluation of the myocardial perfusion in different parts of the heart muscle and identification of the viable and nonviable myocardium tissues. The basic presumption is that the radiopharmaceutical used in myocardial imaging will be taken up by muscular heart tissue in proportion to blood flow. In healthy individuals, blood flow increases during physical exertion. In patients with coronary artery disease, blood perfusion at rest is normal but the ability to provide a larger blood supply during exercise may be compromised. By comparing SPECT images acquired in states of rest and physical stress, a physician can distinguish between normal, underperfused, and necrotic tissue. For example, necrotic tissue demonstrates reduced perfusion in both stress and rest studies but reversible ischemia demonstrates decreased uptake only in the stress scans. The most popular radiotracers used in these studies include $^{99m}$Tc-sestamibi, $^{99m}$Tc-tetrofosmin, and $^{201}$Tl-thalium chloride.

Other common SPECT applications include imaging of lung, for example, in the diagnosis of pulmonary emboli. Inhaled radio-pharmaceuticals such as $^{133}$Xe gas can be used to detect airway obstructions known as pulmonary emboli. These radio-pharmaceuticals bind to the collagen present in an embolus. The resulting SPECT scan provides information about the size and location of the embolus.

In bone imaging, not only tumor metastasis but also small fractures and fissures in bone tissue can be visualized using phosphate complexes (for example, $^{99m}$Tc-phosphate). Phosphate is more highly metabolized in the formation of new bone tissue, so an increase in distribution of this radiotracer allows physicians to detect abnormal bone growth. Renal plasma flow and glomular filtration rates can also be investigated with the use of SPECT imaging or dynamic planar nuclear medicine scans. Low renal blood flow and filtration rates can be an early indicator of kidney damage.

In the past, PET imaging has been used mainly in the research settings. This was due to the relatively high costs and complexity of the equipment and the support infrastructure, such as cyclotrons and radiochemistry laboratories. This has all changed; and presently, because of advances in technology, PET and PET/CT systems are widely used in clinical oncology, cardiology, and neurology to improve our understanding of disease pathogenesis, to aid with diagnosis, and to monitor disease progression and response to treatment.

The main clinical applications of PET [41] are related to oncology, where $^{18}$F-FDG tracer is used for the diagnosis, staging, therapy monitoring, and prognosis of patients with cancer. FDG is a glucose analogue that penetrates cells membranes using sodium-glucose transporters and specific glucose membrane transporters. Since normal physiological distribution of FDG in the body reflects the cellular and tissue consumption of glucose, PET images showing pathological increase in glucose consumption indicate the presence of tumor cells. The most aggressive tumors require greater glucose consumption to maintain their accelerated growth; therefore, $^{18}$F-FDG uptake is being used to differentiate between benign and malignant tumors and show necrotic tissue.

Other PET applications include neurology studies with measurements of such important parameters as regional blood flow, glucose and oxygen metabolism, and receptor binding. In cardiology, PET may be used to assess (a) regional myocardial perfusion defects with $^{82}$Rb for the diagnosis of CAD or (b) heart metabolism with $^{18}$F-FDG. Researchers in neuropsychology and psychiatry study different psychiatric disorders, substance abuse, and schizophrenia using numerous compounds, which bind selectively to brain neuroreceptors and may be radio-labeled with C-11 or F-18.

## REFERENCES

1. S. Webb, ed. *The Physics of Medical Imaging*, Institute of Physics Publishing, Bristol, CT, 1988.

2. Z.-H. Cho, J. P. Jones, and M. Singh, *Foundations of Medical Imaging*, John Wiley & Sons, New York, 1993.

3. S. R. Cherry, J. A. Sorenson, and M. E. Phelps, *Physics in Nuclear Medicine*, W. B. Saunders, Elsevier Science, Philadelphia, PA, 2003.

4. M. N. Wernick and J. N. Aarsvold, *Emission Tomography, The Fundamentals of PET and SPECT*, Elsevier, Academic Press, Amsterdam, Boston, 2004.

5. R. Y. Tsien, Imagining imaging's future, Supplement to *Nature Rev. Mol. Cell Biol.* **4**, SS16–SS21, 2003.

6. D. R. Piwnica-Worms, Introduction to molecular imaging, Supplement to *J. Am. Coll. Radiol.* **1**(1S), 2–3, 2004.

7. R. Weissleder, Center for Molecular Imaging, http://www.mghcmir.org/.

8. M. Phelps, Crump Institute for Molecular Imaging, http://www.crump.ucla.edu

9. H. N. Wagner, Jr., Z. Szabo, and J. W. Buchanan, *Principles of Nuclear Medicine*, W. B. Saunders, Philadelphia, 1995.

10. G. B. Saha, *Fundamentals of Nuclear Pharmacy*, Springer, Berlin, 2005.

11. A. Rahmim and H. Zaidi, PET versus SPECT: Strengths, limitations and challenges, *Nucl. Med. Commun.* **29**, 193–207, 2008.

12. H. Zaidi, *Quantitative Analysis in Nuclear Medicine Imaging*, Springer, Berlin, 2006.

13. Philips CZT camera, http://imarad.com/CustomersProducts.html

14. Gamma Medica-Ideas, http://www.evproducts.com/PDF/czt_spect.pdf

15. K. Iwata, R. G. Greaves, and C. M. Surko, Gamma-ray spectra from positron annihilation on atoms and molecules. *Phys. Rev. A* **55**, 3586–3604, 1997.

16. M. E. Casey and R. Nutt, A multi-slice two-dimensional BGO detector system for PET. *IEEE Trans. Nucl. Sci.* **NS-33**, 760–763, 1986.

17. M. Defrise, P. E. Kinahan, D. W. Townsend, C. Michel, M. Sibomana, and D. F. Newport, Exact and approximate rebinning algorithms for 3D-PET data, *IEEE Trans. Med. Imag.* **16**, 145–158, 1997.

18. D. J. Rovland and S. R. Cherry, Small-animal preclinical nuclear medicine instrumentation and methodology, *Semin. Nucl. Med.* **38**, 209–222, 2008.

19. K. Vunck, D. Beque, M. Defrise, and J. Nuyts, *IEEE Trans. Med. Imag.* **28**, 36–46, 2008.

20. M. Singh and D. Doria, Single photon imaging with electronic collimation, *IEEE Trans. Nucl. Sci.* **32**(1), 843–847, 1985.

21. C. E. Ordonez, A. Bolozdynya, and W. Chang, Doppler broadening of energy spectra in Compton cameras, Rush-Presbyterian-St Luke's Medical Center.

22. S. Watanabe et al., Development of CdTe pixel detectors for Compton cameras, *Nucl. Instr. Meth. A* **567**, 150–153, 2006.

23. L. Zhang, W. L. Rogers, and N. H. Clinthorne, Potential of a Compton camera for high performance scintimammography, *Phys. Med. Biol.* **49**, 617–638, 2004.

24. Y. Seo, C. Mari, and B. H. Hasegawa, Technological development and advances in single-photon emission computed tomography/computed tomography, *Semin. Nucl. Med.* **38**, 177–198, 2008.

25. B. J. Pichler, M. S. Judenhofer, C. Catana, J. H. Walton, M. Kneilling, R. E. Nutt, S. B. Siegel, C. D. Claussen, and S. R. Cherry, Performance test of an LSO-APD detector in a 7-T MRI scanner for simultaneous PET/MRI, *J. Nucl. Med.* **47**, 639–647, 2006.

26. P. P. Bruyant, Analytic and iterative reconstruction algorithms in SPECT, *J. Nucl. Med.* **143**(10), 1343–1358, 2002.

27. J. A. Parker, *Image Reconstruction in Radiology*, CRC Press, Boston, 1990.

28. L. A. Shepp and Y. Vardi, Maximum likelihood reconstruction for emission tomography, *IEEE Trans Med Imaging* **M1-1**, 113–122, 1982.

29. K. Lange and R. Carson, EM reconstruction algorithms for emission and transmission tomography, *J. Comput. Assist. Tomogr.* **8**, 306–316, 1984.

30. H. M. Hudson and R. S. Larkin, Accelerated image reconstruction using ordered subsets of projection data, *IEEE Trans. Med. Imag.* **13**, 601–809, 1994.

31. M. Hudon, D. Lyster, W. Jamieson, A. Qayumi, M. Keiss, L. Rosado, A. Autor, C. Sartori, H. Dougan, and J. V. D. Broek, Efficacy of $^{123}$I-iodophenyl pentadecanoic acid (IPPA) in assessment myocardial metabolism in a model of reversible global ischemia, *Eur. J. Nucl. Med.* **14**, 594–599, 1988.

32. http://en.wikibooks.org/wiki/Basic_Physics_of_Nuclear_Medicine/Dynamic_Studies_in_Nuclear_Medicine

33. C. D. Russell, Fitting linear compartmental models by a matrix diagonalization method, *Nucl. Med. Commun.* **22**, 903–908, 2001.

34. R. L. Furenlid, D. W. Wilson, Y. Chen, H. Kim, P. J. Pietraski, M. J. Crawford, and H. H. Barrett, FastSPECT II: A second-generation high-resolution dynamic SPECT imager, *IEEE Trans. Nucl. Sci.* **51**, 631–635, 2004.

35. M. O'Connor and D. Cho, Rapid radiotracer washout from the heart: Effect on image quality in SPECT performed with a single-headed gamma camera system, *J. Nucl. Med.* **33**, 1146–1151, 1992.

36. A. Smith, G. Gullberg, P. Christian, and F. Datz, Kinetic modeling of teboroxime using dynamic SPECT imaging of a canine model, *J. Nucl. Med.* **35**, 484–495, 1994.

37. D. Berman, J. Ziffer, S. Gambhir, M. Sandler, D. Groshar, D. Dickman, T. Sharir, M. Nagler, S. Ben-Haim, and B. H. Shlomo, D-SPECT: A novel camera for high speed quantitative molecular imaging: Initial clinical results, *J. Nucl. Med.* **47**(Suppl. 1), 131P, 2006.

38. B. W. Reutter, G. T. Gullberg, and R. H. Heusman, Direct least-squares estimation of spatiotemporal distributions from dynamic SPECT projections using a spatial segmentation and temporal B-splines, *IEEE Med. Imag.* **19**, 434–450, 2000.

39. T. Farncombe, A. Celler, C. Bever, D. Noll, J. Maeght, and R. Harrop, The incorporation of organ uptake into dynamic SPECT (dSPECT) image reconstruction, *IEEE Trans. Nucl. Sci.* **48**, 3–9, 2001.

40. R. J. Gibbons, Finding value in imaging; what is appropriate?, *J. Nucl. Cardiol.* **15**, 178–185, 2008.

41. P. E. Valk, D. L. Bailey, D. W. Townsend, and M. N. Maisey, *Positron Emission Tomography: Basic Science and Clinical Practice*, Springer, Berlin, 2003.

# 6 Low-Noise Electronics for Radiation Sensors

GIANLUIGI DE GERONIMO

## 6.1. INTRODUCTION: READOUT OF SIGNALS FROM RADIATION SENSORS

Radiation sensors detect and convert radiation into electric signals; the radiation of interest may include charged particles (e.g., fast electrons and heavy charged particles) and neutral particles (e.g., neutrons, X-rays, and gamma rays). In all cases, the radiation energy must be high enough to cause ionization in the sensing material. The resulting charges can be electron–ion pairs (e.g., gas sensors) or electron–hole pairs (e.g., semiconductor sensors). The amount of charge, $Q$, generated by an ionizing event is proportional to energy deposited in the sensor. Almost all applications require either discriminating or measuring $Q$ and, in some cases, its time of arrival. The sensor has at least two electrodes, but frequently has a larger number, especially on one of its sides (e.g., pixelated, strip, coplanar grid, or semiconductor drift sensors [1–6]). An electric field is generated in the sensor, most often by applying a voltage between (a) the pixelated (or segmented) electrodes on one side and (b) a common electrode on the opposite side. Under this electric field, the ionized charge, $Q$, moves toward one or more electrodes inducing a charge flow in each. This induction can be modeled by associating with each electrode both a current generator $I_Q = Q \cdot s(t)$ and a parallel capacitor $C_{DET}$. It is initially assumed that the charge, $Q$, is induced in an infinitesimally short time—that is, $s(t) = \delta(t)$—although the induction time and shape strongly depend upon the material's transport properties, the configuration of the electrodes, and the distribution of the ionizing event within the sensor. An induction time comparable to the time constant of the filter can introduce an error in the measurement of charge, also known as ballistic deficit.

The sensor may have a continuous component of current, as is the case in semiconductor materials. This component is often referred to as leakage current, $I_{LK}$, which has

*Medical Imaging: Principles, Detectors, and Electronics*, edited by Krzysztof Iniewski
Copyright © 2009 John Wiley & Sons, Inc.

an associated noise of typical unilateral power spectral density $2 \cdot q \cdot I_{LK}$. Mostly, this current, which is proportional to the volume of the sensor, reflects generation-recombination in the sensing material, and it can be reduced by decreasing the sensor's temperature.

Depending upon the type of sensor and the particular application, the charge can be read out event by event (e.g., in energy-resolving imagers and spectroscopic systems for research, security, safety, and medical applications) or it can be integrated from several events in the capacitance of the electrode and read out at a later time (e.g., in CCD or CMOS sensors for imaging and particle tracking). The ionizing events, or photons, occur in most cases randomly in time or, more rarely, during specific intervals, as in some high-energy physics experiments. Invariably, reading out the signals from radiation sensors requires highly specialized electronics, usually referred to as "front-end" electronics. This electronics usually entails stringent requirements in terms of the signal-to-noise ratio, dynamic range, linearity, stability, and the like. In this chapter, the design techniques and the circuit solutions adopted in designing state-of-the-art front-end electronics for radiation sensors are introduced. The sensors considered here are charge-sensing devices and can be modeled as capacitive sources. Other types of sensors, such as microbolometers and superconducting transition-edge, are beyond the scope of this work. Throughout the rest of this chapter, the sensor electrodes of interest will be referred to as "pixels." In most cases, radiation sensors are characterized by a high density of pixels frequently combined with high rate of events per pixel. Furthermore, the requirements on density and rate have tended to increase over the years. Hence, the front-end electronics almost always is developed in the form of an application-specific integrated circuit (ASIC) characterized by many densely packed front-end channels operating in parallel. The input of each channel is connected to one pixel. The channel provides low-noise amplification and processing of the signals from the pixel. The measurements stored in the channels are read out by efficient multiplexing schemes (a description of them is beyond the scope of this chapter).

Figure 6.1 is a block diagram of a typical front-end channel. It is composed of three fundamental blocks: the low-noise amplifier, the filter (usually referred to as the "shaper"), and the extractor. The low-noise amplifier reduces the relative noise

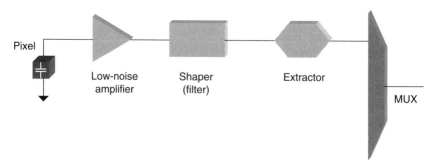

**Fig. 6.1.** Block diagram of a typical front-end channel.

contribution from the next blocks (i.e., the shaper and extractor) to negligible levels. The shaper is required to optimize the signal-to-noise ratio, while the extractor provides discrimination, measurement, and storage depending on the application (e.g., waveform discrimination and counting, waveform peak and/or timing, and periodic sampling).

## 6.2. LOW-NOISE CHARGE AMPLIFICATION

The current signal $I_Q$ induced in the sensing electrode (or pixel) can be integrated in the pixel capacitance and read out with a high input impedance stage, which amplifies the resulting voltage $\sim Q/(C_{det} + C_{amp})$ at the pixel node (e.g., a voltage amplifier, emitter, or source follower wherein $C_{amp}$ is the amplifier's input capacitance), or it can be read out directly with a low-input impedance stage, which amplifies the charge $Q$ (or the current $I_Q$) and keeps the pixel node at a virtual ground, such as a charge or current amplifier. The latter is usually the preferred choice since, among its other advantages, it stabilizes the sensing electrode by keeping its voltage constant during the measurement and/or the readout.

In both cases, the low-noise amplification is required to reduce the noise contribution from the processing electronics (such as the shaper, extractor, and ADC) to negligible amounts; good design practice dictates maximizing this amplification while avoiding overload of subsequent stages. Also, in both cases, low-noise amplification would provide either a charge-to-voltage conversion (e.g., source follower, charge amplifier) or a direct charge-to-charge (or current-to-current) amplification (e.g., charge amplifier with compensation, current amplifier). Depending upon this choice, the shaper would be designed to accept a voltage or a current, respectively, as its input signal.

In a properly designed low-noise amplifier, the noise is dominated by processes in the input transistor. Assuming that CMOS (complementary metal-oxide semiconductor) technology is employed in the design, the input transistor is referred as the "input MOSFET," although the design techniques here discussed can easily be extended to other types of transistors, such as the JFET, the bipolar transistor, and the heterojunction transistor [2, 6, 7]. The design phase, which consists of sizing the input MOSFET for maximum resolution, which essentially consists of maximizing the amount of signal charge that controls the MOSFET current, is called "input MOSFET optimization."

### 6.2.1. Input MOSFET Optimization

The resolution of front-end electronics is usually expressed in terms of equivalent noise charge (ENC), corresponding to the amount of charge that the pixel must deliver to the front-end electronics to have, at the output, a signal whose amplitude is equal to the rms noise. Figure 6.2 is a simplified schematic of the front-end electronics showing the components needed to calculate the ENC [2, 3, 8–10].

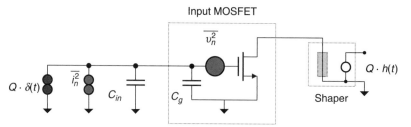

**Fig. 6.2.** Schematic of the front-end electronics showing the components needed to evaluate the ENC. From [10] with permission.

The parallel $\overline{i_n^2}$ and series $\overline{v_n^2}$ noise generators are characterized by unilateral power spectral densities $S_{in}(f)$ and $S_{vn}(f)$, respectively. The input capacitance has the component $C_{in}$, independent of the MOSFET (e.g., pixel, feedback, and interconnect parasitics) along with the component $C_g$, which depends on the MOSFET (e.g., $C_{gs}$ and $C_{gd}$). It is here assumed that the charge, $Q$, is induced in a very short time and that the front end implements a time-invariant filter (shaper) with pulse response $Q \cdot h(t)$, pulse peak amplitude $Q \cdot h(t)|_{max}$, and pulse peaking time $\tau_p$ (calculated from 1% of the peak amplitude to the peak amplitude).

The ENC can be determined in either the frequency domain or time domain. Here, the former approach is discussed and is more suitable for time-invariant filters. The latter, more befitting time-variant filters, such as gated integrators, is described elsewhere [2, 3, 11–16].

In the frequency domain, the ENC can be calculated as

$$
\text{ENC}^2 = \frac{\dfrac{1}{2\pi}\displaystyle\int_0^\infty \left[ S_{in}(\omega)|H(j\omega)|^2 + (C_{in} + C_g)^2\omega^2 S_{vn}(\omega)|H(j\omega)|^2 \right] d\omega}{h(t)|_{max}^2} \tag{6.1}
$$

where $\omega = 2\pi f$, and $H(s)$ is the Laplace transform of the pulse response $h(t)$. Solving Eq. (6.1) requires, along with the expression for $H(s)$, the appropriate modeling of $S_{in}(f)$, $S_{vn}(f)$, and $C_g$. Designers of front-end electronics frequently adopt the following classical model [17–21]:

$$
S_{vn}(f) = \frac{K_f}{C_{ox}WLf} + \frac{2}{3}\frac{4kT}{g_{mw}(J_d, L)W}, \qquad C_g = C_{ox}WL, \qquad S_{in}(f) = 2qI_{LK} \tag{6.2}
$$

where $C_{ox}$ is the oxide capacitance per unit area, $k$ is Boltzmann's constant ($\approx 1.3806 \times 10^{-23}\ \text{m}^2\ \text{kg s}^{-2}\ \text{K}^{-1}$), $T$ is the absolute temperature, $-q$ is the electron charge ($\approx 1.6022 \times 10^{-19}\ \text{C}$), $g_{mw}$ is the MOSFET's transconductance per unit of gate width $W$ (dependent on the density of the drain current, $J_d = I_d/W$, and on gate length, $L$), and $K_f$ is the $1/f$ noise coefficient (a technology-dependent parameter). The contribution from parallel noise typically is associated with the pixel leakage current, $I_{LK}$ (shot noise $2qI_{LK}$), but excess terms related to the charge amplifier's DC feedback and to the detector bias circuit (for AC-coupled pixels) can be added; here, for simplicity, they are assumed to be negligible.

It should be noted that in very deep submicron technologies the contribution from parallel noise also would encompass that associated with the MOSFET gate leakage current, which would depend on $W$, $L$, and $J_d$. Here, this term is neglected, but the optimization process can be easily extended to cover it.

Substituting the models (6.2) into (6.1), the ENC can be simplified:

$$\text{ENC}^2 = (C_{in} + C_{ox}WL)^2 \left[ a_w \frac{1}{\tau_p} \frac{2}{3} \frac{4kT}{g_{mw}(J_d, L)W} + a_f \frac{2\pi K_f}{C_{ox}WL} \right] + a_p \tau_p 2qI_{LK} \quad (6.3)$$

wherein the three coefficients $a_w$, $a_f$, and $a_p$ depend only on the type of shaper [2, 8, 10], as discussed later in Section 6.3.1. Typical values of $a_w$, $a_f$, and $a_p$ for a good time-invariant unipolar shaper are $a_w \approx 1$, $a_f \approx 0.5$, and $a_p \approx 0.5$. Since the second term in (6.3) related to the parallel noise is independent of the input MOSFET, it is effectively ignored in optimization.

The input MOSFET optimization consists of calculating the gate size ($W$, $L$) and the operating point ($J_d$) that, under the given design constraints, specifically minimize [Eq. (6.3)] the input capacitance $C_{in}$, the peaking time $\tau_p$, and the maximum power $P_{d\_max}$ allocated to the input MOSFET branch. To deal with the drain-to-source voltage $V_{ds}$, the designer assumes that the MOSFET will be operated slightly above its saturation voltage, considering the limited requirement on the voltage swing of the drain node in cascode configurations. Accordingly, defining $J_d$ corresponds to fully defining the operating point ($V_{gs}$, $V_{ds}$) of the input MOSFET, apart from the scaling factor $W$ that sets the drain current, $I_d$.

The input capacitance $C_{in}$ is defined by the pixel and its interconnect, while the maximum power $P_{d\_max}$ depends on system-level constraints, and the peaking time $\tau_p$ is set to satisfy requirements on the rate and/or induction time in the pixel. A further constraint may be placed on the peaking time when there is a non-negligible contribution of parallel noise. Using a single supply, $V_{DD}$, imposes a limit on the drain current $I_{d\_max} = P_{d\_max}/V_{DD}$, and consequently, constrains the product $J_dW = I_{d\_max}$. Hence, having defined $C_{in}$, $\tau_p$, and $P_{d\_max}$, the ENC depends only on $W$ (or $J_d$), $L$, and the polarity ($n$ or $p$ channel) of the input MOSFET.

The function $g_{mw}(J_d, L)$ can be extracted either from simple or advanced models or by using numerical results from advanced simulators, such as the Berkeley–Short channel IGFET model (BSIM) in the general-purpose circuit simulation program (SPICE). The coefficients $C_{ox}$ and $K_f$ can be obtained either from data provided by foundries or from direct measurements. Having them, the designer can plot (6.3) for given values of length, $L$. For example, assuming the minimum value $L_{min}$ for $L$ compatible with the technology, the designer would plot (6.3) as a function of $W$ and find the value $W_{opt}$ corresponding to the minimum $\text{ENC}_{min}$. Iterating for larger values of $L$ presumably would give a poorer minimum (lower $g_m$). Consequently, the input MOSFET optimization is complete, returning the design values for $L$, $W$, and $J_d$ as $L_{min}$, $W_{opt}$, and $I_{d\_max}/W_{opt}$.

The process described relies on the accuracy of the classical models (6.2). In entering deep submicron technologies, front-end designers revised these models, thereby generating a more reliable estimate of the achievable ENC and a more accurate optimization process. Front-end designers now would adopt the following enhanced models for $S_{vn}(f)$ and $C_g$ [10]:

$$S_{vn}(f) = \frac{K_f(L)}{C_{ox}WLf^{\alpha_f}} + \alpha_w n\gamma(J_d, L)\frac{4kT}{g_{mw}(J_d, L)W}, \qquad C_g = C_{gw}(J_d, L)W \quad (6.4)$$

where $C_{gw}$ is the MOSFET gate capacitance per unit of gate width, $W$ (dependent on $J_d$ and $L$), $\gamma$ is a dimensionless coefficient ranging between $1/2$ and $2/3$, and the typical values for the excess noise factor, $\alpha_w$, and the subthreshold slope coefficient, $n$, are 1 and 1.25, respectively. The new expression for the ENC can be approximated by [10]

$$\text{ENC}^2 = \left[C_{in} + C_{gw}(J_d, L)\cdot W\right]^2 \left[\frac{a_w}{\tau_p}\frac{\alpha_w n\gamma(J_d, L)4kT}{g_{mw}(J_d, L)\cdot W} + a_f(\alpha_f)\frac{K_f(L)}{C_{ox}WL}\frac{(2\pi)^{\alpha_f}}{\tau_p^{1-\alpha_f}}\right]$$
$$+ a_p\tau_p 2qI_{LK} \qquad (6.5)$$

where the coefficient $a_f(\alpha_f)$ has a typical value of $\approx 0.5$ (discussed in Section 6.3.1).

The most relevant improvements in this equation consist of modeling (a) the dependence of the gate capacitance on $J_d$ and (b) the low-frequency noise with the dependence of the coefficient $K_f$ on $L$ and with the slope coefficient, $\alpha_f$. Again, the function $C_{gw}(J_d, L)$ can be extracted either from simple or sophisticated models or by using numerical results from advanced simulators (e.g., the BSIM in SPICE); similarly, the coefficients $K_f(L)$ and $\alpha_f$ can be taken either from data provided by foundries or from direct measurements.

Figure 6.3 illustrates simulations of $C_{gw}(J_d, L)$ for $n$- and $p$-channel devices with different channel lengths, $L$, in a CMOS 0.25-μm technology. It reveals that $C_{gw}$ increases with the density of the drain current, and the impact of this dependence on the optimization process can be considerable. For example, in cases where the low-frequency noise dominates, using the maximum available power, $P_{d\_max}$, results in a far less satisfactory ENC, since increasing the power $P_d$ (i.e., $J_d$) increases $C_g$ without reducing the contribution from low-frequency noise [10].

Figure 6.4 shows the measured spectral densities of noise, $S_{vn}(f)$, for $n$- and $p$-channel devices with different channel lengths, $L$, in a CMOS 0.25-μm technology. Two observations can be made. First, the slope of the low-frequency component departs from $1/f$ and can be modeled with a noise slope coefficient, $\alpha_f$, that differs from unity [see (6.4)]. Second, the ratio between low-frequency spectra at any given frequency departs from the ratio between sqrt($L$) and can be modeled with a coefficient $K_f$ dependent on $L$. From these measurements, the designer can extract the values required for optimization (Table 6.1).

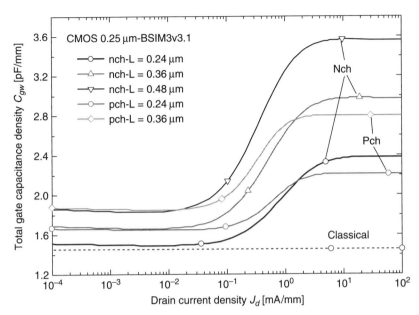

**Fig. 6.3.** Simulated $C_{gw}$ versus $J_d$ for $n$- and $p$-channel devices with different length, $L$, in a CMOS 0.25-μm technology. The classical model for the $n$-channel device at minimum $L$ also is depicted. From [10] with permission.

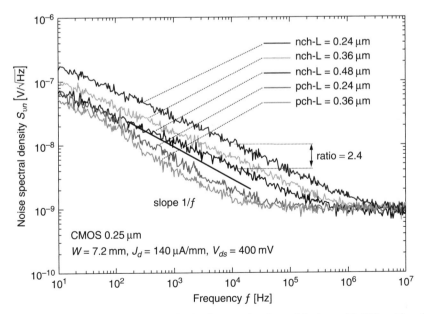

**Fig. 6.4.** Measured spectral densities of noise for $n$- and $p$-channel devices with different length, $L$, in a CMOS 0.25-μm technology. The $1/f$ slope also is shown. From [10] with permission.

**TABLE 6.1. Values for the Coefficients $K_f$ and $\alpha_f$ Extracted from the Measurements in Fig. 6.4. From [10] with permission**

|            | nch | | | pch | |
|------------|-----|---|---|-----|---|
|            | $L = 0.24\ \mu m$ | $L = 0.36\ \mu m$ | $L = 0.48\ \mu m$ | $L = 0.24\ \mu m$ | $L = 0.36\ \mu m$ |
| $K_f$      | $2.71 \times 10^{-24}$ | $1.4 \times 10^{-24}$ | $0.97 \times 10^{-24}$ | $0.60 \times 10^{-24}$ | $0.50 \times 10^{-24}$ |
| $\alpha_f$ | 0.85 | 0.85 | 0.85 | 1.08 | 1.08 |

*Abbreviations*: nch, *n*-channel; pch, *p*-channel.

By combining (6.5) with specific design constraints (e.g., input capacitance $C_{in}$, peaking time $\tau_p$, and maximum power $P_{d\_max}$ allocated to the input MOSFET branch), the designer can select the optimum power $P_{d\_opt}$, size ($W_{opt}$, $L_{opt}$), and operating point ($J_d$) of the input MOSFET that give the minimum ENC = $ENC_{min}$. Furthermore, by calculating $ENC_{min}$ versus $P_{d\_max}$, $C_{in}$, or $\tau_p$, the ENC achievable can be estimated for a given technology under different design constraints. For example, assuming the CMOS 0.25-μm technology discussed here, with a good time-invariant shaper and the constraints $\tau_p = 1\ \mu s$ and $P_{d\_max} = 1\ mW$, the results depicted in Fig. 6.5 will be obtained, where in the $ENC_{min}$ and the corresponding $P_{d\_opt}$ are shown as functions of $C_{in}$ for *n*- and *p*-channel devices with different channel lengths, $L$.

Figure 6.5 demonstrates that the lowest $ENC_{min}$ is not always achieved when using the minimum value $L_{min}$ compatible with the technology due to the dependence of the coefficient $K_f$ on the gate length, $L$. Furthermore, a reduction in the optimum power,

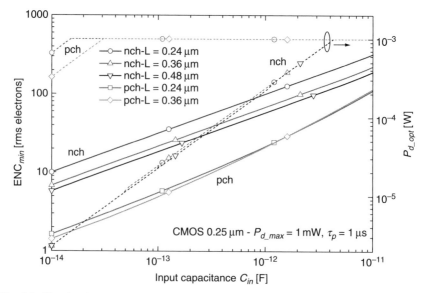

**Fig. 6.5.** Simulated $ENC_{min}$ (solid line, left scale) and $P_{d\_opt}$ (dashed line, right scale) versus $C_{in}$ for *n*- and *p*-channel devices with different length, $L$, in a 0.25-μm technology. From [10] with permission.

$P_{d\_opt} < P_{d\_max}$, is evident in this figure for small values of $C_{in}$, particularly with $n$-channel devices; this is a consequence of the dependence of the gate capacitance on the density of the drain current that pushes the input MOSFET toward weak-to-moderate inversion under dominant low-frequency noise.

The results in Fig. 6.5 correspond to the theoretical limits achievable with the technology, assuming that the noise is dominated by processes in the input MOSFET. Practically, the estimate should cover all the parallel noise contributions and, in some designs, the contributions from other components in the amplifier and the processing electronics. One frequent non-negligible contribution comes from the parallel noise associated with the pixel leakage current, $I_{LK}$, and with the charge amplifier feedback network, as discussed in the next section.

### 6.2.2. Adaptive Continuous Reset

The configuration of the low-noise amplifier most widely adopted by front-end designers is the charge amplifier with compensation, schematized in Fig. 6.6, wherein $C_i = C_{in} + C_g$. The charge amplifier consists of a negative high-gain voltage amplifier with a feedback capacitor, $C_f$. Additional feedback circuitry, referred to as the *reset* network, provides DC stabilization of the amplifier, discharge of the feedback capacitor after each event, and, when applicable, a DC path for the pixel leakage current, $I_{LK}$. The compensation network, *comp*, located between the output of the charge amplifier and the input of the shaper, consists of a capacitor $C_o$ and additional circuitry. This network typically is designed so that the charge injected in the shaper is an amplified replica of the charge $Q$ from the pixel with a charge gain, $G_Q$, given by the ratio $-C_o/C_f$. The shaper provides a virtual ground and accepts current as an input signal. It is represented in Fig. 6.6 as a voltage amplifier with feedback, where a resistor $R_s$ is put in for noise analysis. The series noise from the shaper's voltage amplifier contributes to the spectral density $S_{vn}(f)$ in (6.1) with a factor $(C_i/C_f)^2$; and since $C_i \gg C_f$, it can easily be rendered negligible.

The *reset* can be realized using a time-variant solution—for example, a switch or its MOSFET equivalent—that periodically, or after each event, discharges $C_f$, or a time-invariant solutions (e.g., a resistor or MOSFET equivalent), thereby providing a continuous discharge path. Here the latter is considered since, among other advantages, it minimizes the digital activity in the proximity of the front-end input. Ideally, a

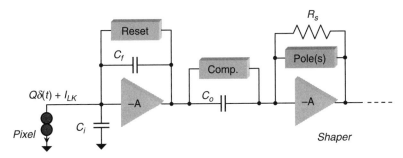

**Fig. 6.6.** Block diagram illustrating the charge amplifier with compensation.

simple resistor $R_f$ in parallel to $C_f$ could be adopted, and a pole-zero compensation could be realized by using, in parallel to $C_o$, a resistor $R_o$ with value $R_o = R_f C_f / C_o$. The resistor $R_f$ would stabilize the amplifier, ensuring the discharge of $C_f$ after each event with time constant $R_f C_f$ (this pole is compensated by the zero from $R_o C_o$), and also a continuous path for $I_{LK}$. The resistors $R_f$, $R_o$, and $R_s$ in Fig. 6.6 would contribute to the noise spectral density $S_{in}(f)$ in (6.1) with the white term $S_{in\_R}$:

$$S_{in\_R} \approx 4kT \left( \frac{1}{R_f} + \frac{1}{R_o G_Q^2} + \frac{1}{R_s G_Q^2} \right) = \frac{4kT}{R_f} \left( 1 + \frac{1}{G_Q} + \frac{R_f}{R_s G_Q^2} \right) \approx \frac{4kT}{R_f} \quad (6.6)$$

where $G_Q = -C_o/C_f = -R_f/R_o$ is the charge gain, assuming $G_Q \gg 1$ and $R_s \gg R_f/G_Q^2$. To make (6.6) negligible with respect to the shot-noise contribution, $2qI_{LK}$, from the pixel, the following design constraint must be applied:

$$\frac{4kT}{R_f} \ll 2qI_{LK} \longrightarrow R_f \gg \frac{4kT}{2q} \frac{1}{I_{LK}} \approx \frac{2V_t}{I_{LK}} \left( V_t = \frac{kT}{q} \approx 25\,\text{mV at } T \approx 300\text{K} \right) \quad (6.7)$$

Equation (6.7) shows that $R_f I_{LK} \gg 50\,\text{mV}$, also known as the "50 mV rule." Considering, for example, the typical pixel leakage current of semiconductor sensors $I_{LK} = 1\,\text{pA} - 1\,\text{nA}$ [ranging from silicon (Si) to cadmium zinc telluride (CZT)], it follows from (6.7): $R_f \gg 50\,\text{M}\Omega - 50\,\text{G}\Omega$. Due to constraints in technology and layout, these resistor values are far from practical in integrated circuit design, where those on the order of few hundred kilohms, at most, can be realistically considered. Hence, front-end designers implement the charge amplifier feedback and the compensation network using active devices; an overview of common solutions is given in reference 22. Here, a solution based on a single MOSFET [20, 23] is introduced, as shown in Fig. 6.7. The MOSFET $M_f$ has source and drain connected, respectively, to the amplifier output and input, and gate connected to a fixed bias voltage $V_G$.

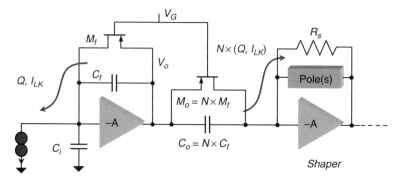

**Fig. 6.7.** Schematic illustrating the charge amplifier with compensation based on a single MOSFET.

The drain voltage of $M_f$ is set by the amplifier's virtual ground. The pixel leakage current, $I_{LK}$, flows through $M_f$ and sets the source voltage (i.e., the amplifier's output voltage) to a value $V_o = V_G + V_{GS}$. In this configuration, $M_f$ is *adaptive* to $I_{LK}$, meaning that via small changes in $V_{GS}$ it adapts to a wide range of values of $I_{LK}$ (sub-picoamperes to hundreds of nanoamperes) without requiring an external control. In terms of noise, $M_f$ contributes to the spectral density $S_{in}(f)$ in (6.1) with the term $S_{in\_Mf}$. The designer defines the gate voltage, $V_G$, and the gate size $(W, L)$ of $M_f$, so that the $M_f$ operates at saturation $(V_{DS} > V_{DS\_SAT})$ and, to the greatest possible extent, in strong inversion $(L \gg W)$. The first constraint makes $M_f$ less sensitive to voltage changes at the amplifier's input, and the second constraint minimizes $S_{in\_Mf}$. Also, when $M_f$ operates in saturation at a given drain current, $I_{LK}$, its noise contribution $S_{in\_Mf}$ decreases, moving from weak to strong inversion [20, 23]. In weak inversion, it results in $S_{in\_Mf} \approx \frac{1}{2} n 4kT g_m \approx 2q I_{LK}$ (where $g_m \approx I_{LK}/nV_t$ and $V_t \approx kT/q$), while in strong inversion it results in $S_{in\_Mf} = \frac{2}{3} 4kT g_m$ (where $g_m \ll I_{LK}/nV_t$). Comparable to the resistor, $M_f$ provides a discharge path to $C_f$ after each event; but, due to the nonlinear dependence of the drain current on $V_{GS}$, the amplifier's signal response, $V_o$, now is nonlinear. To ensure a linear charge amplification, the compensation network shown in Fig. 6.7 can be used, where $C_o$ and $M_o$ are, respectively, the $N$-times replicas of $C_f$ and $M_f$. The voltage amplifier in the shaper is a scaled-down copy of the front-end amplifier and, thus, maintains the input virtual ground at the same voltage.

Accordingly, the charge injected in the shaper reproduces the charge $Q$ from the pixel, with a charge gain $G_Q = -N$. Due to the mirror effect from $M_f$ and $M_o$, which share the same source, drain, and gate voltages, linearity is recovered. As in (6.6), the relative noise contribution from $M_o$ can be rendered negligible with $N \gg 1$.

The configuration in Fig. 6.7, using $p$-channel devices, applies when the negative charge $Q$ must be amplified. In the case of positive charge $Q$, the $n$-channel devices must be used, always keeping the sources connected to the front-end amplifier's output. Typical values of $N$ depend on the required dynamic range and can be as high as several tens. Should it be necessary to decrease the contribution from $R_s$, larger charge gain values will be needed, and the designer can place two or more charge gain stages in a cascade (gains $N_1$, $N_2$, ...); then, the type of MOSFET ($n$- or $p$-channel devices) also must be changed according to the polarity of the signal in each stage. The requirement on the gain, $N$, follows from the design constraint in (6.7):

$$\frac{4kT}{R_s N^2} \ll 2q I_{LK} \quad \longrightarrow \quad N \gg \sqrt{\frac{4kT}{2q} \frac{1}{I_{LK} R_s}} \approx \sqrt{\frac{50\,\mathrm{mV}}{I_{LK} R_s}} \qquad \text{(at } T \approx 300\mathrm{K)} \qquad (6.8)$$

Assuming, for example, $R_S = 100\,\mathrm{k\Omega}$, if $I_{LK} \approx 1\,\mathrm{nA}$ (typical for CZT), then it follows that $N \gg 22$, which could be achieved in a single stage; but if $I_{LK} \approx 1\,\mathrm{pA}$ (as is typical for Si), then $N \gg 700$, requiring a minimum of two stages [24].

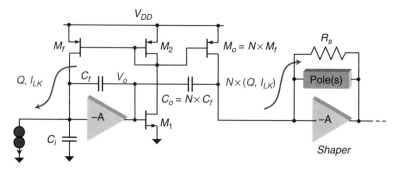

**Fig. 6.8.** Schematic illustrating an alternative configuration for the charge amplifier with compensation.

When designing the reset configuration in Fig. 6.7, problems may occur with the dynamics of the output voltage $V_o$. For example, referring to the case in Fig. 6.7 and using a $p$-channel device as an input MOSFET (threshold voltage $V_{THP}$), $V_o$ would be limited to a dynamic $V_{DD} - V_{THP} + V_G + V_{GS}$ that can be a very few hundreds of millivolts, especially when designing low-voltage technologies. This problem can be resolved via an alternative configuration that exhibits comparable performance without the same limit as the dynamic range. This configuration is shown in Fig. 6.8 [25], where the charge gain $N$ is given by the ratio between the compensation elements, $C_o$ and $M_o$, and the feedback elements, $C_f$ and $M_f$. The sources of $M_f$ and $M_o$ are connected to $V_{DD}$ and the gates are controlled by $M_1$ and $M_2$. This arrangement decouples the bias point of $V_o$ from the virtual ground of the amplifier, thereby offering the much larger dynamic range, $V_{DD} - V_{THN}$. Like the previous configuration, it is adaptive to the pixel leakage current, $I_{LK}$, and even if the signal response $V_o$ is nonlinear, the linearity in the charge amplification is recovered through this compensation. The noise contributions and design guidelines for $M_f$ and $M_o$ are comparable to the ones from the previous configuration. The noise contributions from $M_1$ and $M_2$ are canceled due to the mirror action from $M_f$ and $M_o$ [25].

In general, the solution in Fig. 6.8 is either an alternative or a complement to the solution in Fig. 6.7 and can be considered either as input stage or as second stage in designs that require multiple charge gain stages.

## 6.3. SHAPING AND BASELINE STABILIZATION

The low-noise amplifier is typically followed by a filter, frequently referred to as a shaper, most often responding to an event with a pulse of defined shape and finite duration ("width") that depends on the time constants and number of poles in the transfer function. The shaper's purpose is twofold: First, it limits the bandwidth to maximize the signal-to-noise ratio; second, it restricts the pulse width in view of processing the next event. Extensive calculations have been made to optimize the shape, which depends on the spectral densities of the noise in (6.1) and on additional constraints (e.g., available power and signal rate) [2, 3, 12, 26, 27]. Optimal shapers are difficult to realize, but they can be approximated, with results within a few percent from the optimal, with either analog or digital processors, the latter requiring analog-to-digital conversion of the charge amplifier signal (anti-aliasing filter may be needed). In the analog domain, the shaper can be realized using time-variant solutions (e.g., gated integrator) that limit the pulse width by a switch-controlled return to baseline or via time-invariant solutions (e.g., a semi-Gaussian shaper) that restrict the pulse width using a suitable configuration of poles. The latter, which, among other advantages, minimizes digital activity in the front-end channel, is discussed here.

### 6.3.1. High-Order Shaping

In a front-end channel, the time-invariant shaper responds to an event with an analog pulse, the peak amplitude of which is proportional to the event charge, $Q$. The pulse width, or its time to return to baseline after the peak, depends on the bandwidth (i.e., the time constants) and the configuration of poles. The most popular unipolar time-invariant shapers are realized either using several coincident real poles ($r$-shapers) or with a specific combination of real and complex-conjugate poles ($c$-shapers). The number of poles, $n$, defines the order of the shaper. Designers sometimes prefer to adopt bipolar shapers, attained by applying a differentiation to the unipolar shapers (the order of the shaper now is $n - 1$). Bipolar shapers can be advantageous for high rate applications, but at expenses of a worse signal-to-noise ratio [10, 28]. The transfer functions $T(s)$ of the most widely employed unipolar $r$- and $c$-shapers [29], apart from an amplitude factor, can be written as follows:

$$T(s) = \frac{1}{(s+p)^n} \qquad\qquad r\text{-shapers, } n = 2, 3, 4, \ldots$$

$$T(s) = \frac{1}{(s+p_1) \prod_{i=2}^{(n+1)/2} \left[(s+r_i)^2 + c_i^2\right]} \qquad c\text{-shapers, } n = 3, 5, 7, \ldots \qquad (6.9)$$

$$T(s) = \frac{1}{\prod_{i=1}^{n/2} \left[(s+r_i)^2 + c_i^2\right]} \qquad c\text{-shapers, } n = 2, 4, 6, \ldots$$

where $n$ is the order of the shaper, and $r_j$ and $c_j$ are the real and complex-conjugate parts of the roots of the equation

$$\frac{1}{0!} - \frac{s^2}{1!} + \frac{s^4}{2!} - \frac{s^6}{3!} + \cdots + \frac{s^{2n}}{n!} = 0$$

The time-domain representations of the shapers in (6.9) can be easily calculated as inverse Laplace transform using the partial fractions for real and complex-conjugate roots. Apart from an amplitude factor, we obtain

$$S(t) = \frac{1}{(n-1)!} t^{n-1} \exp(-tp) \qquad \qquad r\text{-shapers}, n = 2, 3, 4, \ldots$$

$$S(t) = K_1 \exp(-tp_1) + \sum_{i=2}^{(n+1)/2} 2|K_i| \exp(-tr_i)\cos(-ti_i + \angle K_i) \quad c\text{-shapers}, n = 3, 5, 7, \ldots$$

$$S(t) = \sum_{i=1}^{n/2} 2|K_i| \exp(-tr_i)\cos(-ti_i + \angle K_i) \qquad \qquad c\text{-shapers}, n = 2, 4, 6, \ldots$$

$$(6.10)$$

where the coefficients $K_i$ for the $c$-shapers are given by

$$K_1 = \frac{1}{\displaystyle\prod_{k=2}^{(n+1)/2} [-p_1 + r_k^2 + i_k^2]} \qquad \qquad c\text{-shapers}, n = 3, 5, 7, \ldots$$

$$K_i = \frac{1}{[-r_i - ji_i + p_1] \displaystyle\prod_{k=2, k \neq i}^{(n+1)/2} [-r_i - ji_i + r_k^2 + i_k^2] - 2ji_i} \qquad i > 1, c\text{-shapers}, n = 3, 5, 7, \ldots$$

$$K_i = \frac{1}{\displaystyle\prod_{k=1, k \neq i}^{n/2} [-r_i - ji_i + r_k^2 + i_k^2] - 2ji_i} \qquad \qquad c\text{-shapers}, n = 2, 4, 6, \ldots$$

$$(6.11)$$

The shapers in (6.9), or (6.10), are referred to as "semi-Gaussian" since as $n$ increases the shape tends to approximate a Gaussian curve. These shapers are characterized by specific time parameters, typically the peaking time $\tau_p$ (the time required to go from 1% of the peak amplitude to the peak) and the width $\tau_w$ (the time required to go from 1% back to 1% of the peak amplitude).

To a first order, the peaking time $\tau_p$ (i.e., $\approx$ the rise time) defines the shaper's bandwidth. As seen from (6.5), the lower the peaking time $\tau_p$ (that is, the higher the bandwidth), the higher the contribution of the white series noise to the ENC. Furthermore, the higher the peaking time, $\tau_p$, the higher the contribution of parallel noise to the ENC. For a given design, a peaking time that minimizes (6.5) invariably can be found.

To the first order, the pulse width, $\tau_w$, defines the time required for the pulse to return to baseline, which is important to the accuracy in processing of the next event. For a given peaking time, $\tau_p$, the higher the order of the shaper, the lower the pulse width, $\tau_w$. At equal order, $n$, and peaking time, $\tau_p$, $c$-shapers offer a lower width $\tau_w$ than the $r$-shapers [10].

As discussed in Section 6.2.1, the three coefficients $a_w$, $a_f$, and $a_p$ in (6.5) depend only on the shaper's type and order. Figure 6.9 shows the waveforms, the values of the coefficients $a_w$, $a_f(\alpha_f)$, and $a_p$, and the ratios $\tau_w/\tau_p$ for several $r$- and $c$-shapers, both

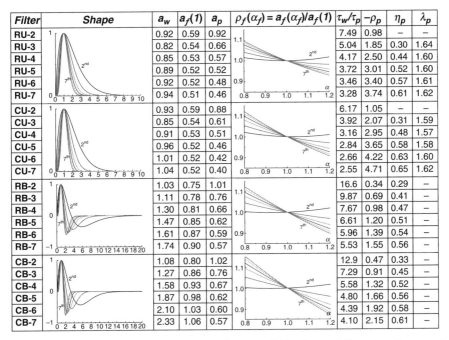

| Filter | Shape | $a_w$ | $a_f(1)$ | $a_p$ | $\rho_f(\alpha_f) = a_f(\alpha_f)/a_f(1)$ | $\tau_w/\tau_p$ | $-\rho_p$ | $\eta_p$ | $\lambda_p$ |
|---|---|---|---|---|---|---|---|---|---|
| RU-2 | | 0.92 | 0.59 | 0.92 | | 7.49 | 0.98 | – | – |
| RU-3 | | 0.82 | 0.54 | 0.66 | | 5.04 | 1.85 | 0.30 | 1.64 |
| RU-4 | | 0.85 | 0.53 | 0.57 | | 4.17 | 2.50 | 0.44 | 1.60 |
| RU-5 | | 0.89 | 0.52 | 0.52 | | 3.72 | 3.01 | 0.52 | 1.60 |
| RU-6 | | 0.92 | 0.52 | 0.48 | | 3.46 | 3.40 | 0.57 | 1.61 |
| RU-7 | | 0.94 | 0.51 | 0.46 | | 3.28 | 3.74 | 0.61 | 1.62 |
| CU-2 | | 0.93 | 0.59 | 0.88 | | 6.17 | 1.05 | – | – |
| CU-3 | | 0.85 | 0.54 | 0.61 | | 3.92 | 2.07 | 0.31 | 1.59 |
| CU-4 | | 0.91 | 0.53 | 0.51 | | 3.16 | 2.95 | 0.48 | 1.57 |
| CU-5 | | 0.96 | 0.52 | 0.46 | | 2.84 | 3.65 | 0.58 | 1.58 |
| CU-6 | | 1.01 | 0.52 | 0.42 | | 2.66 | 4.22 | 0.63 | 1.60 |
| CU-7 | | 1.04 | 0.52 | 0.40 | | 2.55 | 4.71 | 0.65 | 1.62 |
| RB-2 | | 1.03 | 0.75 | 1.01 | | 16.6 | 0.34 | 0.29 | – |
| RB-3 | | 1.11 | 0.78 | 0.76 | | 9.87 | 0.69 | 0.41 | – |
| RB-4 | | 1.30 | 0.81 | 0.66 | | 7.67 | 0.98 | 0.47 | – |
| RB-5 | | 1.47 | 0.85 | 0.62 | | 6.61 | 1.20 | 0.51 | – |
| RB-6 | | 1.61 | 0.87 | 0.59 | | 5.96 | 1.39 | 0.54 | – |
| RB-7 | | 1.74 | 0.90 | 0.57 | | 5.53 | 1.55 | 0.56 | – |
| CB-2 | | 1.08 | 0.80 | 1.02 | | 12.9 | 0.47 | 0.33 | – |
| CB-3 | | 1.27 | 0.86 | 0.76 | | 7.29 | 0.91 | 0.45 | – |
| CB-4 | | 1.58 | 0.93 | 0.67 | | 5.58 | 1.32 | 0.52 | – |
| CB-5 | | 1.87 | 0.98 | 0.62 | | 4.80 | 1.66 | 0.56 | – |
| CB-6 | | 2.10 | 1.03 | 0.60 | | 4.39 | 1.92 | 0.58 | – |
| CB-7 | | 2.33 | 1.06 | 0.57 | | 4.10 | 2.15 | 0.61 | – |

**Fig. 6.9.** Waveforms and coefficients for unipolar and bipolar semi-Gaussian shapers of different order. From [10] with permission.

unipolar (*ru*, *cu*) and bipolar (*rb*, *cb*). The coefficient $a_f(f)$ can be calculated as $a_f(1)\rho_f(\alpha_f)$, where $\alpha_f$ is the slope of the low-frequency noise in $S_{vn}(f)$.

For each shaper, Fig. 6.9 shows three other coefficients: the relative slope at zero crossing $\rho_p$, the relative time at zero crossing $\eta_p$, and the relative peak amplitude $\lambda_p$ (for unipolar only). For unipolar pulses, these coefficients are calculated on the derivative of the pulse. In all cases, the coefficients are normalized to the peaking time, $\tau_p$, and the peak amplitude, $V_p$. The coefficient $\rho_p$ is calculated for unipolar pulses as $v''(\tau_{p0}) \times \tau_p^2/V_p$; thus it is a measure of the curvature at the peak or of the slope of the derivative at zero crossing where $v''$ is the second derivative of the waveform and $\tau_{p0}$ is the time at which the unipolar pulse peaks (equal to the time at which the derivative $v'$ of the unipolar pulse crosses zero). Comparably, for bipolar pulses it is $v'(\tau_{z0}) \times \tau_p/V_p$ (i.e., the slope at zero crossing), where $v'$ is the first derivative of the waveform and $\tau_{z0}$ is the time at which the pulse crosses zero. The coefficient $\eta_p$ is calculated for unipolar pulses as $\tau_{pb}/\tau_p$, where $\tau_{pb}$ is the time for the derivative of the unipolar pulse to go from 1% of its peak amplitude to zero crossing, and for bipolar pulses as $\tau_p/\tau_z$, where $\tau_z$ is similarly defined. The coefficient $\lambda_p$ is calculated as $\tau_p V'_p/V_p$, where $V'_p$ is the peak amplitude of the derivative. These coefficients are useful when evaluating the performance of shapers in terms of ballistic deficit or timing resolution, as discussed later.

When designing without constraints on the peaking time, it is easily verified that higher-order unipolar shapers perform better than lower-order ones in terms of

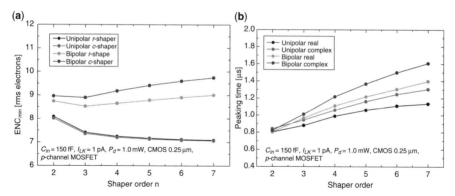

**Fig. 6.10.** (a) Calculated $ENC_{min}$ for semi-Gaussian shapers of different order assuming $C_{in} = 150$ fF, $I_{LK} = 1$ pA, $P_d = 1$ mW, and $p$-channel input MOSFET in a 0.25-$\mu$m technology. (b) The corresponding peaking times.

$ENC_{min}$. For example, in Fig. 6.10a the $ENC_{min}$ is calculated for a $p$-channel input MOSFET in a 0.25-$\mu$m technology and $C_{in} = 150$ fF, $I_{LK} = 1$ pA, $P_d = 1$ mW; Fig. 6.10b shows the corresponding peaking times. Furthermore, after first decreasing, the $ENC_{min}$ of bipolar shapers increases with the order. This change reflects the negative lobe of the bipolar shape that increases with the order, thus raising the level of noise more than the decrease due to the positive lobe's higher symmetry.

It should be stressed that here the $ENC_{min}$ is calculated by applying the MOSFET optimization process (Section 6.2.1) without any constraint on the peaking time $\tau_p$ or width $\tau_w$. In most applications, this is not so, and constraints are imposed on $\tau_p$ and/or $\tau_w$, typically from the event rate and/or the ballistic deficit (e.g., the charge induction time in the pixel).

As an example, it is considered a front-end electronics for high-rate (some megahertz) photon counting applications [30] operating with a pixelated sensor affected by a relatively long (few tens of nanoseconds) charge induction time (e.g., CdZnTe). Due to the small pixel effect, the charge induction time is (to a first order) independent of the depth of the ionizing interaction [31, 32]. The choice of the shaper results from combining good energy resolution, low ballistic deficit, and high-rate operation. In this analysis, four unipolar shapers are compared: two $r$-shapers of the second order (also known as RC-CR) and the fourth order (also known as RC$^3$-CR), and two $c$-shapers of the fifth and ninth order. Figure 6.11 depicts these four shapers, assuming an equal 1% pulse width, $\tau_w$. This comparison is justified considering that, to a first order, pulses with equal width should exhibit a comparable pile-up effect (pulse peak amplitude affected by the previous event). Figure 6.11 reveals that the difference in the four shapers mainly lies in three parameters: the pulses' peaking time $\tau_p$, their curvature in the proximity of the peak, and their symmetry. In particular, the ninth-order shaper exhibits the longest peaking time and the highest symmetry, while the fifth-order shaper shows the lowest peak curvature. The impact of these differences on the ENC, ballistic deficit, and pile-up can be analyzed.

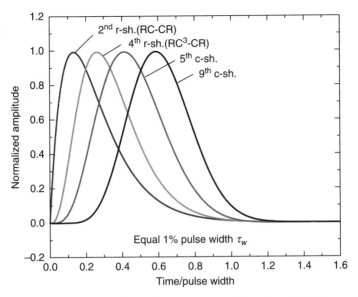

**Fig. 6.11.** Comparison between four shapers at equal 1% pulse width. From [30] with permission.

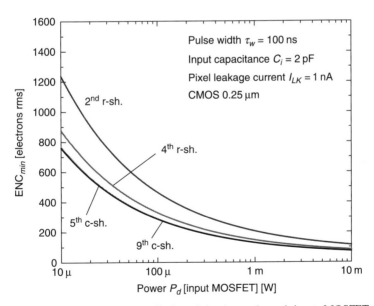

**Fig. 6.12.** $ENC_{min}$ versus power $P_d$ dissipated in the $p$-channel input MOSFET branch in a CMOS 0.25-$\mu$m technology for the four shapers in Fig. 6.11, assuming $\tau_w = 100$ ns, $C_i = 2$ pF, and $I_{LK} = 1$ nA. From [30] with permission.

When comparing the shapers in terms of $ENC_{min}$ under the pulse width constraint, it is here assumed that the contribution of white series noise from the input MOSFET is the dominant component, as is typical in high-rate designs. Figure 6.12 is a plot of the $ENC_{min}$ versus the power $P_d$ dissipated in the input MOSFET branch, assuming a pulse width $\tau_w = 100$ ns, an input capacitance $C_i = 2$ pF, and a pixel leakage current $I_{LK} = 1$ nA. The figure demonstrates that the higher-order shapers offer a better resolution due to their higher peaking time $\tau_p$ (i.e., lower bandwidth) that effectively reduces the white series noise contribution. Their advantage can be somewhat attenuated if the power dedicated to the additional poles is taken into account. On the other hand, their advantage is even higher when the ballistic deficit is considered, as discussed below.

In comparing the shapers in terms of ballistic deficit under the pulse width constraint, it must be considered that the charge induced on the pixel electrode is characterized by a finite time. For simplicity, it is also assumed that the charge is uniformly distributed in time and that the total charge is fixed and independent of the induction time; these suppositions represent some approximation for pixelated CZT sensors. Figure 6.13a compares the four shapers' responses for charge collection times ranging from negligible (0) to half of the pulse width (0.5). In Fig. 6.13b the corresponding peak amplitudes are compared. The fifth-order shaper offers the lowest ballistic deficit due to its lower curvature in the proximity of the peak. In general, the performance of the semi-Gaussian shapers in terms of ballistic deficit can be evaluated from the product $\rho_p(\tau_w/\tau_p)^2$ from Fig. 6.9, which gives the normalized curvature at equal pulse width, $\tau_w$. It is readily verified that the unipolar $c$-shapers provide the lowest curvature, with a minimum around the fifth-order. The ballistic deficit effectively attenuates the amplitude of the peak of each detected event. It appears reasonable to assume that the shot noise generated in the sensor's bulk is somewhat affected by this deficit, since the associated charge behaves like the signal charge. On the other hand, the electronic noise from the charge amplifier (i.e., series noise and noise from continuous reset $S_{in\_Mf}$) is not so affected. Hence, the effective ENC observed by the user

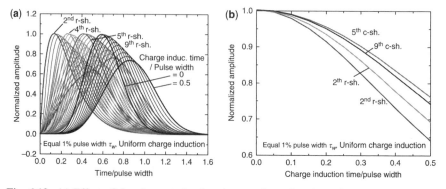

**Fig. 6.13.** (a) Effect of the charge induction time on the pulse shape for uniform induction ranging from small (0) to half (0.5) of the 1% pulse width $\tau_w$ for the four shapers in Fig. 6.11. (b) Corresponding effect on the peak amplitude. From [30] with permission.

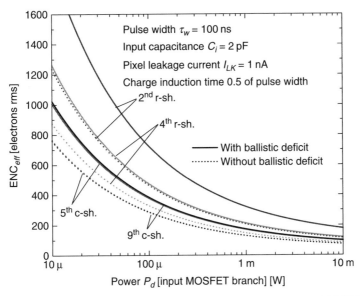

**Fig. 6.14.** Effective resolution $ENC_{eff}$ versus power $P_d$ dissipated in the input MOSFET branch for the four shapers in Fig. 6.11, assuming $\tau_w = 100$ ns, $C_i = 2$ pF, and $I_{LK} = 1$ nA, with ballistic deficit corresponding to a charge induction time half of the 1% pulse width (0.5 in Fig. 6.13). For comparison, the cases without ballistic deficit from Fig. 6.12 are shown as dashed lines. From [30] with permission.

increases in an amount that depends upon the ballistic deficit [30]. In Fig. 6.14 the effective ENC is simulated for the four shapers in Fig. 6.11, assuming a charge induction time of half of the pulse width; undoubtedly, the ballistic deficit further emphasizes the advantages of high-order shapers. The previous results suggest that the fifth-order $c$-shaper is as a good candidate in terms of resolution and ballistic deficit. However, when operating at high rate, the effect of the pile-up on the shaper under a pulse width constraint also must be investigated.

Accordingly, it is initially considered a sequence of two pulses characterized by a given time delay, also assuming a discrimination window centered at the peak amplitude, and a wide 10% of the peak amplitude. Figure 6.15a compares the four shapers for a sequence of two pulses with time delay equal to half of the pulse width $\tau_w$. Figure 6.15b, correspondingly, illustrates the piled-up amplitude (peak amplitude of the second pulse) versus the time delay, ranging from half of the pulse width to one pulse width.

The ninth-order $c$-shaper performs better in terms of pile-up, being the only one to properly count two events in the discrimination window; this advantage is due to the higher pulse symmetry and lower pulse width in the proximity of the peak. Further analysis of pile-up can be undertaken assuming the more realistic case of a Poisson distribution of time delays from the first to the second pulse, thereby demonstrating that the ninth-order $c$-shaper could offer the lowest spectral distortion and, consequently, the highest throughput in high-rate photon counting applications [30].

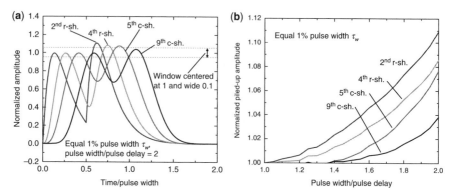

**Fig. 6.15.** (a) Effect of the pile-up of a sequence of two pulses, characterized by a time delay equal to half of the pulse width $\tau_w$, on the four shapers in Fig. 6.11. A window discrimination centered at the peak amplitude and with width 10% of the peak amplitude also is shown. (b) Piled-up amplitude for a sequence of two pulses characterized by a time delay ranging from half to one pulse width for the same shapers. From [30] with permission.

### 6.3.2. Output Baseline Stabilization—The Baseline Holder

The DC component of the shaper from which the signal pulse departs is referred to as the output baseline. Since most extractors process the pulses' absolute amplitude, which reflects the superposition of the baseline and the signal, it is important to properly reference and stabilize the output baseline. Nonstabilized baselines may fluctuate for several reasons, like changes in temperature, pixel leakage current, power supply, low-frequency noise, and the instantaneous rate of the events. Nonreferenced baselines also can severely limit the dynamic and/or the linearity of the front-end electronics, as in high gain shapers where the output baseline could settle close to one of the two rails, depending on the offsets in the first stages. In multiple front-end channels sharing the same discrimination levels, the dispersion in the output baselines can limit the efficiency of some channels.

A baseline holder (BLH) circuit can assure a referenced and stabilized baseline [33]. The BLH, shown in Fig. 6.16, consists of a feedback loop applied from the analog output of the channel (e.g., the output of the shaper) to a suitable node at the input of one of the first stages of the channel (e.g., the shaper's input).

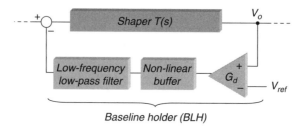

**Fig. 6.16.** Block diagram illustrating the baseline holder (BLH).

The BLH is composed of three stages: a difference amplifier, a nonlinear buffer, and a low-frequency low-pass filter. The difference amplifier compares the output voltage $V_o$ to a reference voltage $V_{ref}$, while the filter attenuates the response of the feedback circuit in the frequency range of interest and provides the dominant pole required to stabilize the loop. The nonlinear buffer suppresses the response of the feedback loop in the presence of signals.

The role of each block is discussed, starting with the assumption that the nonlinear buffer is not present. It is shown that the nonlinear buffer is relevant for stabilization when the channel operates at a high rate with unipolar pulses.

The difference amplifier subtracts $V_{ref}$ from the output voltage and amplifies the difference with a gain, $G_d$, of about a few tens. The feedback loop forces the output voltage to settle at a DC level close to $V_{ref}$. A stable and accurate $V_{ref}$ can be obtained using, for example, a bandgap referenced circuit [34, 35]. The low-frequency low-pass filter makes the feedback loop response to a single pulse essentially negligible. The transfer function of the system in Fig. 6.16 can be approximated with [33]

$$F(s) \approx \frac{T(s)}{G_{loop}(0)} \frac{1 + s\tau_F}{1 + s\dfrac{\tau_F}{G_{loop}(0)}} \qquad (6.12)$$

where $T(s)$ is the transfer function of the shaper, $\tau_F$ is the time constant of the low-frequency filter, and $G_{loop}(s)$ is the loop gain. From (6.12) it can be observed that, assuming $G_{loop}(0) \gg 1$, the feedback introduces into the shaper's response a unity-gain high-pass filter composed of a low-frequency zero with time constant, $\tau_F$, and a pole with time constant $\tau_F/G_{loop}(0)$. It can be verified that stability is reached when $\tau_F/G_{loop}(0) \gg \tau_p$, where $\tau_p$ is the shaper's peaking time. The small signals of the shaper's response with feedback are similar to those obtained by introducing an AC-coupling with time constant $\tau_F/G_{loop}(0)$ after the shaper.

By referencing the output voltage to a stable $V_{ref}$ and strongly attenuating the gain of the channel at DC and low frequency, the BLH provides a referenced and stabilized output baseline with almost no impact on the channel's single pulse response. The baseline dispersion of multiple channels can be strongly reduced, being limited by the offset in the difference amplifier. Unfortunately when such a configuration, characterized by an AC-coupling in the signal path, operates with unipolar pulses, it generates a baseline shift that depends on the pulse rate [2]. The shift can be intuitively understood by considering that an AC-coupling imposes a zero area on the output waveforms. The purpose of the nonlinear buffer is to suppress the undesired baseline shift. By nonlinearly attenuating the pulses' amplitude, the buffer strongly attenuates the area of the pulses processed by the BLH filter, and hence the baseline is much less dependent on the rate. The nonlinear buffer can be realized via a slew-rate limited follower, $M_B$, as shown in Fig. 6.17, whose response is controlled by the current $I_B$ and the capacitance $C_B$, which introduce a limit $I_B/C_B$ on the slew-rate of the output $V_{Bo}$.

For slow and small signals the output, $V_{Bo}$, is not affected by the slew-rate limit and the follower's response is linear. Thus, low-frequency fluctuations flow through the BLH without being affected by nonlinear attenuation, maximizing the BLH's

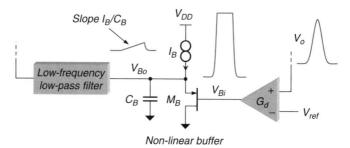

**Fig. 6.17.** Slew-rate-limited follower.

stabilizing efficiency. For large and fast signals, the output $V_{Bo}$ is slew-rate limited and the follower displays a nonlinear response. As Fig. 6.17 reveals, the unipolar pulse $V_o$, after being amplified, $V_{Bi}$, by the gain, $G_d$, is strongly attenuated by the slew-rate limited response, $V_{Bo}$. Consequently, the area of the pulses processed by the BLH filter is much smaller than that in the linear case, resulting in a baseline much less sensitive to the rate.

The design criteria for the slew-rate-limited follower must encompass the stability of the feedback loop and the minimization of the baseline shift at a high rate [33]. Good practice adopts the lowest possible gate size for $M_B$, since this minimizes the parasitic capacitance between the gate and the source. The follower introduces a pole at the source node with time constant, $\tau_B$. Operating $M_B$ in weak inversion, the time constant then can be approximated with $\tau_B \approx C_B V_t / I_B$, where $V_t = kT/q$. Assuming that the feedback loop is stable without a nonlinear buffer, it can be verified that stability is maintained when $\tau_B \ll \tau_F / G_{loop}(0)$. By equating the areas at the output, the maximum baseline shift, $\delta V_B$, can be approximated with the following [33]:

$$\delta V_B \approx -\frac{\tau_p R_t}{G_d} 2 V_{DD} K_a \approx -\frac{\tau_p R_t}{G_d} 2 V_t \qquad (6.13)$$

where $R_t$ is the rate, and $K_a$ is the ratio between the areas at the gate and at the source of $M_B$. In (6.13) the condition $\tau_B > 6\tau_p$ was applied, resulting in a minimum for $K_a$, given by $K_a \approx V_t / V_{DD}$ [33].

In Fig. 6.18, a possible realization of the BLH is shown, wherein the signal is fed back as current, $I_{Fo}$, to a shaper accepting input currents, as in Fig. 6.6. The nonlinear buffer is based on the slew-rate limited follower in Fig. 6.17 and includes a mirror that restricts the current through $M_B$ to twice $I_B$. The limiter prevents $C_B$ discharging to ground after negative pulses, thus protecting the circuit from unwanted memory effects. Typical values of $I_B$ and $C_F$ are, respectively, 1 nA and 200 fF. The low-frequency low-pass filter is conceived using a similar configuration, but $M_F$ is designed with a long channel and operates with low current, $I_F$, and large capacitor, $C_F$, typically tens of picoamperes and tens of picofarad, respectively. The filter also

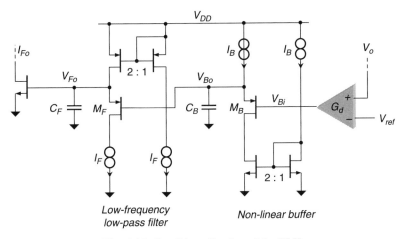

**Fig. 6.18.** Possible realization of the BLH.

includes a protection mirror. Figure 6.19 plots the measured response of a channel implementing a shaper with a BLH. In Fig. 6.19a a small signal transfer function is shown, where the poles of a fifth-order $c$-shaper and the high-pass filter from the BLH can be observed. Figure 6.19b illustrates the response to a sequence of unipolar pulses with peaking time $\tau_p = 400$ ns and rate increasing from 20 kHz to 400 kHz; it is compared to the response of an AC-coupling with high-pass time constant equal to that of the BLH. There is a visible shift of the baseline with the rate in the latter. The shift is negligible with the BLH due to the limiting effect of the nonlinear buffer.

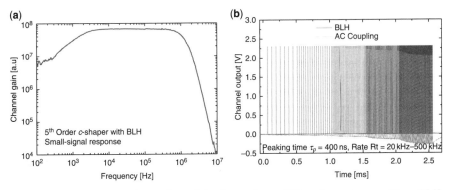

**Fig. 6.19.** Measurements for a channel implementing a fifth-order $c$-shaper with a BLH: (a) Small-signal transfer function; (b) response to unipolar pulses with peaking time $\tau_p = 400$ ns and rate increasing from 20 kHz to 500 kHz, compared to an AC-coupling with high-pass time constant equal to that of the BLH. From [33] with permission.

## 6.4. EXTRACTION

A shaped pulse carries information on the amplitude, timing, and shape of an ionizing event. The information of interest depends on the application; it typically covers the pulse' peak amplitude (a measure of the event's energy) and/or the pulse timing (a measure of the event's timing relative to a trigger). Some cases necessitate having different timing information, like (a) the pulse rise time to correct for a ballistic deficit or (b) the pulse time-over-threshold to discriminate piled-up events [36, 37]. Other cases require periodically sampling the waveform, as in time projection chambers with high multiplicity of events.

Accurately extracting such information requires employing specialized circuits. Here, three solutions are discussed: the discriminator, which provides either single- or multiamplitude discrimination; the multiphase peak detector, which provides accurate measurements of the peak amplitude and/or peak timing; and the current-mode peak detector digitizer, which simultaneously detects and digitizes the peak at moderate resolution.

### 6.4.1. Single- and Multiamplitude Discrimination

The comparator is a widely adopted circuit for discriminating amplitude. It is a mixed-signal (analog-to-digital) circuit composed of (a) a high-gain differential amplifier with hysteresis and (b) a digital output stage. One of the two inputs is connected to a threshold voltage, $V_{TH}$, while the other is connected to the output of the shaper, $V_i$. When the output, $V_i$, exceeds the threshold $V_{TH}$ (and the hysteresis), the comparator changes the state of the digital output, thus discriminating between below- and above-threshold amplitudes.

Hysteresis can be obtained by introducing a positive feedback at the output of the differential stage, and its design value results from considering of the effects of noise, dynamic range, and baseline dispersion on $V_i$. A small hysteresis is required to pick up very low amplitude events, setting the threshold voltage $V_{TH}$ very close to the baseline. A large hysteresis is needed to compensate for baseline dispersion and avoid multiple firing of the comparator due to noise associated with the waveform. Since the latter can severely compromise the performance of the front-end electronics, designers tend to adopt wide safety margins using large hysteresis, thereby putting a lower limit on the discrimination of the amplitudes (i.e., restricting the dynamic range). In systems wherein measurement and readout are two separate phases, the parasitic injections associated with the readout phase can trigger the comparator immediately before the measurement phase. If a simple edge-sensitive logic follows the comparator, this may lock the front-end channel into a triggered state, thus blocking the processing of valid events. For all these reasons, a state-of-the-art discriminator includes, with the comparator, additional analog and digital functions that allow the comparator to operate with a small hysteresis without these undesired effects. Figure 6.20 is a block diagram of a complete discrimination circuit.

The comparator in Fig. 6.20 is characterized by a low hysteresis, set according to the minimum expected noise; it includes a digital control signal, $s$. When $s$ is high,

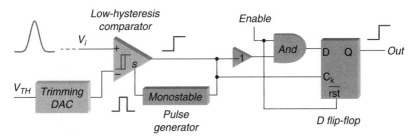

**Fig. 6.20.** Block diagram of a discrimination circuit.

the comparator is held in its current state independent of the input signal. The trimming DAC modifies the threshold to compensate for the output baseline dispersion and comparator offset. The discriminator is controlled by the *Enable* signal. When *Enable* is low, the *D* flip-flop is kept in a reset state and the discriminator is inactive, independent of the comparator's activity. Conversely, when *Enable* is high, the *D* flip-flop is sensitive to the comparator's output. Positive threshold crossing triggers the discriminator's output and starts a pulse from the monostable, which through *s* holds the comparator in its state preventing multiple firing.

Figure 6.21 is a schematic of a comparator implementing the hold function; in this configuration, digital feedback is provided though the two MOSFETs $M_{s+}$ and

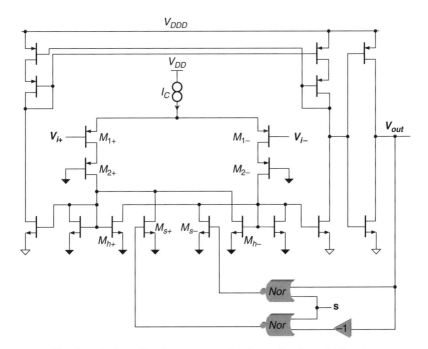

**Fig. 6.21.** Schematic of a comparator implementing the hold function.

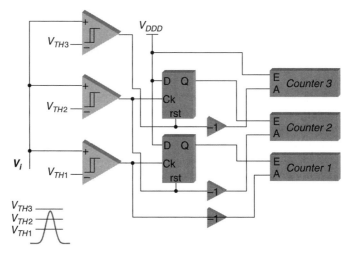

**Fig. 6.22.** A multiamplitude discrimination circuit (with three windows).

$M_{s-}$, the two *nor* gates and the *inverter*. The two MOSFETs, $M_{2+}$ and $M_{2-}$, provide protection against the comparator's kickback, while $M_{h+}$ and $M_{h-}$ provide positive feedback for the hysteresis. To lower mixed-signal noise, all MOSFETs in the circuit's digital section are connected to a dedicated digital supply, $V_{DDD}$, and to ground (open triangles). The $p$-channel MOSFETs in the circuit's digital section are cross-connected to minimize the current in the digital supply.

The edge of the discriminator can be used to trigger a flag for the external electronics (DAQ), to increment a counter (as in photon-counting applications), or to initiate processing of the pulse (e.g., peak and/or time detection).

In some photon-counting applications, discrimination is attained using two or more energy windows. For a small number of windows, this can be achieved using multiamplitude discrimination logic, as demonstrated for the three windows in Fig. 6.22. Each window has an associated counter with an enable control (level sensitive) and an advance one (positive edge sensitive). The rising edge of the analog input pulse controls the counter's enable control. At the first threshold crossing, the first counter is enabled through the flip-flop. Should the second threshold be crossed, the first counter is disabled and the second one enabled, and so on. The falling edge of the analog input pulse triggers the counter's advance, and at the crossings only the enabled counter increases its count.

### 6.4.2. Peak- and Time-Detection: The Multiphase Peak Detector

The peak detector measures and stores in an analog memory the peak voltage of the input pulse $V_i$ and, if needed, it releases a "peak-found" digital signal corresponding to the peak. The stored peak voltage can be routed to an analog-to-digital converter (ADC) later during the readout. In Fig. 6.23, the CMOS peak detector is shown in its classical configuration. The analog pulse, $V_i$, feeds the negative input of the

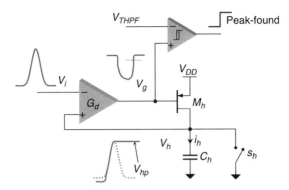

**Fig. 6.23.** The classical configuration of the CMOS peak detector.

difference amplifier, while the positive input is connected to the drain of $M_h$ and to the hold capacitor, $C_h$, that is kept discharged by the reset switch, $s_h$. The edge of the discriminator can be used to start peak detection by opening $s_h$. Due to the difference at the inputs of the amplifier, the output voltage, $V_g$, sharply decreases from $V_{DD}$, enabling the flow of a current $i_h$ from $M_h$ that charges $C_h$. Because of the feedback loop and high gain, $G_d$, the voltage $V_h$ on $C_h$ tracks the input pulse, while the charging current, $i_h$, is proportional to the derivative of the input signal $V_i$ according to $i_h \approx C_h$ $(dV_i/dt)$. When the signal $V_i$ approaches the peak, the current $i_h$ nears zero and the voltage, $V_g$, equals $V_{DD} - V_{THh}$, where $V_{THh}$ is the threshold voltage of $M_h$. Right after the peak, the signal $V_i$ starts falling, but, since there is no discharge path for $C_h$, the voltage $V_h$ holds the absolute peak value $V_{hp} = V_{ip}$, where $V_{ip}$ is the peak voltage of $V_i$. It is noted that the value of the absolute peak includes the baseline. Due to the high gain, $G_d$, the output voltage $V_g$ sharply increases toward $V_{DD}$ disabling the current source, $M_h$. The peak voltage $V_{hp} = V_{ip}$ is now held in $C_h$, which behaves as an analog memory a for the next readout, while the comparator can generate a peak-found digital signal if $V_{DD} - V_{THh} < V_{THPF} < V_{DD}$. The transistor level design of the configuration in Fig. 6.23 should take into account the finite bandwidth of the amplifier and the nonideality of the readout circuit, which may impose significant limits on its stability and accuracy. First, during pulse tracking, the time constant, $\tau_h$, of the pole from $C_h$ can change by orders of magnitude, and in combination with the poles from the amplifier, can readily destabilize the feedback loop. Second, the readout of the peak voltage $V_{hp}$ requires a high impedance, high accuracy buffer to minimize error and loss of charge. In 1994, Kruiskamp and Leenaerts introduced the configuration in Fig. 6.24 that addressed these two limits [38]. They replaced the difference amplifier with an operational transconductance amplifier (OTA) so that the amplifier's dominant pole is located at node $V_g$ with time constant $\tau_0$ defined by the parasitic capacitance and the resistance from the diode connected MOSFET $M_c$. During pulse tracking, the current through $M_c$ increases, effectively reducing the loop gain, decreasing $\tau_0$, and moving the OTA pole to higher frequency. The dynamic change in $\tau_0$ tracks the change in the dominant pole, $\tau_h$, resulting in an increased stability. The follower $M_f$ in the feedback loop allows the readout of the

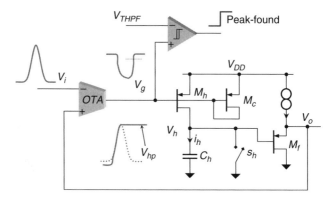

**Fig. 6.24.** The CMOS peak detector in the Kruiskamp and Leenaerts configuration.

peak voltage $V_{hp}$ without loss of charge. A thorough analysis of the configuration in Fig. 6.24, however, reveals that the nonidealities from the OTA and $M_h$ can severely affect accuracy and dynamic range. It can be shown that the error, $\delta V_p = |V_{hp} - V_{ip}|$, on the peak value can be approximated with [39]

$$\delta V_p \approx \frac{2.5\, V_{DD} L_h^2 V'_{i,max}}{\mu_h (V_{DD} - V_{THh})^2} + V_{OFF} + \frac{V_{THh} - V_{Acm}}{A_0} + \frac{V_{DD} - 2V_{ip}}{2\mathrm{CMRR}_A}$$
$$+ \frac{V_{DD} C_{ox} W_A L_A}{1.4\, C_h} + (V_{DD} - V_{THh}) \frac{V'_i}{V'_{g,max}} \tag{6.14}$$

The first term is due to capacitances ($C_{gd}$, $C_d$) and channel charge injection from $M_h$, where $L_h$ is the gate length, $\mu_h$ is the mobility, $V_{THh}$ is the threshold voltage, and $V'_{i,max}$ is the maximum slope of the input pulse $V_i$. All other terms are from nonidealities in the OTA, as follows. The second term is due to the offset at the input of the OTA and, for a simple differential stage, can be approximated with $V_{OFF} \approx A_{VT}/(W_A L_A)^{1/2}$ [40], where $L_A$ and $W_A$ are the gate length and width of the OTA's input MOSFETs, and $A_{VT}$ is the technology-dependent mismatch coefficient. The third and fourth terms are due to the finite gain for the OTA, where $V_{A,CM}$ is the common-mode voltage reference, $A_0$ is the DC differential voltage gain, and $\mathrm{CMRR}_A$ is the common-mode rejection ratio. The fifth term is due to the OTA's input capacitances, while the sixth term is related to the slew-rate limit, $V'_{g,max}$, at the output of the OTA.

By using a minimum value for $L_h$, the first term can be contained within a negligible fraction of the percent of $V_{DD}$ for pulses with a peaking time of a few tens of nanoseconds. With suitable design of the output stage of the OTA, the sixth term also can be made negligible. The design values of $L_A$ and $W_A$ reflect the compromise between the second term (offset) and the fifth term (capacitance), and the minimum error can be about of a percent of the $V_{DD}$. Depending on the values of $V_{A,CM}$, $A_0$, and $\mathrm{CMRR}_A$, the three corresponding contributions also can be non-negligible.

Equation (6.14) can be simplified as follows:

$$\delta V_p \approx V_{OFF} + \frac{V_{THh} - V_{Acm}}{A_0} + \frac{V_{DD} - 2V_{ip}}{2\mathrm{CMRR}_A} + \frac{V_{DD}C_{ox}W_A L_A}{1.4C_h} \qquad (6.15)$$

It is observed that while the first contribution, $V_{OFF}$, here is usually independent of $V_{ip}$ (for a simple differential stage), the last three contributions may exhibit a non-negligible nonlinear dependence on it. Assuming that a fixed error is acceptable, the values of $L_A$ and $W_A$ can be chosen to ensure that the fourth contribution in (6.15) is negligible.

The dynamic range of the configuration in Fig. 6.24 also is limited by the threshold voltage of the follower, $M_f$, and by the input stage of the OTA, assumed to be a simple differential stage achieved by using $p$-channel input MOSFETs. To increase dynamic range, the designer should replace the follower with a rail-to-rail buffer, carrying, in turn, penalty in terms of power and constraints on the loop stability. The designer should also adopt a rail-to-rail differential stage at the OTA input by combining two $n$-channel and $p$-channel differential stages, with a penalty of a total offset, $V_{OFF}$, that depends on $V_{ip}$.

The circuit shown in Fig. 6.25 addresses the limits of accuracy and dynamic range of the previous version by operating the peak detector in two phases, the peak-detect phase and the readout phase [41]. During the former phase the switches $S_{i1}$ and $S_{g1}$ are closed and the switches $S_{f2}$ and $S_{g2}$ are open. Then, the circuit operates like the one in Fig. 6.24, tracking the pulse, finding the peak, and releasing the associated peak-found signal. The peak-found signal switches the configuration into the readout by opening the switches $S_{i1}$ and $S_{g1}$ and closing the switches $S_{f2}$ and $S_{g2}$. In this phase, the amplifier is reconfigured as a unity-gain buffer and the peak voltage $V_{hp}$ stored on $C_h$ is made available via the buffer at the output $V_o$.

By reconfiguring the amplifier as a buffer, the contributions from $V_{OFF}$, $V_{A,CM}$, and $\mathrm{CMRR}_A$ in (6.15) can be shown to be canceled and the error $\delta V_p = |V_{hp} - V_{ip}|$ in the

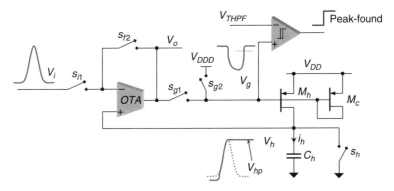

**Fig. 6.25.** The two-phase CMOS peak detector.

peak value is reduced to [41]

$$\delta V_p \approx \frac{V_{THh} - V_{ip}}{A_0} + \frac{V_{DD}C_{ox}W_A L_A}{1.4C_h} \tag{6.16}$$

The values of $L_A$ and $W_A$ can be chosen to make the second contribution negligible independent of $V_{OFF}$, whose contribution now is canceled. Also, the first term can be reduced by increasing the DC gain, $A_0$. Then, the OTA can be designed with a rail-to-rail differential stage at the input: Because of the cancelation of its contribution, an offset $V_{OFF}$ dependent on $V_{ip}$ will not affect accuracy. The OTA can be configured with a rail-to-rail output stage, thus increasing the dynamic range. Finally, reusing the amplifier as a buffer makes the output impedance of the peak detector at readout small, without increasing the amount of power dissipated.

The two-phase peak detector in Fig. 6.25 gives the designer an instrument that can process pulse peaks with high absolute accuracy. Two or more peak detectors can be combined in a configuration where several channels share a much smaller number of peak detectors, as discussed [37, 41–45]. This self-triggered, sparsifying, and derandomizing configuration greatly relaxes the requirement on the readout and on the analog-to-digital conversion rate, which then can approach the average rate of the input events with negligible impact on efficiency. The configuration in Fig. 6.25, integrated into a single channel, can be bettered by introducing a third tracking phase in which the circuit tracks the input waveform without peak detection, as shown in Fig. 6.26. During this phase, the switch, $S_{n0}$, is closed while the switch, $S_{n1}$, is open. The peak detector tracks the waveform (i.e., the baseline and below threshold events) until an above threshold event initiates the peak detection phase by opening the former, and closing the latter. It is noted that $V_g$ now starts from a value close to $V_{DD} - V_{THh}$ rather than from $V_{DD}$. This softer transition can improve performance when processing very low amplitude pulses.

The peak-found signal generated by the peak detector can be used to measure the timing of the event with respect to a trigger [45, 46]. Compared to the

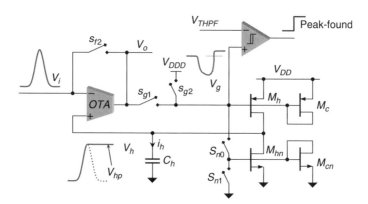

**Fig. 6.26.** The three-phase CMOS peak detector.

measurement-based one threshold crossing of $V_i$, advantageously the peak-founds is to a first order independent of the peak amplitude. In principle, the timing performance of the peak-found signal is equal to the one of a zero-crossing on a bipolar pulse. The current flowing through $C_h$ during peak detection equals the derivative of the input pulse according to $i_h = C_h\, dV_i/dt$. When the signal $V_i$ approaches the peak, the current $i_h$ approaches zero, and the voltage $V_g$ equals $V_{DD} - V_{THh}$. Immediately after the peak the signal, $V_i$, starts falling and, due to the high gain, $G_d$, the output voltage $V_g$ sharply increases toward $V_{DD}$ crossing the comparator threshold, $V_{THPF}$, set at $V_{DD} - V_{THh} < V_{THPF} < V_{DD}$. The timing resolution $\sigma_t$ can be approximated with

$$\sigma_t \approx \frac{\sigma_{i_h}}{i_h'(\tau_{p0})} \approx \frac{i_{hp}}{Q}\frac{\mathrm{ENC}_b(\tau_{pb})}{i_h'(\tau_{TH})} \approx \frac{V_p\,\lambda_p}{Q}\frac{\mathrm{ENC}_b(\tau_p\eta_p)}{\tau_p}\frac{\tau_p^2}{\rho_p}\frac{\tau_p^2}{V_p}$$

$$= \frac{\mathrm{ENC}_b(\tau_p\eta_p)\tau_p\lambda_p}{Q\rho_p} \tag{6.17}$$

where $\sigma_{i_h}$ is the noise on the current signal $i_h$, and $i_h'$ is the derivative calculated at the time, $\tau_{TH}$, of the threshold crossing of the comparator, where $i_h(\tau_{TH}) = i_{hTH}$ is the current. It is pointed out that $i_{hTH}$ is set by the threshold $V_{THPF}$; and for $\tau_{TH}$ approaching the time $\tau_{p0}$ at which $V_i$ peaks, it follows that $i_h'(\tau_{TH}) \approx i_h'(\tau_{p0})$. In (6.17), the noise $\sigma_{ih}$ is assumed to be dominated by that on the input pulse, $V_i$, from the first front-end stage, and it is calculated in (6.17) as $\mathrm{ENC}_b$ multiplied by the gain from $Q$ to the current $i_h$ in $C_h$ (i.e., $i_{hp}/Q$, where $i_{hp}$ is the peak value of $i_h$). Therefore, the term $\mathrm{ENC}_b$ is calculated assuming a bipolar shape (current $i_h$) with peaking time $\tau_{pb}$. As a next step, the normalized coefficients in Fig. 6.9 can be adopted, yielding the final expression in (6.17) that only depends on $Q$ and on the peaking time $\tau_p$. This expression assumes that contributions to noise from the shaper, the OTA, the comparator, and other parts are negligible. In practice, overlooking these other contributions can severely limit the timing resolution.

For example, particular attention must be paid to the high-frequency noise components from the shaper and the OTA. As already observed, the signal, $i_h$, decreases with the peaking time, and its slope $i_h'(\tau_{p0})$ declines with the square of the peaking time. On the other hand, these noise components, along with their emphasis by the derivative, are to a first order independent of peaking time. The contribution to the timing resolution is proportional to the square of the peaking time and can be approximated with $\sigma_{ih\_hf}/i_h'(\tau_{p0})$, where $\sigma_{ih\_hf}$ is the corresponding rms current noise. A resistor in series with the capacitor $C_h$, while helping to stabilize the peak detector loop, also would filter such high-frequency noise.

The noise from the comparator can be evaluated as $\sigma_{vc}/\psi_{v_g}$, where $\sigma_{vc}$ is the rms voltage noise at the input of the comparator, and $\psi_{v_g}$ is the slope of $V_g$. The slope can be approximated from the Taylor series of the signal, $V_i - V_{ip}$, right after the peak, yielding $\psi_{v_g} \approx A\cdot(2i_h'(\tau_{p0})V_{ip}/C_h)^{1/2}$, where $A$ is the difference voltage gain. Again, this contribution is proportional to the square of the peaking time.

Assuming that the timing signal is converted into a voltage using a time-to-amplitude converter (TAC), ( such as a voltage ramp), the series noise from the TAC and its readout should be included. The TAC contribution can be approximated with $[(S_{iTAC}T_{TAC})/(\psi_{TAC}C_{TAC})]^{1/2}$, where $S_{iTAC}$ is the spectral noise power density of the current flowing into a capacitor $C_{TAC}$ to generate the ramp, $T_{TAC}$ is the duration of the ramp, and $\psi_{TAC}$ is the slope. The contribution from the readout can be approximated with $\sigma_{vro}/\psi_{TAC}$, where $\sigma_{vro}$ is the rms voltage noise from the readout stage.

### 6.4.3. Current-Mode Peak Detector and Digitizer

The multiamplitude discrimination discussed in Section 6.4.1 can yield an efficient solution for applications requiring discrimination and counting among a very small number of energy levels. Complexity and power dissipation can be contained due to the few comparators, logic, and associated counters.

The peak detector discussed in Section 6.4.2, followed by a high-resolution ADC, can be an effective approach to applications requiring accurate measurements of energy and/or timing. For each event, it would convert only the necessary information, namely, peak and/or timing. For applications requiring discrimination among (or digitization with) a moderate number of energy levels ($\sim$few tens), the current-mode peak detector and digitizer (PDD) [46] might be selected because it simultaneously provides low-power and clock-less peak detection and analog-to-digital (A/D) conversion in real time.

In general, the PDD is a type of current-mode flash converter consisting of $2^n - 1$ cells, each containing a unit current sink and a set of switches and associated logic. The input current first is compared to the unit current in the first cell. If it exceeds one unit current, then the second cell current is added and the sum is again compared to the input current. The process is repeated sequentially until the total sink current equals the input current within one less significant bit (LSB). A thermometer-to-binary converter encodes the result into an $n$-bit word.

For a time-varying input current, the circuit can have peak-detecting behavior. Additional unit currents are added to the comparison current as long as the input current exceeds the comparison one. Once the input current begins to decrease, the digitized maximum value is maintained until the circuit is reset. In this mode, the PDD also generates a "conversion-done" pulse when the current derivative changes its sign.

The operating principle of the PDD is illustrated in Fig. 6.27. Before entering the PDAD circuit, the shaped voltage pulse undergoes a voltage-to-current conversion (not shown in Fig. 6.27) resulting in a shaped current, $I$. The PDD circuit is composed of $n - 1$ cells, where $n$ is the number of desired energy levels. Each $j$ cell (where $j = 1$ to $n - 1$) implements two switches $s_{aj}$ and $s_{bj}$, a logic port dsc$_j$ (e.g., an inverter or a Schmitt trigger), and a current source $i_j$. In cases where the levels must be equally spaced, all current sources, $i_j$, can be matched; and $i_j n = I_{max}$, where $I_{max}$ is the maximum expected current.

After a circuit reset, all $s_{aj}$ switches initially are open and all $s_{bj}$ switches are closed. All voltages, $v_j$, must be equal to 0 V. Processing of the pulse starts with the switch $s_{a1}$ closed. The shaped current, $I$, is compared to the unit current, $i_1$, of the first cell, corresponding to one LSB. As long as the former is lower than the latter, the voltage, $v_1$, remains at 0 V. If the shaped current exceeds the unit current (one LSB), the voltage $v_1$,

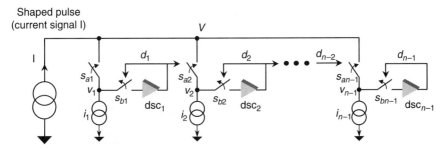

**Fig. 6.27.** Schematic diagram illustrating the operating principle of the peak detector and digitizer (PDD). From [46] with permission.

which is now equal to the common voltage, $V$, rises from $0$ V until the port $dsc_1$ changes state. At that time, the switch $s_{b1}$ is open and the switch $s_{a2}$ is closed, thereby adding one unit of current $i_2$. The voltage $v_2$ rapidly increases toward the voltage $V$. The current $I$ is now compared to the sum of $i_1$ and $i_2$ and, depending on the result, $v_2$ (that is, $V$) will decrease to $0$ V or will increase, forcing the next port, $dsc_2$, to change state. Accordingly, a new unit current, $i_3$, will be added for the current comparison, and so on. This enabling of further units of current will proceed, tracking the rising edge of the current pulse, $I$. After reaching its peak, it starts decreasing. Since the switches $s_{bj}$ of all the enabled cells are open, the corresponding unit currents cannot be disabled and the quantized peak is preserved and converted into a discrete level according to the last enabled unit current. The voltage $V$ decreases due to the comparison between the quantized peak current and the input current, $I$. This decrease generates the peak-found and end-of-conversion signal.

The result of detecting and converting the peak can be used to increment on-chip counters or it can be encoded into an $m$-bit digital word where $n = 2^m$. A simple encoding approach, based on a two-dimensional matrix configuration, has been described, and the PDAD design criteria have been discussed [46]. The accuracy on the threshold voltage of the dsc elements in Fig. 6.27 can be shown to satisfy the condition [46] (quite relaxed compared to the case of voltage-mode flash converters)

$$\sigma_{V_{th}} < \frac{1}{3} V_{th} \frac{1}{1 + (n-1)\frac{C_j}{C}} \tag{6.18}$$

where $C$ and $C_j$ are, respectively, the parasitic capacitance at nodes $V$ and $v_j$ in Fig. 6.27. The maximum conversion time can be approximated with [46]

$$T_{max} \approx \frac{V_{th} C_j n}{I_{max}} \sum_{j=1}^{n-1} \frac{1}{n-j} + (n-1)\delta t \tag{6.19}$$

where $I_{max}$ is the maximum expected peak value and $\delta t$ is the digital switching time typical for the technology. For example, assuming $C_j = 20$ fF, $C = 1$ pF, $n = 256$, $V_{th} \approx 1.25$ V, $I_{max} = 256\ \mu A$ ($1\ \mu A/cell$), and $\delta t = 100$ ps, then $T_{max} \approx 175$ ns.

In general, the power dissipated by the PDD depends on the pulses' rate and amplitude. Assuming that after the peak the digital word is transferred to a memory and the PDD is reset, the worst-case dissipated power $P_{d_{max}}$ can be approximated by

$$P_{d_{max}} \approx \frac{1}{2} V_{DD} n i_j \times \text{peaking time} \times \text{rate} \tag{6.20}$$

Compared to other solutions, the PDD has the advantage of combining in the same circuit both peak detection and A/D conversion at low power (sub-milliwatts, depending on the resolution) and without needing a clock signal. Specifically, compared to the available voltage-mode flash ADCs, the requirement on voltage accuracy is more relaxed [see (6.18)]. Furthermore, the PDAD dissipates less power than current-mode flash ADCs, but at the price of a limited conversion speed. However, because peak detection and conversion occurs during the transition of the shaped pulse, the converted peak becomes available right after the peak occurs, thus effectively minimizing the processing time and consequently maximizing efficiency. Thus, by comparison with other non-flash clocked and clock-less solutions, it has the advantage of offering peak detection capability.

## 6.5. CONCLUSIONS

This work was aimed at introducing the reader to some of the state-of-the-art practices in formulating low-noise electronics for radiation sensors. The design concepts and circuit solutions presented here were successfully adopted in several high-resolution and high-counting-rate radiation detection systems. The rapid evolution of such systems, mainly associated with the progress in the technology for integrated circuits, continues to generate an extensive literature about developments in this area.

In all cases, it is important not to underestimate the considerable effort required to translate these solutions into practical realizations that the final user will find reliable and convenient. This effort frequently involves system- and transistor-level design concurrent with extensive simulations, specialized physical layout, and thorough experimental characterization.

## ACKNOWLEDGMENTS

The author wishes to express his deep gratitude to Veljko Radeka for his continuous support and encouragement, as well as for frequent and constructive discussions. He is also indebted to Avril D. Woodhead for assistance with editing.

## REFERENCES

1. G. F. Knoll, *Radiation Detection and Measurement*, 3rd ed., John Wiley & Sons, New York, 2000.

2. E. Gatti and P. F. Manfredi, Processing the signals from solid-state detectors in elementary particle physics, *Riv. Nuovo Cimento* **9**, 1–147, 1986.

3. V. Radeka, Low noise techniques in detectors, *Annu. Rev. Nucl. Part. Sci.* **38**, 217–277, 1988.

4. G. Lutz, *Semiconductor Radiation Detectors*, Springer-Verlag, Berlin, 1999.

5. H. Spieler, *Semiconductor Detector Systems*, Oxford University Press, New York, 2005.

6. L. Rossi, P. Fisher, T. Rohe, and N. Wermes, *Pixel Detectors*, Springer-Verlag, Berlin, 2006.

7. G. Bertuccio, A. Pullia, and G. De Geronimo, Criteria of choice of the front-end transistors for low-noise preamplification of detector signals at sub-microsecond shaping times for X- and γ-ray spectroscopy, *Nucl. Instrum. Methods A* **380**, 592–595, 1996.

8. E. Gatti, P. F. Manfredi, M. Sampietro, and V. Speziali, Suboptimal filtering of $1/f$-noise in detector charge measurements, *Nucl. Instrum. Methods A* **297**, 467–478, 1990.

9. P. Seller, Noise analysis in linear electronic circuits, *Nucl. Instrum. Meth. A* **376**, 229–241, 1996.

10. G. De Geronimo and P. O'Connor, MOSFET optimization in deep submicron technology for charge amplifiers, *IEEE Trans. Nucl. Sci.* **52**, 3223–3232, 2005.

11. T. H. Wilmshurst, *Signal Recovery*, 2nd ed., Adam Hilger, Bristol, England, 1990.

12. V. Radeka, Optimum signal-processing for pulse-amplitude spectrometry in the presence of high-rate effects and noise, *IEEE Trans. Nucl. Sci.* **15**, 455–470, 1968.

13. V. Radeka, Trapezoidal filtering of signals from large germanium detectors at high rates, *Nucl. Instrum. Methods A* **99**, 525–539, 1972.

14. A. Pullia, Impact of non-white noise impulse amplitude measurements: A time domain approach, *Nucl. Instrum. Methods A* **405**, 121–125, 1998.

15. M. Feser, B. Hornberger, C. Jacobsen, G. De Geronimo, P. Rehak, P. Holl, and L. Struder, Integrating silicon detector with segmentation for scanning transmission X-ray microscopy, *Nucl. Instrum. Methods A* **565**, 841–854, 2006.

16. W. Buttler, B. J. Hosticka, and G. Lutz, Noise filtering for readout electronics, *Nucl. Instrum. Methods A* **288**, 187–190, 1990.

17. W. M. C. Sansen and Z. Y. Chang, Limits of low noise performance of detector readout front ends in CMOS technology, *IEEE Trans. Circ. Syst.* **37**, 1375–1382, 1990.

18. Y. Hu, J. D. Berst, and W. Dulinski, Semiconductor position-sensitive detectors, *Nucl. Instrum. Methods A* **378**, 589–593, 1996.

19. C. G. Jakobson and Y. Nemirovsky, CMOS low-noise switched charge sensitive preamplifier for CdTe and CdZnTe X-ray detectors, *IEEE Trans. Nucl. Sci.* **44**, 20–25, 1997.

20. G. De Geronimo and P. O'Connor, A CMOS detector leakage current self-adaptable continuous reset system: Theoretical analysis, *Nucl. Instrum. Methods A* **421**, 322–333, 1999.

21. T. H. Lee, G. Cho, H. J. Kim, S. W. Lee, W. Lee, and H. Han, Analysis of $1/f$ noise in CMOS preamplifier with CDS circuit, *IEEE Trans. Nucl. Sci.* **49**, 1819–1823, 2002.

22. G. De Geronimo, P. O'Connor, V. Radeka, and B. Yu, Front-end electronics for imaging detectors, *Nucl. Instrum. Methods A* **471**, 192–199, 2001.

23. G. De Geronimo and P. O'Connor, A CMOS fully compensated continuous reset system, *IEEE Trans. Nucl. Sci.* **47**, 1458–1462, 2000.

24. G. De Geronimo, P. O'Connor, R. H. Beuttenmuller, Z. Li, A. J. Kuczewski, and D. P. Siddons, Development of a high-rate, high-resolution detector for EXAFS experiments, *IEEE Trans. Nucl. Sci.* **50**, 885–891, 2003.

25. G. De Geronimo, J. Fried, P. O'Connor, V. Radeka, G. C. Smith, C. Thorn, and B. Yu, Front-end ASIC for a GEM based time projection chamber, *IEEE Trans. Nucl. Sci.* **51**, 1312–1317, 2004.

26. J. Llacer, Optimum filtering in presence of dominant $1/f$ noise, *Nucl. Instrum. Methods* **130**, 565–570, 1975.

27. E. Gatti, M. Sampietro, and P. F. Manfredi, Optimum filters for detector charge measurements in presence of $1/f$ noise, *Nucl. Instrum. Methods A* **287**, 513–520, 1990.

28. E. Fairstein, Bipolar pulse shaping revisited, *IEEE Trans. Nucl. Sci.* **44**, 424–428, 1997.

29. S. Ohkawa, M. Yoshizawa, and K. Husimi, Direct synthesis of the Gaussian filter for nuclear pulse amplifiers, *Nucl. Instrum. Methods* **138**, 85–92, 1976.

30. G. De Geronimo, A. Dragone, J. Grosholz, P. O'Connor, and E. Vernon, ASIC with multiple energy discrimination for high-rate photon counting applications, *IEEE Trans. Nucl. Sci.* **54**, 303–312, 2007.

31. U. Lachish, Driving spectral resolution to the noise limit in semiconductor gamma detector arrays, *IEEE Trans. Nucl. Sci.* **48**, 520–523, 2001.

32. F. Mathy, A. Gliere, E. G. D'Aillon, P. Masse, M. Picone, J. Tabary, and L. Verger, A three-dimensional model of CdZnTe gamma-ray detector and its experimental validation, *IEEE Trans. Nucl. Sci.* **51**, 2419–2426, 2004.

33. G. De Geronimo, P. O'Connor, and J. Grosholz, A CMOS baseline holder (BLH) for readout ASICs, *IEEE Trans. Nucl. Sci.* **47**, 818–822, 2000.

34. P. R. Gray and R. G. Meyer, *Analysis and Design of Analog Integrated Circuits*, 3rd ed., John Wiley & Sons, New York, 1993.

35. D. Johns and K. Martin, Analog integrated circuit design, 3rd ed., John Wiley & Sons, New York, 1997.

36. C. G. Ryan, D. P. Siddons, G. Moorhead, R. Kirkham, P. A. Dunn, A. Dragone, and G. De Geronimo, Large detector array and real-time processing and elemental image projection of X-ray and proton microprobe fluorescence data, *Nucl. Instrum. Methods B* **260**, 1–7, 2007.

37. A. Dragone, G. De Geronimo, J. Fried, A. Kandasamy, P. O'Connor, and E. Vernon, The PDD ASIC: Highly efficient energy and timing extraction for high-rate applications, in *Proceedings, 2005 Nuclear Science Symposium*, Puerto Rico, Vol. 2, October 2005, 914–918.

38. M. W. Kruiskamp and D. M. W. Leenaerts, A CMOS peak detect sample and hold circuit, *IEEE Trans. Nucl. Sci.* **41**, 295–298, 1994.

39. G. De Geronimo, P. O'Connor, and A. Kandasamy, Analog CMOS peak detect and hold circuits—Part 1. analysis of the classical configuration, *Nucl. Instrum. Methods A* **484**, 533–543, 2002.

40. M. J. M. Pelgrom and M. Vertregt, CMOS technology for mixed signal ICs, *Nucl. Instrum. Methods A* **395**, 298–305, 1997.

41. G. De Geronimo, P. O'Connor, and A. Kandasamy, Analog CMOS peak detect and hold circuits—Part 2. The two-phase offset-free and derandomizing configuration, *Nucl. Instrum. Methods A* **484**, 544–556, 2002.

42. G. De Geronimo, A. Kandasamy, and P. O'Connor, Analog peak detector and derandomizer for high rate spectroscopy, *IEEE Trans. Nucl. Sci.* **49**, 1769–1773, 2002.

43. P. O'Connor, G. De Geronimo, and A. Kandasamy, Amplitude and time measurement ASIC with analog derandomization: First results, *IEEE Trans. Nucl. Sci.* **50**, 892–897, 2003.

44. P. O'Connor, G. De Geronimo, and A. Kandasamy, Amplitude and time measurement ASICs, *Nucl. Instrum. Methods A* **505**, 352–357, 2003.

45. P. O'Connor, G. De Geronimo, J. Grosholz, A. Kandasamy, S. Junnarkar, and J. Fried, Multichannel energy and timing measurements with the peak detector/derandomizer ASIC, in *Proceedings, 2004 Nuclear Science Symposium*, Rome, Italy, 2004.

46. G. De Geronimo, J. Fried, G. C. Smith, B. Yu, E. Vernon, W. L. Brian, C. L. Britton, L. G. Clonts, and S. S. Frank, ASIC for small angle neutron scattering experiments at the NSLS, *IEEE Trans. Nucl. Sci.* **54**, 541–548, 2007.

# PART III
## Ultrasound Imaging

# 7 Electronics for Diagnostic Ultrasound

ROBERT WODNICKI, BRUNO HAIDER, and KAI E. THOMENIUS

## 7.1. INTRODUCTION

Medical ultrasound, which has never been a very static field, is undergoing particularly interesting and exciting changes today. The most dramatic of these changes involves extensive miniaturization from a large (e.g., refrigerator/washing machine-sized) scanner to a laptop, tablet computer, or handheld device. Surprisingly, much of this reduction has been accomplished without significant loss in performance. In fact, additional capabilities such as real-time 3D imaging have become available as part of this miniaturization process.

As often happens, a dramatic change such as the one just described introduces many unexpected benefits. Foremost among these is a migration of the ultrasonic imager from its traditional home in radiology and cardiology departments to the hands of specialists. One can even envision migration to the hands of the primary care physician.

A key driver behind these changes is a continued reduction in power and size of electronic devices due to Moore's Law. Not only has this enabled a reduction in the physical size of the devices themselves, it has also had a secondary impact through migration of functionality from hardware to software. All of this has yielded an unprecedented amount of flexibility to the systems designer and promises to transform the field in a fundamental way.

Medical ultrasound is an ideal target for miniaturization and further integration, since the spatial requirements of the physical sensor array are particularly small, especially when compared to the room-sized gantries required for computed tomography (CT), magnetic resonance imaging (MRI), positron emission tomography (PET), and related fields. Ultrasound transducer arrays being used today are layered structures a few millimeters thick with contact surfaces on the order of a few centimeters by about 1 cm. The remainder of the system is largely comprised of processing electronics

*Medical Imaging: Principles, Detectors, and Electronics*, edited by Krzysztof Iniewski
Copyright © 2009 John Wiley & Sons, Inc.

that are highly amenable to miniaturization. Therefore, as opposed to other imaging modalities, ultrasound stands to benefit greatly from current and future trends in electronic miniaturization.

This chapter gives the technical background needed to appreciate the processes just described. Section 7.2 covers the general principles involved in forming ultrasonic images. Sections 7.3 and 7.4 discuss ultrasound systems and probes, respectively, and give the reader a background for appreciating the role of large-scale integration in this context. Sections 7.5 through 7.7 cover the current state-of-the-art of electronics. Finally, Section 7.8 is dedicated to describing current approaches to miniaturization in widespread use throughout the industry.

## 7.2. ULTRASOUND IMAGING PRINCIPLES

Ultrasound machines form images of biological tissue by transmitting focused beams of sound waves into the body and then listening for the reflected sound waves to determine the structure of the tissue being imaged.

This process is illustrated in Fig. 7.1: A handheld probe containing an array of transducer elements is applied to the body. The ultrasound machine creates a focused beam of sound using the array of transducer elements whose outputs are phased relative to each other. The sound waves are created by applying a transmit voltage to the transducer elements, which generates a related acoustic pressure at the face of the transducer. Sound waves propagate out from the face of the transducer and add constructively at a point distant from the plane of the transducer inside the body.

Depending on the acoustic impedance of the tissue along the path of the acoustic beam, a proportionate level of sound is reflected and travels back to the face of the transducer. The received echoes are phased relative to one another and added up in order to preferentially focus the attention of the system along the line of sight of the receive beam.

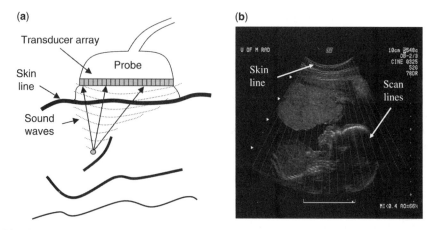

**Fig. 7.1.** Illustration of ultrasound imaging. (a) Schematic of the imaging process. (b) Image of fetus in the womb showing scan lines.

Unlike other medical imaging devices, ultrasound images reflections that arise from changes in acoustic impedance as sound waves travel from tissue of one type to another. One important consequence of this is that structures which are oblique to the path of the ultrasound beam are often weakly imaged [1]. In addition, strong reflectors will tend to shadow deeper structures by preventing sound waves from reaching them. Another source of information is backscatter from small reflectors which give tissues such as liver a speckle-like structure (as seen in Fig. 7.3).

### 7.2.1. Ultrasound Scanning

Imaging using ultrasound consists of building up a 2D or 3D representation of the subject using multiple scan-lines that are acquired adjacent to each other within the body. This is illustrated in Fig. 7.1b, where a number of individual scan lines (white lines) are highlighted. While they are shown separated in the figure, in reality they are contiguous and build up the image. The transducer transmits ultrasound into the body, echoes representing tissue structure are received, and these echoes are then processed and displayed by the ultrasound machine.

Figure 7.2 shows the progression in the development of probe architectures. The first ultrasound probes consisted of a single element that was manually moved along the patient in order to generate an image (Fig. 7.2a). These were soon replaced by mechanically scanned probes in which a single element is translated using a motor in order to build up an image. Some mechanically scanned probes are still used most

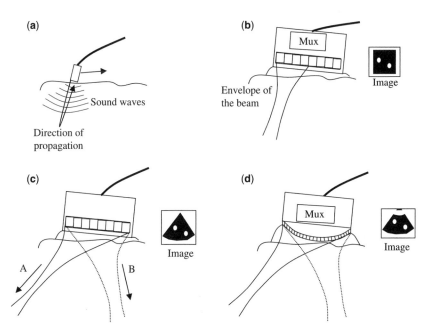

**Fig. 7.2.** Ultrasound scanning probe types. (a) Manual scan with single element. (b) Electronically scanned 1D linear array. (c) Electronically scanned 1D phased array showing two different steering directions (A and B). (d) Electronically scanned 1D curved array.

notably in applications where electronic scanning is impractical such as high-frequency probes and some 3D probes. Mechanically scanned probes historically suffered from significant reliability problems and limited beam agility.

Today, the majority of ultrasound products use electronically scanned probes. These can be further divided into *linear scan* and *sector scan* types. In linearly scanned probes, the beam is translated horizontally along the face of the transducer also known as the *aperture* (Fig. 7.2b). Translation is accomplished by successively selecting different groups of transducers electronically. In sector scan probes (also called *phased array* probes), the beam is scanned radially with the transducer probe as the pivot for the fan beam (Fig. 7.2c). Radial scanning is done by relative phasing of channel delays in order to steer the beam. The group of elements that are active at any given time is known as the *active aperture*. Finally, curved array probes are linear probes that implement sector scanning due to their geometry (Fig. 7.2d).

The choice of probe type is dependent on the requirements for access to the tissue being imaged as well as the nature of the structures themselves. For example, in cases where large structures (such as a fetal head) need to be imaged from a small aperture, a curved or phased array probe might be used. In cases where access is not as much of an issue, but large imaging area near the probe is required (such as for carotid artery imaging), linear probes would be used.

***7.2.1.1. Sector Scan Probes.*** Sector scan probes are used for applications in which a small probe must image a large volume such as when imaging the heart or when imaging a fetus. In all such cases, the imaging area is larger than the transducer aperture. While sector probes provide increasing image width with increasing depth, they lack width close to the skin. Due to the requirement for a transformation from polar to Cartesian coordinates, optimal image resolution may not be obtained using these probes. The line density is in fact much higher in the near field than in the far field, which leads to nonuniform resolution over the image. The smaller aperture also leads to inferior image resolution when compared to linear array probes which may have larger active apertures.

***7.2.1.2. Linear Scan Probes.*** In a linear scan probe, the ultrasound beam is translated linearly along the full face of the transducer. These probes translate the ultrasound beam across the imaging aperture by reconfiguring a multiplexer that changes how the ultrasound system channels are connected to the transducer array. Linear scan probes are preferred in applications where larger imaging area close to the transducer is important. These applications include imaging of carotid arteries and cysts in the liver. They are generally used for vasculature (carotid artery and leg veins) and small parts imaging (breast and testicles). In these cases, access to the tissue being imaged is not an issue, so a linear array can be used.

***7.2.1.3. Curved Array Probes.*** In abdominal imaging, curved arrays are used to create a curved, trapezoidal image. The curved probe geometry provides a naturally phased array without the requirement for explicit steering of the beam. Curved array probes provide larger contact area (to image a fetus head, for example), which allows a larger volume of coverage.

*7.2.1.4. Compound Imaging.* Compound imaging combines beam-steering and linear array scanning in order to obtain improved contrast. Compound imaging reduces noise by imaging at multiple angles and combining the resulting images to a single lower noise image.

## 7.2.2. Understanding Ultrasound Images

Ultrasound images are reviewed by trained radiologists and cardiologists. The images can sometimes be difficult to interpret without significant prior experience. In this section some of the more common types of images are presented for reference.

*7.2.2.1. Ultrasound Tissue Phantom.* Figure 7.3 shows a typical ultrasound image. In this case the subject is what in diagnostic imaging is called a *phantom* [1, 2]. This is an insulated box that contains tissue mimicking materials arranged in a well-defined way so that they can be used to test the resolution and sensitivity of a particular imaging device. These include anechoic (non-echo-producing) cysts, spherical targets with different tissue densities, and highly echogenic (echo-producing) wire targets as shown. The image appears shaped as a fan because it is created by a beam that is effectively steered from one side of the image to the other. The top of the image represents the probe face that is pressed up against the body, while the bottom of the image is the deepest part of the body that is displayed by the system for the particular probe. *Penetration depth* is limited by the noise floor of the imager; past a certain depth, noise becomes evident in the image and so these data are typically not displayed. Ultrasound is a coherent imaging modality. This causes a specific image artifact known as *speckle*, which is shown in Fig. 7.3b.

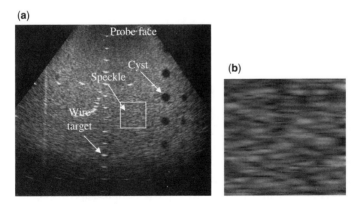

**Fig. 7.3.** Ultrasound fan-beam image of standard acoustic phantom. (a) Cysts are the dark circles; wire targets are the bright (white) dashed lines. (b) Close-up view of speckle. (Adapted from reference 6.)

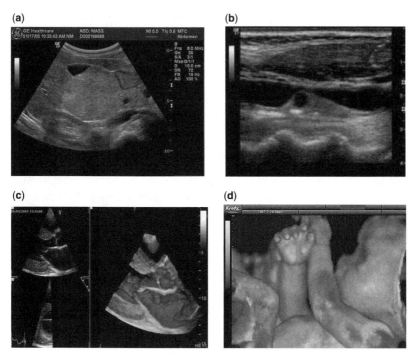

**Fig. 7.4.** Examples of typical ultrasound images. (a) Cysts in the liver. (b) Carotid artery blood flow. (c) Heart imaged in 3D. (d) Fetus in the womb.

**7.2.2.2. Diagnostic Images.** Ultrasound is most commonly used for imaging abdominal organs, blood flow, and vasculature, for example in imaging cysts in the liver (Fig. 7.4a), blood flow (Fig. 7.4b) and anatomical structures within the heart (Fig. 7.4c), and imaging the fetus in the womb (Fig. 7.4d). It also has an important use in treatment of breast cancer where it is used to assist in needle biopsy.

A variety of ultrasound machines are currently offered by the principal manufacturers in the industry. These include portable units as well as console imagers specialized for cardiology, obstetrics, and radiology. Images are produced in real-time providing both 2D and 3D images of structures in the body.

### 7.2.3. Ultrasound Beam Formation

Early ultrasound scanners were mechanically translated single element transducers. The availability of arrays of transducer elements that could be steered and focused by electronic *beam formation* made the mechanically scanned devices obsolete. Beam formation can be divided into the simultaneous operations of *focusing* and *steering* [3]. In addition, certain systems use multiplexing to accomplish aperture translation.

**7.2.3.1. Focusing and Steering.** Focusing is accomplished by time delay of the transmit channels such that the acoustic waves add constructively at the focal point

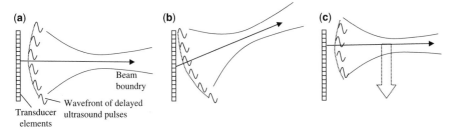

**Fig. 7.5.** Beam formation. (a) Focused beam. (b) Focusing and steering. (c) Aperture translation. (Adapted from reference 41.)

as shown in Fig. 7.5a. Steering refers to changing the beam angle while the aperture remains fixed in space. Steering is accomplished by delaying the transmit channels so that the focal point is moved at an angle relative to the normal center of the active aperture. This is shown in Fig. 7.5b, where the beam is steered to a focal point. Note that in this case both focusing and steering are used.

***7.2.3.2. Translation of the Aperture.*** Translation of the aperture is accomplished by successively switching in different groups of transducer elements to accomplish a linear scan of the active beam across the face of the array as shown in Fig. 7.5c. Steering and linear scanning can be combined, as is the case with spatial compounding [2].

***7.2.3.3. Transmit Beam Formation.*** Time delay of the transmit signals is accomplished in the system beam-forming electronics by shifting the transmit control signals relative to each other in time so that the resulting acoustic waves are transmitted with the required phase delay. The delay is typically implemented using digital delay registers or an array of synchronized and programmable clocks. The extent of phase delay required on each channel is calculated in real-time by the beam-forming electronics based on the required instantaneous location of the focused ultrasound beam for the given scan. For focusing, the delay is less than 5 μs, whereas for beam steering it can be higher than 15 μs [3].

***7.2.3.4. Receive Beam Formation.*** Beam formation is also done during reception in a similar fashion as on transmit. In this case it can be thought of as focusing the attention of the receive electronics such that only echoes originating at the focal point achieve the maximum antenna gain at the particular instant. In practice, this is implemented digitally using a FIFO, but it can also be done using analog delay lines (see Section 7.7 for further discussion). The beam-forming configurations can be different on transmit and receive to independently optimize the transmit and receive beams. The beamwidth at the focus can be approximated by [4]

$$bw = \lambda(f\#)$$

The *f#* (or *f*-number) is a measure of the degree of focusing and is defined as [5]

$$f\# = \frac{\text{focus}}{D}$$

where *focus* refers to the location of the axial focus, and $D$ is the aperture size.

### 7.2.4. Ultrasound Transmit/Receive Cycle

Ultrasound machines build images one line at a time by multiple successive transmit and receive cycles. Each of these cycles forms an ultrasound beam in a particular direction to acquire the respective line. This process is illustrated in Fig. 7.6a.

A diagram of the ultrasound transmit/receive cycle is shown in Fig. 7.6b. The cycle as shown is known as a *B-mode* cycle and is the most commonly used cycle today. Other cycles are used particularly for Doppler imaging discussed in Section 7.2.5.5. The transmit cycle is repeated each time that a new line is to be acquired. Typically between 50 and 400 lines are acquired where each one takes approximately 100–300 μs. The *pulse repetition frequency* (PRF) depends on the application. For example, cardiology requires PRF rates as high as 10 kHz while abdominal imaging is typically done at 3- to 4-kHz PRF rates.

The cycle begins with reconfiguration of the electronics including any beam-specific control data for the mux switches and the pulsers. Reconfiguration is generally avoided during the receive cycle in order to prevent any digital noise from coupling into the receivers and causing image artifacts. Therefore, given a fixed frame rate set by the expectations of the user, the time for reconfiguration takes away from the available receive time. This reduces the penetration capability of the system and is therefore undesirable. For this reason, reconfiguration must be made as short as possible (typically less than 10 μs).

Next, the transmit cycle begins during which all active pulsers in the array are fired each according to its assigned phase delay. The cycle ends after the last pulse has completed its transmission.

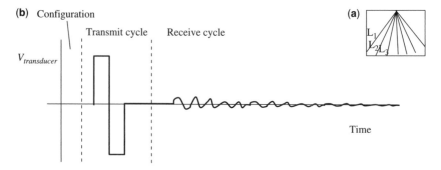

**Fig. 7.6.** Ultrasound beam acquisition. (a) Successive lines used to build up the image. (b) Transmit/receive cycle in terms of the voltage as seen at the transducer.

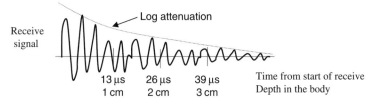

**Fig. 7.7.** Signal level attenuation with depth in the body. Note that the return signal frequency is shown much lower than in reality for illustration.

Immediately following the transmit cycle, the receive cycle begins and lasts until all of the echoes have returned from deep within the body. As shown in Fig. 7.7, a graph of the receive cycle has the important property that depth in the body is linearly related to increasing time along the $x$-axis. This can best be understood as follows: Immediately after the first transmit pulses occur, echoes begin to come in from tissue in the body. In fact, the transmit pulses first hit the skin line, then hit the tissue immediately underneath the skin, then the tissue following that, and so on, until they have propagated on through, very deep into the body. At each point the tissue being stimulated by transmit pulses returns an echo. These echoes, in turn, propagate back through the tissue to the transducer array where they are then converted into electrical energy for the receive signal. With the assumption of a constant speed of sound, one can convert the time interval between echos to distance in tissue as shown in Fig. 7.7.

Another important property of Fig. 7.7 is the log magnitude signal attenuation that echoes experience when traveling through tissue in the body (discussed further in Section 7.2.6.2). Echoes that return sooner are coming from structures that are close to the transducer and therefore experience much less attenuation than echoes that return later. This attenuation must be corrected for in order to display a useful image with uniform grayscale levels at all pixels.

### 7.2.5. Imaging Techniques

A significant amount of design work is required to optimize the beam patterns such that adequate resolution for the application can be obtained at all points along the beam line. A tradeoff is reached where some areas achieve better resolution at the expense of others. A number of special techniques have been developed to maximize the benefits of this tradeoff.

*7.2.5.1. Apodization or Weighting.* Spatial weighting of the aperture is known as *apodization* and is analogous to time-domain weighting in signal processing. Apodization is sometimes used for transmit and receive beam formation. In this case the amplitudes of the output waveforms across the aperture are modified using a windowing function (for example, a Hamming window). This is typically implemented using multilevel pulse circuits (discussed in Section 7.5). Apodization is used in order to reduce the presence of imaging artifacts caused by sidelobes of the ultrasound

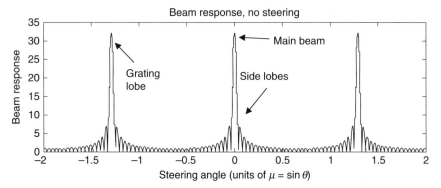

**Fig. 7.8.** Simulated beam pattern providing an illustration of side lobes and grating lobes.

beam [6]. *Side lobes* are smaller lobes of transmit energy on either side of the main beam which are generated naturally as part of the beam-formation process as illustrated in Fig. 7.8. *Grating lobes* are similar to sidelobes in that they generate acoustic energy outside the desired main lobe. Their origin stems from the underlying sampling of the acoustic array. Grating lobes generate image artifacts that typically show up as blurred duplicate structures or added haze some angular distance away from the source [2]. The design of the aperture is optimized to reduce side and grating lobe energy to the point where they do not create image artifacts.

***7.2.5.2. Dynamic Focusing.*** One advantage of using an electronically configured array transducer is that it is possible to optimize the focus over a much longer axial path. During the receive phase it is possible to continuously adjust the focal point of the array by changing the time delay relationships of the active beam-forming channels. This is illustrated in Fig. 7.9 where the overlapping beam patterns represent different configurations of the beam-forming delays [1]. This process is known as *dynamic focusing* on receive and is standard in most current products. The individual beam configurations are referred to as *focal zones* since the composite axial beam pattern is essentially divided into multiple segments with each one obtaining optimal focus over a short distance. The combined beam will have reasonably improved focus at the expense of more sophisticated beam-forming electronics. Dynamic focusing essentially moves the focus at the speed of sound in order to track the traveling wavefront.

It is worth noting that dynamic focusing is straightforward to implement on receive but cannot be implemented in real-time on transmit. This is due to the fact that forming a unique transmit focus requires sound waves to be transmitted into the body and waiting for the echoes to be received. However, it is possible to form multiple focal zones on transmit by repeated acquisitions with varying focal locations. Forming multiple focal zones on transmit will reduce the PRF rate significantly. In cardiac imaging systems, the high frame rate precludes the use of multiple transmit focal zones, however this technique is widely used in abdominal imaging.

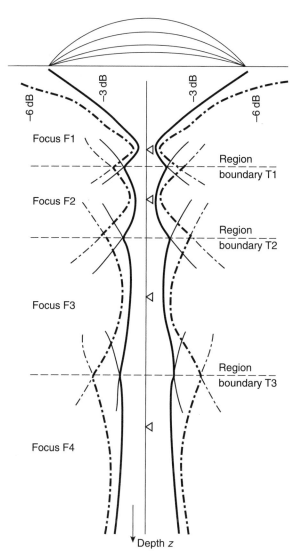

**Fig. 7.9.** Dynamic receive. Multiple focal zones are formed along the receive beam scan line in order to increase the focal depth of the imager. (Adapted from reference 1.)

**7.2.5.3. Multiline Acquisition.** *Multiline acquisition* (MLA) is a special-purpose imaging mode in which a single broad transmit beam is generated, and then a number of more tightly focused receive beams are simultaneously processed. This mode of imaging can be accomplished simply by additional processing of the same receive data [2, 7].

Frame rate is fundamentally limited by the speed of sound in the body. The relatively slow speed of sound is a basic limiter of 3D imaging in ultrasound. For each line that is *transmitted*, the ultrasound machine must wait for the sound to return from the farthest point that is being imaged before it can move onto the next line. The advantage of MLA is that it increases the frame rate since it becomes possible to acquire multiple receive points simultaneously. MLA requires the digital beam former to be larger since it needs to process more beams on receive simultaneously. It often is used to facilitate real-time imaging in 3D [7].

***7.2.5.4. Codes.*** The Food and Drug Administration (FDA) in the United States regulates the peak transmit power for ultrasound machines in basic imaging modes. The peak transmit voltage is also limited by the breakdown voltage of the semiconductor process used to fabricate the transmit electronics. In some applications, it is necessary to improve penetration of the ultrasound beam into the body while still satisfying these constraints on peak transmit power. This can be accomplished using *coded excitation*, which essentially improves the signal-to-noise ratio (SNR) by increasing the average transmitted acoustic power without exceeding the limits on transmit voltage. In this case, the transmit sequence consists of a predefined series of phase-coded transmit states which implement specific codes such as Golay codes [6]. The receive beam former contains a filter that compresses the code sequence to a length of one transmit state. In this way, range resolution is equivalent to the resolution of a noncoded system while penetration depth is increased. Typical code lengths are on the order of 8 or more transmit cycles. Longer codes are not used since they can contribute to imaging artifacts.

***7.2.5.5. Doppler Imaging.*** Doppler imaging is used to measure the speed of blood flow or tissue vibration in the body. For example, it is used to measure blood flow in the carotid artery where any perturbations in the flow can signal possible atherosclerotic plaque formation [2]. Measurements are performed on a volume of blood and displayed as a distribution of Doppler frequencies representing the range of velocities of the blood constituents in the vessel. Doppler imaging is also used in cardiology to assess the performance of the cardiac valves.

Doppler imaging takes advantage of the fact that sound reflected by a moving object (relative to the stationary observer) incurs a Doppler frequency shift. The amount of the Doppler shift can be measured and used to determine the velocity of the object being imaged. The Doppler frequency $\Delta f$ is given by [1]

$$\Delta f = \frac{2f_0 v_d}{c} \cos \phi$$

where $f_0$ is the transmitted frequency and $\phi$ is the angle of incidence, $v_d$ is the velocity of the moving objects (e.g., red blood cells), and $c$ is the speed of sound in the medium.

Doppler imaging constitutes the most stringent imaging mode for ultrasound in terms of noise floor [8]. It requires very low noise receive amplifiers as well as careful attention to transmit electronics including the clock generation and the high-voltage transmitters. Required jitter specification on clocking of transmit and A/D circuitry is on the order of tens of picoseconds root mean square (rms).

Different types of Doppler are used including *pulse wave (PW) doppler* with range localization used to localize the echoes from the Doppler signal and *continuous-wave (CW) doppler*, which cannot provide any location information [2], but is capable of measuring increased flow velocities. *Color flow doppler* is an imaging mode where B-mode data is acquired simultaneously to pulsed Doppler data by interleaving of B-mode and PW Doppler pulse trains in subsequent transmit cycles.

***7.2.5.6. Harmonic Imaging.*** Propagation of ultrasound in the body is associated with the formation of harmonic energy due to nonlinear properties of the tissue. This harmonic energy can be imaged and has a number of important uses: First, the transmit beam that is formed due to harmonic energy is more tightly focused because the beam shape drops off with the square of the pressure as opposed to linearly for fundamental imaging. Second, harmonic imaging rejects echoes from near-field subcutaneous tissue structures since the harmonic beam forms deeper in the body. Harmonic imaging requires higher transmit power and therefore transmitters capable of generating higher transmit voltages must be used. Since higher voltage circuitry consumes more area and power, it may be somewhat challenging to implement good harmonic imaging performance in newer systems that are compact and portable.

### 7.2.6. Image Quality Performance Parameters

Image quality in ultrasound images is directly affected by a number of parameters that are discussed below.

***7.2.6.1. Reflection.*** Significant losses can be incurred when the acoustic wave propagates from one tissue type to another. This is, in general, not an issue because waves propagate between layers of soft tissue in which water is a major constituent; however, two types of reflectors do transmit the majority of sound back. These are gas in the lungs and gastrointestinal tract and bone. Therefore when imaging through these tissues, partial or complete occlusion of other surrounding structures may occur.

Reflections are important because they are the basis of the detection of structure in the body using ultrasound; however, they represent a problem since they make it difficult to image certain parts of the body effectively. For example, imaging the heart is particularly challenging since much of the acoustic window for this part of the body is hidden by the rib cage or the lungs.

Sound is also strongly reflected at the transition between the transducer and the body. In fact, direct contact is needed since the attenuation going from the body to air would be too great to be useful. Typically a special acoustic gel is used to ensure that a good acoustic coupling is maintained between the transducer probe face and the body.

***7.2.6.2. Absorption.*** Absorption, $\alpha$, is measured in $dB/cm/MHz$, where $\alpha$ varies from 0.5 for fat to 2.0 for muscle [1]. Round-trip attenuation is given by $2\alpha$, so for typical tissue (fat) it is approximately $1\,dB/cm/MHz$.

For example, imaging at a depth of 10 cm at 4 MHz yields a round-trip attenuation of 40 dB, whereas imaging at a depth of 20 cm yields round-trip attenuation of 80 dB. The ability of a particular probe to image to a certain depth within the body is defined as *penetration depth* and varies significantly depending on application.

***7.2.6.3. Resolution.*** Spatial resolution in ultrasound imaging is determined by three interacting effects as can be seen in Fig. 7.10 [9]. *Axial* (or *range* or *longitudinal*) *resolution* is the resolution along the direction of travel of the transmitted ultrasound beam. This resolution is limited by the length of the transmitted ultrasound pulse. *Elevational* and *azimuthal resolutions* are determined by how well the ultrasound beam can be focused in these two orthogonal directions. Elevational focusing in 1D arrays is often done using a fixed focus acoustic lens, while focus in azimuth is done electronically. Circular apertures (such as annular arrays, which are composed of concentric ring transducers [10]) are also used, in which case there is no difference between elevational and azimuthal focus.

Probes that image finer detail do so using higher frequencies (5–10 MHz), and therefore they achieve much poorer penetration than probes that image at lower detail at lower frequencies (1–4 MHz) due to absorption as described in Section 7.2.6.2.

In the usual nomenclature for characterization of acoustic fields, $\lambda$ is the wavelength of the ultrasound signal given by

$$\lambda = \frac{c}{f}$$

where $f$ is the frequency of the transmitted signal, and $c$ is the speed of sound in the medium (nominally 1540 m/s for soft tissue).

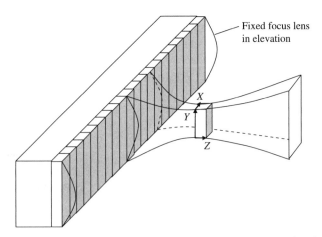

**Fig. 7.10.** Illustration of ultrasound beam generated using a linear 1D array showing elevational ($Y$), axial ($Z$) and ($X$) azimuthal resolutions at the beam focus. The array is focussed in elevation using a fixed focus lens. (Adapted from reference 9.)

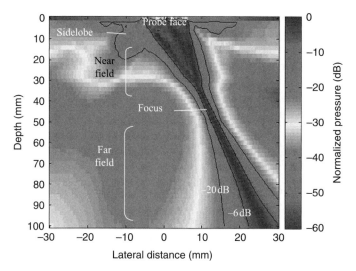

**Fig. 7.11.** Plot of annular array beam profile cross section showing side lobe and −6-dB and −20-dB contours [11].

Figure 7.11 [11] shows a cross sectional plot of an ultrasound beam through the azimuthal plane for an annular array aperture. Here the −6-dB beamwidth contour line is plotted and shown to be varying along the axial beam line. In fact the beam can be broken into three distinct regions as shown in the figure: the focal region appearing near the center of the beam, the near field, a region of poor focusing which appears close to the transducer; and the far field, a region of poor focus distant from the transducer. The term "strength of focusing" is sometimes used to describe arrays. A strong focus is associated with a narrow beamwidth at the focus and a short depth for the focal region; operating conditions with lower $f$-numbers are considered more strongly focused. The array shown in Fig. 7.11 is relatively strongly focused; hence a short but narrow axial "waist" to the beam. A good approximate expression for depth of focus (DOF) is given by [4, 5]

$$DOF = 7\lambda(f\#)^2$$

***7.2.6.4. Dynamic Range.*** Signal dynamic range is limited by the maximum transmit pressure and the minimum detectable return signal. The transmit pressure is directly related to the required pulsing voltage. The maximum transmit pressure is strictly regulated by the FDA in the United States in order to reduce the chances of adverse events such as heating of tissue and mechanical bioeffects such as cavitation [2]. The minimum receive voltage is technology-dependent and is currently limited by the noise in the front-end preamplifier. However, assuming that this noise can be reduced, the noise in the sensor itself as well as the noise due to Brownian motion in the body will then set the minimum detectable signal [1]. Given these considerations, the total round-trip signal dynamic range is typically on the order of 120 dB for today's products.

For example, to image at a resolution of 100 $\mu$m requires the transmit frequency to be on the order of 15 MHz for an $f\# = 1$ transducer. At this frequency, nominal round-trip attenuation of about 15 dB/cm can be expected. Given a typical front-end dynamic range of about 120 dB and required instantaneous dynamic range of 60 dB for image display, the probe would only be able to view objects which are at most 4 cm deep into the body. Conversely, the same system imaging at 3 MHz would be able to view objects 20 cm deep into the body. Therefore there exists a natural tradeoff between penetration and resolution which implies that high-resolution probes are only used to image shallow depths while imaging much deeper into the body can only be done at low resolution.

***7.2.6.5. Speckle.*** As can be seen in the Fig. 7.3b, a key feature of ultrasound images is a faint pattern in the background of the image which is known as *speckle*. Speckle may provide diagnostically useful information to the clinician [12]. However, speckle also degrades image contrast, so there is a need to reduce its effects. Speckle reduction is an active area of research and can be achieved by techniques such as spatial compounding [2].

### 7.2.7. Ultrasound Imaging Modalities

The physical limitations described above have split the ultrasound market among naturally occurring lines of application. For example, vascular probes image at high resolution to visualize plaques and blood flow, but they generally are used on arteries (such as the carotid) that are close to the surface of the skin. Conversely, abdominal probes are used to image structures deep inside the body such as cysts in the liver or the fetus in the womb, and these structures yield diagnostically relevant information at much lower imaging resolutions. Table 7.1 summarizes the current applications of ultrasound imaging and associated key imaging parameters.

**TABLE 7.1. Comparison of Currently Available Ultrasound Imaging Modalities**

| Application | Steered | Probe | 2D/3D | Depth | Resolution | Examples of Clinical Use |
|---|---|---|---|---|---|---|
| Cardiology | Yes | 2D | 2D/3D | 2–16 cm | 2–3 mm | Characterize heart function |
| OB/Gyn | No | Curved | 2D/3D | 2–20 cm | 2–3 mm | Imaging fetus in 3D |
| Intravascular (IVUS) | Yes | Linear | 2D | <1 cm | 0.3 mm | Assess severity of plaques |
| Intracardiac (ICE) | Yes | Linear | 2D | 1–5 cm | 0.5 mm | Characterize heart function |
| General imaging | Yes | Linear | 2D | 2–20 cm | 2–3 mm | Locate cysts in liver/kidney |
| Vascular/ small parts | No | Linear | 2D/3D | 1–5 cm | 0.3 mm | Assess carotid plaques, breast tumors |

**TABLE 7.2. Comparison of Currently Available Medical Imaging Modalities**

| Imaging Type | Ionizing | Image Quality | 2D/3D | Real-Time | Portable | Cost |
|---|---|---|---|---|---|---|
| MR | No | High | 2D/3D | No | No | High |
| CT | Yes | High | 2D/3D | No | No | High |
| US | No | Medium/High | 2D/3D | Yes | Yes | Low |
| X-ray | Yes | High | 2D | Yes | No | Medium |
| PET/Nuclear | Yes | Low | 2D | No | No | High/Medium |

It is useful also to understand ultrasound imaging in view of the general diagnostic imaging market. Today, there exists a spectrum of imaging modalities that are used to provide diagnostic information in medicine. The most commonly used of these are shown in Table 7.2. Ideally, only a single imaging method would be used to cover all possible diagnoses; however, as is evidenced by the broad distribution of devices in use today, there is no one method that is capable of providing all required diagnostic information. In practice, each device covers a range of specific conditions for which it is optimally suited. For example, X-ray computed tomography (CT) is ideally suited to imaging anatomical structure in fine detail but is not as sensitive to soft tissue, while magnetic resonance imaging (MRI) is much more sensitive to soft tissue but it is also a more expensive procedure and thus is not as widely accessible.

As an imaging modality, ultrasound is somewhat unique in the sense that it can provide high-quality, real-time imaging at a fraction of the cost of other imaging devices in use today. It is also highly portable and therefore is an attractive option for use in a clinic, or even in emergency care. Ultrasound also has the benefit of not using ionizing radiation, although there are still biological safety issues which must be considered [2].

## 7.3. THE ULTRASOUND SYSTEM

A variety of manufacturers of ultrasound imagers exist today with varying areas of expertise. Around the year 2000, significant consolidations took place with the entry of major imaging companies such as GE, Philips, and Siemens. These large imaging companies acquired ultrasound powerhouses such as ATL, Acuson, Vingmed, and Diasonics. Today, they produce an array of ultrasound equipment including console and portable units. The majority of these imagers follow a standard system architecture that is illustrated in Fig. 7.12. The function of the subunits shown is summarized here and described in detail in the remaining sections.

### 7.3.1. Transducers

As shown in Fig. 7.12, an array of transducers is interfaced to a high-voltage multiplexer over an analog interface cable. The cable consists of a bundle of 64–500 micro-coax cables that are each no more than 300 μm thick. The transducers are

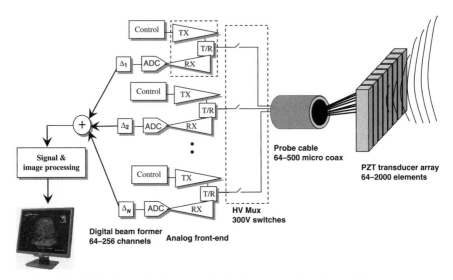

**Fig. 7.12.** Schematic diagram of typical ultrasound imaging architecture.

operated with (a) transmit voltages ranging between 5 Vpp and 300 Vpp and (b) receive voltages ranging between 1 μV and 100 mV. The transducer is applied in contact to the body, transmits ultrasound waves into tissue, and converts the received acoustic echoes to electrical signals which are than sent back to the system for processing. Therefore it is a bilateral device that does both transmit and receive functions.

### 7.3.2. High-Voltage Multiplexer

In a linear or curved array, the high-voltage multiplexer is used to select a series of groups of transducers during a typical image scan in order to translate the beam pattern across the array.

### 7.3.3. High-Voltage Transmit/Receive Switch

The transmit/receive (T/R) switches are high-voltage switches (historically diodes) that are used to protect the low-voltage receive circuitry from the high-voltage transmit pulses.

### 7.3.4. High-Voltage Transmitters

The transmitters (sometimes called *pulsers*) are high-voltage output stages typically operating between 1 MHz and 15 MHz and producing voltages as high as 300 Vpp. They may produce unipolar pulses (swing between ground and +HV), bipolar pulses (swing between ground, +HV and −HV), or multilevel pulses that are typically quantized to about 4–8 bits. Transmit cycles range from 1 or typically 2 cycle pulses for *B-mode* (standard 2D) imaging to as many as 50 pulses in *pulse-wave (PW) doppler* and continuous imaging in *continuous-wave (CW) doppler* modes.

### 7.3.5. Receive Amplifier and Time Gain Control

The receive signal is amplified by a low-noise preamplifier and then scaled in time using a special-purpose programmable gain amplifier known as a *time gain control (TGC)* amplifier. These are typically stand-alone ASICs; however, recently, parts that integrate multiple channels on a single chip have become available [13].

### 7.3.6. Analog-to-Digital Converter and Beamformer

The resulting normalized receive signals are then processed by an array of analog-to-digital converters typically running between 30 and 60 MHz at 10- to 12-bit resolution. The digitized channel signals are fed into digital ASICs that perform beam-forming calculations in real time. These beam-forming calculations consist of phase delays and weighted summation. Control settings for the transmit and the receive beam-forming hardware are calculated as the beam is scanned across the array.

### 7.3.7. Signal and Image-Processing

The summed signals are transformed from polar to Cartesian coordinates by a scan converter and then processed to improve image quality and displayed on a monitor for the operator. The user typically has control of a number of imaging functions in real time during a scan procedure including image contrast, depth of focus, and scaling of receive gain.

## 7.4. TRANSDUCERS

Since their introduction in the 1970s, ultrasound machines have for the most part used piezoelectric materials for transduction. The most popular material is lead zirconate titanate, which is a ceramic that is also known as PZT. Today, the majority of ultrasound transducers are still manufactured by dicing slabs of piezoceramic to form individual resonators as shown in Fig. 7.13a. The transducer pitch varies from 400 μm to as little as 80 μm, depending on the intended application of the probe and resulting required imaging frequency of operation. Aside from PZT, a number of other promising new transducer technologies have been proposed including capacitive micromachined ultrasound transducers (cMUTs) and single-crystal relaxor piezoelectrics and electrostrictive ceramics. Poly-vinylidine fluoride (PVDF) transducers have also been used in the past. Some of these new technologies will be discussed below.

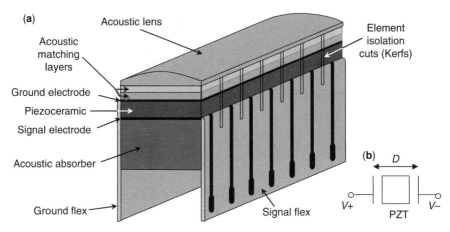

**Fig. 7.13.** Construction of 1D linear PZT ultrasound probe array. (a) Diagram of the array. (b) Schematic representation of a single PZT element.

### 7.4.1. Acoustic Characteristics

The typical PZT transducer is shown in Fig. 7.13b and consists of an acoustic–electric material (PZT) that is sandwiched between two electrodes. When a voltage is applied to the electrodes, the electric field causes the piezoelectric material to deform mechanically. Since the material is coupled to the body through matching and lens layers, acoustic energy in the form of sound waves is transferred from the transducer into the body for imaging. On receive, vibrations from the body are transferred back to the transducer, which causes the transducing material to deform mechanically. These deformations change the distance $D$, thereby generating an electrical signal in response to the acoustic input. Typical values for the receive voltage range between a few microvolts rms and 100 mV rms.

The acoustic lens shown in Fig. 7.13a is a material such as silicone, which can be shaped to focus the beam in elevation. An acoustic matching material is used between the piezoelectric material and the lens in order to improve transfer of acoustic energy from the PZT elements into the body. The backing material is an acoustic absorber such as scatters in an absorbing matrix which damps reverberations and reduces crosstalk. Connection to the elements is typically made using a flexible circuit assembly with very fine line spacing in order to match the pitch of the elements, which can be on the order of 100 μm [14].

Most transducers up until the 1990s were 1D arrays (also called *linear arrays*), meaning that they were only diced in a single direction forming a 1D array of long and thin transducers as seen in Fig. 7.13a.

One-dimensional arrays are focused electronically in the azimuth or image plane. Often a lens is used for elevation focusing in such arrays. This creates limitations in the slice thickness of the imaged plane. This limitation is a result of the fact that the lens provides a fixed focus that cannot achieve optimal focusing along the entire receive path.

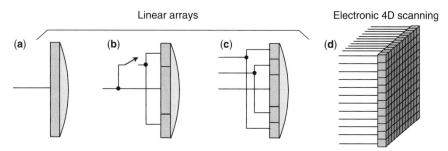

**Fig. 7.14.** Evolution of transducer array topologies showing elevational view of the array elements. (a) 1D array with fixed focus in elevation. (b) 1.25D array with 1 degree of freedom focus in elevation. (c) 1.5D array wth 2 or more degrees of freedom focus in elevation. (d) 2D array for electronically scanned volumetric imaging. (Adapted from reference 15.)

In the 1990s, the problem of poor focus in elevation was addressed by introducing additional dicing along the elevation direction as shown in Fig. 7.14 [15]. These so-called 1.25D and 1.5D arrays are now better focused in the elevation direction during the receive cycle and achieve an improvement in overall imaging resolution.

Ideally the beam profile would be focused with equal strength in elevation as it is in azimuth. This is achieved using an array that is finely diced in both the elevation and azimuth directions as shown in Fig. 7.14d. Development of so-called 2D arrays has been hampered by a number of issues including the sheer complexity of dealing with the large number of signals generated by a two-dimensional transducer as well as manufacturing issues [16].

Further improvements in focus can be achieved by using annular arrays that maintain axisymmetric focus throughout the entire imaging depth [17].

### 7.4.2. Transducer Performance Characteristics

The dimensions of the PZT transducer determine the key performance characteristics of the device. Figure 7.15a shows the typical output signal of a transducer stimulated by a very short transmit pulse. This is the impulse response and is used to derive the frequency response of the device by transformation to the frequency domain as shown in Fig. 7.15b. The frequency domain plot provides information on a number of important transducer performance parameters.

*Bandwidth* can be quoted either as one-way or two-way (pulse-echo) measurements. One-way signifies the transmit or receive bandwidth alone and two-way signifies the transmit bandwidth multiplied by the receive bandwidth. For a single transducer the one-way bandwidth is given by the $-3$-dB frequencies while the two-way bandwidth is given by the $-6$-dB frequencies. In second harmonic imaging, it is important to be able to image the receive signal at higher frequencies so a wider bandwidth is preferable.

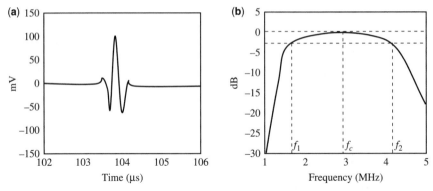

**Fig. 7.15.** Acoustic output of ultrasound transducer. (a) Impulse response in time. (b) Frequency response obtained by transforming impulse response to frequency domain. The center frequency $f_c$ is customarily given as the average of $-3$-dB frequencies, $f_1$ and $f_2$. (Adapted from reference 19.)

*Fractional bandwidth* is a measure of the range of frequencies over which the transducer will operate and is given by

$$FB = (f_1 - f_2)/f_c$$

where $f_1$ is the upper $-3$-dB point, $f_2$ is the lower $-3$ dB point, and $f_c$ is the center frequency as shown in Fig. 7.15b. Typical fractional bandwidth for PZT is on the order of about 70% whereas for cMUTs, it can be as high as 110%. Since higher-resolution applications require higher imaging frequencies, a device with a larger fractional bandwidth has the potential to cover a broader range of applications with a single probe.

Figure 7.16 shows the frequency response of a number of different transducers superimposed, including PZT and cMUT, to show how the bandwidth of these devices

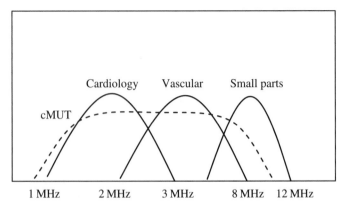

**Fig. 7.16.** Frequency response of some currently available ultrasound probes and potential cMUT-based probes.

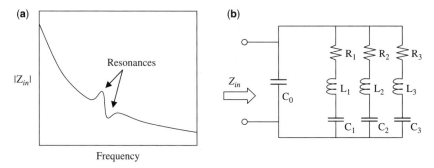

**Fig. 7.17.** Electrical impedance model of PZT transducer. (a) Frequency-domain impedance plot. (b) RLC equivalent circuit.

differs by application. cMUTs are a promising device for general imaging since they can be used to cover a broad range of imaging applications.

*Center frequency* is the frequency at which imaging is nominally performed. *Resonance frequencies* are the natural modes of resonance of the transducer in conjunction with the case capacitance. There are typically two resonance frequencies as shown in Fig. 7.17a. *Output power* is an important parameter and is a measure of the ability of the transducer to convert the driving electrical power into acoustic power. One can form a direct analog to electrical units by noting that volts are analogous to pressure and current is analogous to particle velocity, with the power in both cases measured in watts.

### 7.4.3. Design and Modeling

Due to the highly specialized nature of each ultrasound imaging procedure, transducers are typically designed specifically to meet the requirements of the particular application.

Transducer design takes into account both electrical and acousto-mechanical aspects. Special-purpose finite element modeling (FEM) tools have been developed [18] in order to allow for optimization of the acoustic characteristics of the transducers.

In addition to the PZT itself, these tools also model backing layers that are placed behind the array to damp unwanted resonances as well as lensing material placed in front of the arrays to focus the acoustic beam. The extra layers affect the resonance and bandwidth of the transducers and therefore must be modeled in order to optimize the complete transducer stack-up.

***7.4.3.1. Electrical Impedance Models.*** Once the acoustic design of a transducer is complete, an electrical model can be generated which can be used in development of the system electronics to drive the transducers. Electrical models mimic the electrical impedance looking into the transducers from the electrical port as shown in Fig. 7.17b. They may also be used to model the acoustic output for a given electrical input as represented by a voltage equivalent of the acoustic output power.

A number of different models exist which provide an increasingly accurate representation of electrical impedance and electro-acoustic transfer characteristics. The simplest model of a loaded transducer (neglecting electro-acoustic effects) is that of a parallel plate capacitor, $C_0$, which appears between two plates separated by a distance $d$, each with a plate area $A$ and an insulating material (most often PZT) with dielectric constant $\varepsilon_S$ between them. In this case, $C_0$ is given by [6]

$$C_0 = \varepsilon_S A / d$$

The capacitance $C_0$ is a reasonably accurate first-order approximation of the behavior of the device as a load at frequencies other than resonance and is often referred to as the bulk capacitance.

A more accurate model of the transducer will take into account the resonant behavior that is due to the natural modes of operation of the device shown in Fig. 7.17a. A simple approximation of this behavior can be obtained using linear circuit components as shown in Fig. 7.17b. Here the bulk capacitance $C_0$ also appears and can be calculated using the equation above. In addition, an RLC circuit is added to model the resonance of the transducer. Multiple RLC branches can be added to represent additional resonant modes and obtain increasingly accurate representations of the full behavior of the device.

More sophisticated models of the transducer, such as the KLM model or the Mason model [95], can be used to simulate both the electrical impedance behavior of the device as well as energy transformation from the electrical to the acousto-mechanical domains and vice versa. For the most accurate simulation of transducer behavior, finite element models such as PZFlex can be used [18].

For design of front-end electronics, each of these models can be used to represent the impedance of the transducer in order to calculate the required drive current, as well as ring-down effects and power transfer from the source to the acoustic load. They can also be expanded for receive simulation with the addition of a voltage source that models charge generated by acoustic to electrical transduction.

### 7.4.4. Alternative Transducer Technologies

While PZT remains the gold standard material for ultrasound transducers, it has a number of features that are somewhat undesirable. The requirement to form individual elements by mechanically dicing the material makes it difficult to create large 2D arrays of very small elements for volumetric imaging. In addition, the low fractional bandwidth of PZT makes it difficult to use a single probe to cover a number of imaging applications. Also, while there have been some promising results obtained with single crystal PZT [19, 20], it is difficult to fabricate PZT array elements that operate at high frequencies, which makes it unattractive for use in very high resolution applications. Lastly, the current manufacturing process for PZT makes it difficult to integrate efficiently with associated front-end electronics.

To address some of these issues, a number of groups have been working in recent years on alternative technologies. These include capacitive micromachined ultrasound

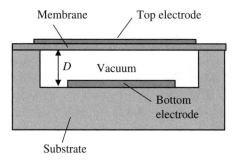

**Fig. 7.18.** cMUT Cross section showing vibrating membrane and top electrode suspended above substrate comprising static bottom electrode.

transducers (cMUTs) [21–25] and piezoelectric micromachined ultrasound transducers (pMUTs) [26–28], as well as polyvinylidene fluoride (PVDF) [29, 30].

cMUTs are micromachined devices that are manufactured using standard semiconductor processing. Figure 7.18 illustrates a cross section of a cMUT device. These operate in a fashion similar to that of the PZT structure shown in Fig. 7.13b except that there is no piezoelectric material. Instead, the electrodes are separated by a cavity that is typically evacuated and sealed. The distance between the electrodes, $D$, is on the order of a few hundred nanometers. This very small separation greatly increases the electric field across the electrodes. In addition, the electrodes themselves are composed of very thin material; typically, the top electrode covers or is comprised of a membrane that is free to vibrate. The large electric field and thin electrodes make it possible to generate sufficient electro-acoustic conversion without the requirement for a piezoelectric material. The operating frequency and output power of the devices is directly determined by the thickness of the cavity and the thickness of the suspended membrane [31].

There are currently two ways to manufacture cMUTs: *surface micro-machined* devices are processed in a thin layer on the surface of the wafer. They are typically manufactured by etching shallow cavities in the bulk silicon wafer and covering these with a silicon nitride membrane to form the top electrode. The bottom electrode can be formed with either a deposited metal layer or by using a highly doped silicon wafer for the substrate.

*Bulk micromachined* devices are processed using silicon on insulator (SOI) bonding techniques [32]. The use of SOI wafers allows the resonant cavities to be precisely defined in the active silicon layer of the first SOI wafer. The resonant membrane is formed out of the active Si layer of the second SOI wafer with the bulk section etched back to reveal the membranes.

cMUTs are readily bonded to supporting electronics using bump-bonding techniques as illustrated in Fig. 7.19. As can be seen in the figure, bonding to supporting electronics requires through-wafer vias to be etched from the front of the wafer to the back side in order to bring the individual transducer connections from the front of the cMUT substrate to the back, where they can be accessed by the front-end processing electronics [22]. Curved arrays have also been shown [33].

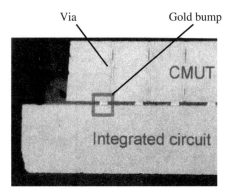

**Fig. 7.19.** Flip-chip integrated cMUT and electronics die. (Adapted from reference 91.)

A persistent challenge with cMUTs has been the inability to bring the round-trip sensitivity to match that of PZT [21]. This issue limits penetration and is exacerbated by the impedance mismatch between the high-impedance transducers and the probe cable capacitance. Preamps in the probe handle can be used to mitigate this problem, but integration has proven difficult.

Therefore, while cMUTs show promise as a next-generation transducer technology, to date this has not been realized in practical applications. In particular, unbuffered transmit and receive sensitivity remains inferior to that of comparable PZT devices, and uniform and consistent fabrication has proven difficult to realize.

For very high frequency imaging, PZT is not typically used. In the 1990s, effort was centered around the development of PVDF, which has been successful in application to ophthalmology where a product is currently available [29]. Hydrophones, which are used in testing of transducers in the lab, are often PVDF devices due to the required wide bandwidth in this application.

## 7.5. TRANSMIT ELECTRONICS

Ultrasound transducers are driven at high voltages in order to increase the output power, causing larger signals to be received from the body. Design of the transmit circuitry must therefore balance minimizing exposure to the patient while still transmitting enough acoustic power to yield acceptable image quality. Transmit electronics comprise the transmitters, multiplexers, and transmit/receive switches, all of which are implemented using special-purpose high-voltage semiconductor processes.

### 7.5.1. High-Voltage CMOS Devices

The double diffused MOSFET (DMOS) device is commonly used in ultrasound high-voltage circuits. A cross section of this device is shown in Fig. 7.20a. The principal distinguishing feature of this structure is the second diffused *p*-well, which is part

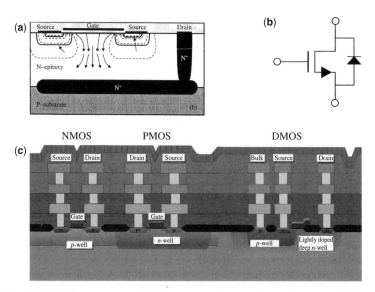

**Fig. 7.20.** DMOSFET structure: (a) Cross section showing flow of current. (Adapted from reference 34.) (b) Schematic symbol showing parasitic drain-source diode. (c) Cross section of a representative high-voltage CMOS process with high-quality co-integrated low-voltage devices. (Adapted from and modified from reference 37.)

of the channel of the device. The area between the *p*-well and the drain connection is called the *drift region*. The drift region provides the high-voltage stand-off capability of the device since the resistance in this path effectively insulates the gate of the device from the high drain voltage [34].

As shown in Fig. 7.20b the additional *p*-well leads to a parasitic diode that is intrinsic to the DMOS device. This diode leads to unwanted effects in switching applications including crosstalk and loading. Crosstalk on transmit leads to bleeding of signal between adjacent channels and can cause a degradation of imaging performance. For example, in cardiac imaging, crosstalk can reduce the range of steering angles that can be achieved from $\pm 45°$ down to $\pm 30°$, contributing to a loss of sensitivity at the edges of the image. Therefore, special circuit architectures must be used in order to mitigate the effects of this and other sources of crosstalk.

Discrete high-voltage *N*-type DMOS devices are typically vertically integrated with current flowing from the drain at the bottom of the device up to the source connection at the top of the device (Fig. 7.20a) [34]. This vertical current flow provides for significant current sourcing capability and very low on resistance which are important for high power applications. In this case, the substrate of the device is effectively its drain. This makes it difficult to integrate multiple devices to build circuits since each device needs to have its own isolated drain substrate.

Lateral DMOS devices do not suffer from this limitation [35]. Since the drain and the source are both isolated from the common substrate, it is possible to integrate multiple high-voltage devices on a single substrate in order to build complex circuits [35, 36].

Recently, a number of vendors have begun to supply specialized hybrid processes in which high-quality lateral DMOS devices have been cointegrated with small-feature-size low-voltage transistors [37–39]. This makes it possible to build high-voltage circuits that are cointegrated with high-quality control logic and high-speed, low-noise, and low-power analog circuitry. These processes are commonly used to implement display drivers and for automotive applications [36]. They are also ideally suited for implementing high-voltage circuits required in ultrasound systems.

Figure 7.20c shows an example cross section of a typical high-voltage CMOS process in which DMOS devices are integrated adjacent to high-quality low-voltage devices [37]. The extra silicon area needed to isolate high-voltage devices from one another leads to significant overhead in the layout, which makes it challenging to integrate a large number of ultrasound channels on a single substrate.

Particular care must be taken to ensure that leak paths between devices are minimized [34] and that parasitic high-voltage devices are suppressed. These parasitic devices can lead to degradation of performance due to crosstalk between adjacent channels or, even worse, could cause complete device failure due to a CMOS latch-up condition taking place. In many cases, these parasitics are not accurately modeled and therefore cannot be anticipated in simulation. Great care is therefore required during the design of these circuits in order to detect and eliminate these parasitics.

### 7.5.2. Transmit/Receive (T/R) Switch

The classical ultrasound T/R switch is typically composed of a diode bridge as shown in Fig. 7.21. The circuit operates as follows: A constant current flows from $+V$ to $-V$ in order to bias the diodes. The on resistance of the diodes is chosen in order to

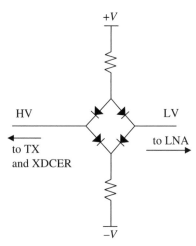

**Fig. 7.21.** Transmit/receive switch implementation in the form of a biased diode bridge circuit.

minimize noise in the device. Typically, the noise is designed to be below $1 \text{ nV}/\sqrt{\text{Hz}}$. Given this requirement, the current flowing in this circuit can often be quite high. Therefore, as the channel count on systems has increased, this type of T/R switch has become less popular due to the excessive power consumption.

Operation of the protection circuit relies on the fact that at 200 Vpp, the transmit voltage transitions above the bias voltage cause the diodes to be reverse-biased and therefore no current flows from the transmitter to the receiver when pulsing. The receive signals are typically below the reverse-bias threshold voltage at about 100-mV maximum. Therefore, the receive signals are passed across the diode while the high-voltage transmit signals are effectively blocked. A diode bridge as shown in the figure is used in order to block bipolar signals. If only unipolar pulsing is used, then a simple back-to-back diode circuit could be sufficient.

It is also possible to use a MOSFET switch in place of diodes [40]; this does not require a static bias current and therefore is much more power efficient. Use of a MOSFET switch requires a control circuit and therefore is more complicated than the simple diode bridge that operates automatically. Therefore the diode bridge is somewhat safer in terms of protection since it does not need to be controlled. Another drawback of the MOSFET T/R switch is that discrete DMOS devices contribute significant parasitic capacitance to load the transducer during the receive cycle. This problem is somewhat mitigated in an ASIC implementation but must still be accounted for.

### 7.5.3. High-Voltage Pulsers

*7.5.3.1. Unipolar and Trilevel Pulsers.* A standard architecture for an ultrasound pulser (sometimes called a transmitter) is shown in Fig. 7.22a. The circuit uses DMOS devices in the output stage in order to implement a half-bridge to charge the transducer load capacitance up to HVP and then down to HVN. In a unipolar

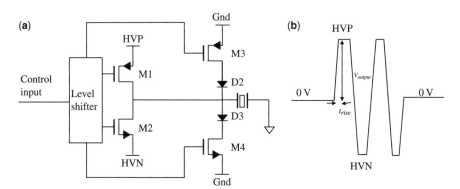

**Fig. 7.22.** Bipolar ultrasound pulser. (a) Representative circuit. (b) Format of the bipolar output pulse.

pulser, the signal swings between HVN and HVP, where HVN in this case is simply ground. The peak-to-peak transmit voltage is typically on the order of 300 V, and in some portable systems it could be closer to 50 V. In a trilevel (or bipolar) pulser, the output voltage swings between HVN and HVP and returns to a ground level when the cycle is completed as shown in Fig. 7.22b.

Return to zero is accomplished using the clamping circuit formed by M3 and M4 in conjunction with D2 and D3. The diodes are used to prevent the clamp from turning on when the voltage swings above and below ground. This could occur if, for example, the gate of M4 was biased at ground (i.e., off state) and its drain was dropped more than 5 V below ground. Since this situation happens each time the output drops to HVN, the diode D3 is inserted to prevent M4 from turning on. In a unipolar pulser, this circuit is not needed.

Control of the output FETs is accomplished using level shifters that take the transmit timing signals generated by a low-voltage circuit referenced around 0 V and shifts them up to the positive high-voltage HVP and down to the negative high-voltage HVN. This circuit could be composed of a resistor capacitor circuit or could also be given by a transmit latch [34] or a current mirror. The level shifter needs to be sized such that it can drive the input capacitances of the large output devices fast enough to switch the devices within the required rise time of the output waveform. This current can often be a significant power draw in itself.

The output devices are sized such that they can drive the output load fast enough to satisfy the rise-time requirements. The load can be approximated to first order as a capacitance. In the case of pulsers in the system, this capacitance consists, in large part, of the cable capacitance $C_c$, which typically is on the order of 100 pF or more. The transducer is modeled to first order as a capacitance $C_d$ which can be as large as 40 pF for a linear probe but is typically closer to 10 pF for a 2D probe.

In addition to the transducer and cable load capacitances, a significant additional parasitic capacitive load is provided by the output devices themselves. These include the drain-to-bulk capacitance $C_{db}$ as well as the drain-to-source parasitic diode capacitance $C_{ds}$. Both the NMOS and PMOS devices must be accounted for. In addition, routing capacitance $C_p$ can be important, depending on the interconnect technology used. The total load consists of each of the above-mentioned capacitances in parallel and is therefore given by

$$C_{load} = C_d + C_c + 2(C_{ds} + C_{db}) + C_p$$

The required output current for rise time, $t_{rise}$, is given by

$$I_{rise} = \frac{C_{load}\Delta V_{output}}{\Delta t_{rise}}$$

The output devices and clamp circuit will therefore be sized in order to drive this current along with some margin to account for variations in DMOS $I_{ds}$ which may be significant.

In second harmonic imaging, the waveform of Fig. 7.22b must be completely symmetric in order to avoid generation of any spurious transmitted signal power at the second harmonic frequency. Therefore, it is important that the pull-up and pull-down devices as well as the clamp be well-matched so that their rise times are as close to identical as possible. Also, in Doppler mode the requirement for very low phase noise means that jitter in the control circuitry must be minimized as much as possible.

Power dissipated by the unipolar pulser is given by

$$P = CV^2 fN$$

where $C$ is the total load capacitance, $V$ is the peak-peak output voltage, $f$ is the pulse repetition frequency (PRF), and $N$ is the number of pulses that occur during one complete transmit cycle. The square law dependence on voltage makes heat dissipation a challenging problem at high transmit voltages.

**7.5.3.2. Multilevel Pulsers.**  Multilevel pulsing can be used in transmit beam forming for apodization of the transmit aperture. Multilevel pulsers are also used for transmitting Gaussian-shaped pulses or chirps. Figure 7.23 shows examples of different multilevel pulse shapes that are typically used. Figure 7.23a shows apodization on multiple parallel channels where channel 1 transmits at a peak voltage of $V_1$, channel 2 transmits at a peak voltage of $2V_1$, and channel 3 transmits at a peak voltage of $3V_1$. Figure 7.23b shows transmission of a shaped pulse, here a Gaussian pulse that has improved acoustic properties. Figure 7.23c shows a series of transmit cycles in

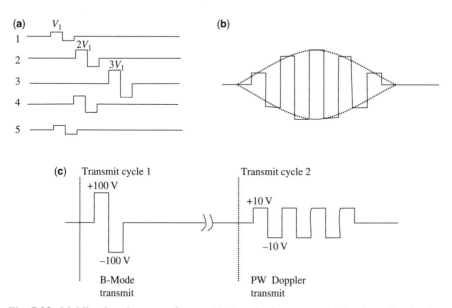

**Fig. 7.23.** Multilevel pulser waveforms. (a) Apodization on multiple channels showing changing amplitude on different channels. (b) Example of temporal apodization on the transmit pulse in the form of a Gaussian pulse. (c) Color flow Doppler pulse sequence.

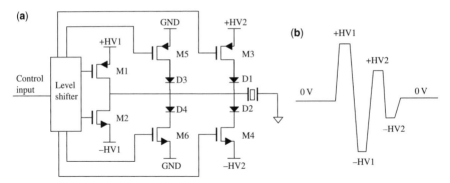

**Fig. 7.24.** Two-level capable pulser circuit. (a) Two-level capable pulser. (b) Output waveform showing two different transmit voltage levels ($\pm$ HV1 and $\pm$ HV2).

which multiple transmit pulses of different voltage are used. With a multilevel pulser it is possible to send different amplitude pulses without the need to quickly switch the supply voltage.

A circuit capable of implementing some of the waveforms in Fig. 7.23 is shown in Fig. 7.24. Two half-bridges are used to drive a single transducer. The first output stage is biased at HV1 while the second output stage is biased at HV2. The voltages HV1 and HV2 can be set with complete freedom by the system on each transmit cycle. The use of two output stages allows for instantaneous switching between pulsing at output voltage HV1 and pulsing at output voltage HV2. For example, to generate the color flow Doppler waveform of Fig. 7.23c, pulser 1 would first transmit a B-mode signal at 200 Vpp followed in the next transmit cycle by a longer pulsed Doppler stream at 20 Vpp by pulser 2. Note that the length of the pulsed Doppler sequence is related to the amount of transmitted acoustic power that is limited by the FDA requirements.

Rapidly switching between the two output stages shown in Fig. 7.24 in order to generate a color flow Doppler sequence requires the diodes D1 and D2 which prevent the output devices in the lower-voltage transmitter from turning on while the higher-voltage pulse is being transmitted.

The waveform of Fig. 7.23a can be generated using multiple channels where each channel has a different transmit voltage supplied by the system. Although these voltages can be updated on each transmit cycle to achieve the desired apodization envelope across the array, switching the supply voltages is usually undesirable due to the high power consumption on the power supplies.

Another way to generate the waveforms of Fig. 7.23 is to use a transmit amplifier as shown in Fig. 7.25 [41]. Here an arbitrary waveform is generated by a DAC operating at a multiple of the transmit frequency. Using this circuit, it is possible to generate any of the waveforms in Fig. 7.23. The penalty for this flexibility is the requirement for a high-quality high-voltage amplifier which consumes a significant amount of current when biased on. The Si area required to implement this circuit can also be prohibitive. Nevertheless, in a high-end ultrasound machine, these costs may be acceptable in

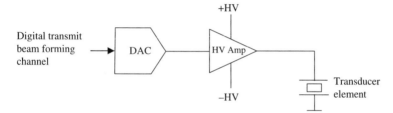

**Fig. 7.25.** Multi-level capable transmit amplifier. Adapted from [42].

order to achieve the excellent imaging performance that is made possible by transmitting flexible multilevel waveforms.

### 7.5.3.3. High-Voltage Multiplexers.

A number of architectures exist for high-voltage multiplexer switches [36, 42–45]. These include circuits where the gate source voltage is kept constant by allowing the gate to float relative to the control circuit [43–45]. These circuits can be used to integrate muxing directly in the probe handle in order to reduce the number of coaxes in the probe cable [8].

All of these switch circuits solve the problem of the source voltage having to pass a large negative and positive excursion of voltages relative to the TTL or CMOS level control voltages. They do so by using DMOS devices in the configuration shown in Fig. 7.26a. Back-to-back MOSFETs are used in order to block bipolar signals in the off-state. These would otherwise be passed through the parasitic source-drain

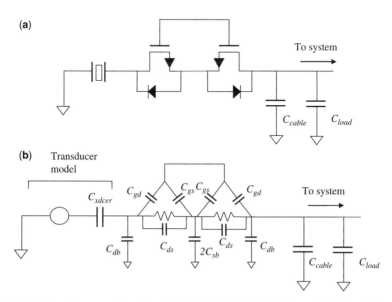

**Fig. 7.26.** High voltage switch using back to back DMOS devices. (a) Circuit topology showing parasitic drain-source diodes. (b) Small signal model showing parasitic capacitances.

diode, which is an inherent feature of DMOS devices. Unipolar pulsers only require a single DMOS device.

The gate-source voltage is kept at a constant voltage (typically 5 V) while the source voltage goes from −HV to +HV as the bipolar transmit signal is passed through the switch [43]. This keeps the switch on throughout the transmit cycle regardless of the source voltage.

$R_{DS}(ON)$ is an important parameter for multiplexer design since it directly affects the noise produced by the device as well as the signal attenuation. The behavior of DMOS devices in the linear region can be approximated using a low-voltage CMOS FET in series with a resistance representing the resistance of the drift region [34]. The resistance of the low-voltage FET can be estimated using the equation for $Ids$ versus $Vds$ for a MOSFET when operating in the triode region [46],

$$Ids = K[2(Vgs - Vt)Vds - Vds^2]$$

An estimate of the low-voltage FET resistance can be found by taking the derivative with respect to the drain-source voltage,

$$\frac{\partial I_{DS}}{\partial V_{DS}} = K(2(V_{GS} - V_t) - 2V_{DS})$$

The second term on the right-hand side of the equation is very close to zero (nominally about 0.2 V in the triode region) and can therefore be neglected. $R_{DS}(ON)$ may then be approximated by

$$R_{DS}(ON) = 1/[2K(Vgs - Vt)] + R_{drift}$$

where $R_{drift}$ accounts for the resistance of the drift region. $R_{DS}(ON)$ is usually quoted in $\Omega \cdot mm^2$ because it is strongly dependent on the device area.

Isolation of the switch depends on the path from the drain to the source when the switch is off. This is given by the off resistance (typically in the megaohms) in parallel with the drain-source parasitic diode capacitance combined with the load seen at the other end of the switch to ground. The parasitic capacitances of the FET are shown in Fig. 7.26b. $C_{gs}$ and $C_{gd}$ are the gate drain and gate source overlap capacitances, and $C_{db}$ is the drain-bulk capacitance. $C_{ds}$ is due to the parasitic diode that is intrinsic to the DMOS device.

In ultrasound, switch isolation is typically kept below −30 dB to reduce the negative effects of crosstalk between neighboring elements. $C_{ds}$ contributes to a degradation in off switch isolation. Also, in the back-to-back DMOS case, $C_{gs}$ and $C_{gd}$ can be problematic. In some architectures, the gate voltage is left floating in order to isolate the switch from its control circuit [43]. In the off-state, signals can then pass from one terminal of the switch to the other across the gate overlap capacitance since they are not damped at the gate by a bias voltage.

Transients that occur when the switch is turned on and off lead to charge injection and feedthrough [47]. These spurious signals can be injected both into the receiver and into the transducer. Charge injection is caused by charge being drawn from the switch terminals on turn-on of the switch to create the channel and then being forced out to the terminals of the switch to collapse the channel when the switch is turned off [47]. Feedthrough, which is a related phenomenon, is caused by charge being stored on the gate-source and gate-drain overlap capacitances being transferred to the terminals of the switch when the gate voltage is toggled on or off.

Injected charge into the receiver must be minimized in order to reduce the chance of saturating the high gain receive amplifier. A saturated amplifier right after the transmit cycle is incapable of imaging the first few millimeters of tissue at the skin line, which is important in certain applications.

Charge injected into the transducer is problematic because it leads to spurious transmitted acoustic waves that are transmitted into the body. Depending on the magnitude of this transmitted energy and the configuration of the array, unwanted image artifacts could result.

***7.5.3.4. Tuning.*** Series inductors are sometimes added in order to cancel the reactance of the ultrasound element as well as the cable. The net effect is to provide a 2× boost in transmit voltage that is advantageous when the available CMOS process does not support the full range of transmit voltages needed. Due to their large size, tuning inductors are more difficult for integration in the probe itself, but they are used extensively on the system side.

## 7.6.  RECEIVE ELECTRONICS

The receive electronics consist of very-low-noise and high-gain amplifiers that are designed to maximize the signal-to-noise ratio at all parts of the receive cycle.

### 7.6.1.  Front-End Receive Signal Chain

Figure 7.27 shows the typical front-end signal chain for the receive configuration. Acoustic pressure signals are converted to electrical signals by the transducer, which then drives the receive chain. The load presented to the transducer consists of the accumulated parasitic capacitances due to the ultrasound probe cable (100–200 pF),

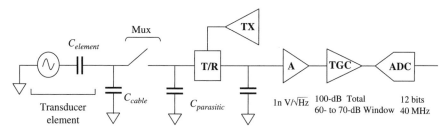

**Fig. 7.27.** Front-end receive signal chain.

the high-voltage mux (100 pF), the T/R switch (100 pF), and the input to the pream-plifier as well as any routing capacitances. The source impedance of the transducer is typically on the order of 40–150 pF for a linear array, and therefore the receive signal experiences significant attenuation from the transducer to the input of the preamplifier. The lost signal can be recovered by moving the preamplifiers to the probe handle where they can be placed right next to the transducers themselves in order to bypass the cable load capacitance [8]. As the transducer elements become smaller, their source impedance grows, and therefore the signal attenuation increases dramati-cally. In 2D arrays, this is especially problematic [48].

The noise that is added to the signal from the element is that due to the series resist-ance of the multiplexer as well as the dynamic resistance of the T/R switch diode bridge. These noise components must be less than the input referred noise of the pre-amplifier, which is typically on the order of $1\,\mathrm{nV}/\sqrt{\mathrm{Hz}}$, but can be higher in some applications.

### 7.6.2. Low-Noise Preamplifier

The preamplifier must be high gain in order to boost the weak signal above the noise of the successive analog processing stages. For a general-purpose imager, it must also have a bandwidth that allows the highest-frequency probe that might be used to inter-face with the system. Typically, the amplifier bandwidth is set between 10 MHz and 40 MHz, but it can be higher in certain specialized applications. Both charge [49, 50] and voltage mode [51] amplifiers are used.

An important consideration in some imaging modes is settling of the receive preamplifiers. Any signal that is fed through the T/R switches to the receive amplifier during transmit pulsing could cause the preamplifiers to be saturated. If the amplifiers cannot come out of saturation immediately following the last transmit pulse, then the tissue that is very close to the skin line will be obscured by the poor response of the amplifier during that period. Therefore, it is important to minimize the amount of feedthrough during the transmit phase as well as to design the amplifiers for fast recovery [8].

### 7.6.3. Time Gain Control Amplifier

The Time Gain Control (TGC) stage is typically implemented as a low-noise amplifier (LNA) followed by a resistor ladder attenuator [51]; however, other architectures have also been proposed [52, 53]. The main requirement of this stage is that it have continu-ously variable and user-control of the input gain over a very large range. Control of the gain should be linear in decibels in order to match the rate of attenuation found in the body (discussed in Section 7.2.6.2). This control is typically done using an analog ramp voltage divided into a series of gain ranges at various preset depths in the body. The value of the gain in each of these ranges is optimized during the system design according to the specific application and probe used. Digital TGC settings may also be used in some systems. The TGC controls are used to equalize the bright-ness in different areas of tissue during the examination.

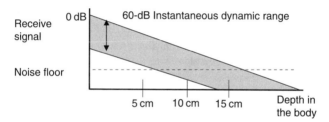

**Fig. 7.28.** Time gain control (TGC) and dynamic range of the ultrasound system. A 60-dB instantaneous dynamic range window slides across the entire 100-dB dynamic range of each system channel as imaging goes deeper into the body.

As is shown in Fig. 7.28, the instantaneous dynamic range of the TGC output constitutes a window on the order of 60 dB. During the receive cycle, this window is slid across the overall dynamic range of the received signal on each channel, which is in excess of 100 dB. The window starts close to 0 dB at the beginning of the receive cycle and drops as time progresses and receive signals become increasingly attenuated. Dynamic range on each channel is augmented by the signal to noise ratio gain due to summation of all 128 channels. This effect provides an additional factor of $\sqrt{128}$ equivalent to roughly 21 dB of dynamic range. Therefore, the overall SNR for the system on receive is about 121 dB. As was discussed in Section 7.2, penetration depth is directly limited by this overall SNR. In practice, the system noise floor is set by thermal and flicker noise in the LNA integrated over the bandwidth of operation which is typically about 3–10 MHz.

### 7.6.4. Analog-to-Digital Converter

The ADC must oversample the input signal in order to relax constraints on subsequent digital processing and filtering. It is therefore typically sampled at 30–60 MHz. In order to achieve even higher timing quantization, digital interpolation can be applied [8]. The tradeoff is therefore a compromise between time resolution and ADC power and complexity. Typical systems quantize to 12 bits accuracy. Alternative structures are also proposed such as 1-bit Sigma–Delta converters [54–56]. There exists a tradeoff between ADC bit depth and TGC gain range: With greater amount of bits in the ADC, smaller gain ranges in the TGC are needed and therefore design complexity and power consumption are transferred from the TGC to the ADC [8].

### 7.6.5. Power Dissipation and Device Integration

In addition to noise considerations, power consumption is often a key design parameter in the overall design of the front-end receiver chain. For certain portable systems the amount of power available for each receive channel is on the order of tens of milliwatts. Given that this must be shared by a high-gain LNA as well as a high-speed ADC, the design of the front-end electronics can be very demanding. Since typical catalog TGC parts dissipate as much as 200 mW/channel [57], these architectures do not lend themselves well to use in low-power portable systems. Many groups are addressing the tradeoff of noise, bandwidth, power, and circuit size with novel architectures such as capacitive [58] and current feedback amplifiers [59].

Efforts to integrate various elements of the receive chain are ongoing and are mostly concentrated along the lines of grouping multiple channels of like functionality rather than grouping each signal chain on a single ASIC [8]. Examples of these include multichannel preamps and TGC amplifiers [60–62] as well as ADC arrays designed specifically for ultrasound applications [54, 63]. ASICs that integrate multiple complete front-end receive chains have also been most recently introduced [13]. These implementations may also yield improvements in per-channel power dissipation.

## 7.7. BEAM-FORMING ELECTRONICS

The beam former combines the element signals (as described in Section 7.2.3) into a single signal for image generation. Initially, this function was implemented entirely using analog electronics. Then in the 1990s, once ADCs with adequate speed and resolution became available, a significant effort was undertaken to transition the beam-forming function to digital circuitry [3]. Today, the majority of beam-forming circuits are implemented in digital ASICs. In large 2D arrays where thousands of beam-forming channels are necessary, it is very difficult to implement a full digital solution. This is mostly driven by the inability to make large arrays of low-power ADCs. Therefore, another possible method is hybrid beam-forming electronics composed of both analog and digital beam-forming sections that are optimized to solve this problem [64]. Each of these will be described in this section in turn.

### 7.7.1. Digital Beam Formers

The digital beam former is typically a very large high-speed dedicated ASIC that processes the receive data in real time [65–67]. Time delays are accomplished using a FIFO register. Unaligned interpolated data are written into the FIFO at the receive clock rate. Data are then read out of the FIFO at the same rate but with the starting point for each channel adjusted such that the relative delays between channels are now equalized. One possible implementation is shown in Fig. 7.29.

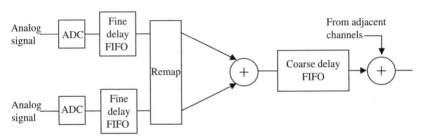

**Fig. 7.29.** Schematic representation of a digital beam former. (Adapted from reference 65.)

Delays are typically broken into two parts: a coarse delay and a fine delay. In certain implementations, fine delays are achieved as a result of interpolation, whereas coarse delays are based on ADC sample steps. Delay quantization of the beam former directly impacts the quality of the image. Insufficiently small quantization steps create background noise due to spatial sidelobes of the receive beam. Typically, the delay quantization of the beam former is on the order of 5–20 ns.

The beam former may also include anti-aliasing filtering as well as bandpass filters that exclude unwanted frequencies in favor of the specific passband of the transducer. In a digital beam former these would be implemented as a DSP filter, typically a FIR or IIR filter.

Other possible implementations include systems using FPGA [68] and DSP [69, 70] processors. The advantage of using an FPGA or DSP is complete reconfigurability of the beamforming process. Additionally, the design cost may not be as high as with an ASIC. The disadvantage of using an FPGA or DSP is increased cost of the electronics in volume, higher power dissipation, and larger overall chip size.

### 7.7.2. Analog Beam Formers

Before A/D converters achieved adequate resolution and sampling rates for ultrasound processing, all beam forming was done using analog circuits. Historically, these analog delays were implemented using lumped constant delay lines [71]. All-pass filters [72, 73] have also been proposed. More recently, a number of time-discretized solutions have been investigated, including analog RAMs and CCD shift registers [74–77], as well as switched capacitor and switched current delay lines [78].

Analog beam formers suffer from at least two major limitations: First there is the issue of maintaining the exact beam-forming coefficients in the face of (a) process variations in the chip manufacture and (b) drift due to temperature; second, there is the challenge of achieving uniform linear delay with respect to frequency over a very long time period. Long time delays are difficult to achieve because of the size of each individual delay element.

The delay in an all-pass filter is not uniform with frequency. The limited bandwidth of operation for most transducers means that over a narrow range of frequencies, the delay characteristic can be approximated as linear; however, the delay error increases significantly outside of this bandwidth. Therefore, the all-pass filter is best suited for more narrow band systems such as for cardiac imaging.

The tapped delay line is better suited for a broader range of frequencies since it achieves linear phase delay that is unrelated to the circuit topology. The main challenge for a tapped delay line is maintaining the voltage sample resolution (especially a problem for switched-current implementations) [66] as well as power dissipation. Power dissipation can be a serious issue for CCDs, which have high-voltage clocks that switch at high rates and therefore dissipate significant power. In a switched capacitor implementation, the resolution is given by the settling time of the capacitor drive amplifier as well as the noise on the capacitor. The capacitors must be sized large enough such that their noise contribution will be above the required level. This dictates both the speed of operation and size of the capacitor bank array.

### 7.7.3. Hybrid Beam Formers

A hybrid beam former consists of an analog beam former followed by a digital beam former as illustrated in Fig. 7.30. This architecture is especially important for 3D beam forming in which thousands of transducer elements are used as is required by volumetric imaging [64]. The analog beam former implements fine delays, while the digital beam former processes coarse delays. The advantage of this arrangement is that the fine delay can be accomplished with a phase rotation that can be done with an analog circuit, whereas the coarse delay, which cannot be easily achieved with the analog circuit, is readily implemented with the digital circuit.

The analog delay and first-stage beam sum are done in the probe handle in order to greatly reduce the number of system channels and coax cables. The beam formation in the probe handle performs beam steering and focuses over a small area of the aperture; as a consequence, only shorter delays are needed. This makes a 2D array possible since it cuts down the number of coaxes in the cable (128 versus 2000), and also it cuts down on the number of required front-end receive circuits (i.e., TGC and ADC). For example, channels can be added together in groups of four, which reduces the number of system channels accordingly from $4 \times 128$ to just 128.

### 7.7.4. Reconfigurable Arrays

A reconfigurable array is a 2D array of transducers in which the elements are connected together using a large distributed switch matrix as shown in Fig. 7.31 [79, 80]. The elements are grouped along iso-delay lines. This takes advantage of the fact that in some imaging applications, regions of essentially constant delay span relatively large areas of the transducer face. As a consequence, the number of system channels can be reduced since only such regions require unique delays. It is desirable to have the switch matrix integrated directly behind the transducers in order to greatly reduce the number of signals from the array to the signal processing circuitry in the system.

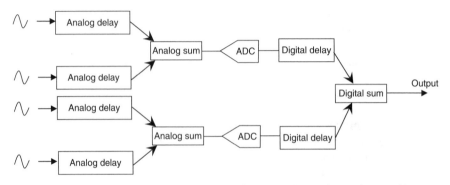

**Fig. 7.30.** Hybrid digital and analog beam former. (Adapted from reference 64.)

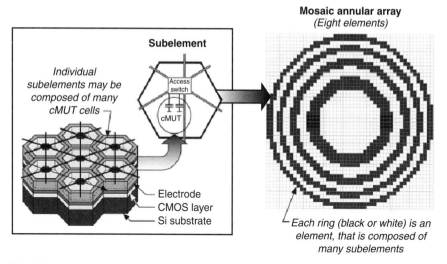

**Fig. 7.31.** Reconfigurable array architecture implementing a mosaic annular array topology. (Adapted from reference 80.)

## 7.8. MINIATURIZATION

A major effort in recent years has begun to miniaturize much of the functionality of the imaging electronics in the ultrasound system. In particular, there has been a trend to combine multiple digital beam-forming channels on a single ASIC [65, 66], as well as multiple ADCs and TGCs on a single mixed signal ASIC [8]. Further work has also been done in merging complete transmit channels on a single device [81]. All of this activity is driving toward a highly miniaturized ultrasound front-end with reduced cost and weight. As a side benefit, miniaturization has also allowed some of the front-end electronics (notably HV muxes, pulsers, and receive amplifiers) to move into the ultrasound probe. The net effect of this is reduction in power consumption and improved signal-to-noise ratio due to the elimination of the cable capacitance from the transmit and receive chain.

Figure 7.32 shows a schematic view of the major functional sections of an ultrasound system. While the basic functions have remained relatively constant since the first systems became available, their implementation has undergone significant migration in recent years. In particular, much of the front-end interface as well as some of the digital processing has been migrating toward increased miniaturization in the form of custom ASICs or implementation in software. At the same time, these functions have also been moving closer to the transducer with key systems today including some level of beam forming in the probe handle itself. With the availability of very fast general-purpose and dedicated computing hardware, there has also been a competing migration of some beam-forming functions to software. All of these trends led to further miniaturization of the ultrasound system, which has broad implications for the quality and availability of this important imaging modality.

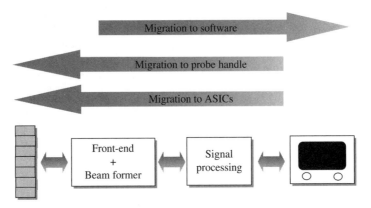

**Fig. 7.32.** Migration of functionality in the ultrasound machine.

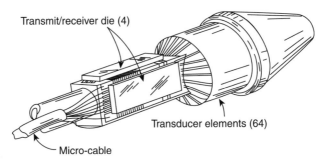

**Fig. 7.33.** Catheter-based side-looking ultrasound probe. (Adapted from reference 83.)

Miniaturization has also become increasingly important in recent years in solving the problem of 3D volumetric imaging in which thousands of transducer channels must be processed in real time. It has also been used in challenging applications such as intravascular ultrasound (IVUS) [82, 83] as shown in Fig. 7.33.

In this section, the importance of miniaturization to future ultrasound systems will be discussed. In particular, portable systems will benefit greatly in the next few years and some of the history and challenges in this regard will be presented. Ultimately, direct integration of the transducer and the front-end electronics can yield a highly compact and energy-efficient solution, and new work in this area will also be discussed.

### 7.8.1. Portable Systems

The trend toward miniaturization began in the 1990s with low-cost portable systems [84–86]. These systems attempted to bring basic ultrasound functionality to a large number of general clinicians. Example applications would be those in which image quality is not a high priority such as emergency medicine and in general practice.

As discussed previously, image quality depends greatly on the number of channels used as well as the transmit frequency. Also important is the noise performance of the front-end amplifiers and the available transmit voltage. All of these have historically been compromised in portable systems in order to realize a high level of miniaturization.

Today, the market for portables is growing quickly and is expected to reach $1 billion by 2010 [87]. A range of products are currently on the market including laptop style units and handheld tablets. Smaller devices are also in the pipeline [88]. Market share of the portable ultrasound segment continues to grow at a much faster rate than traditional console units [89] and is projected to accelerate.

***7.8.1.1. Tablet and Handheld Style Units.*** In lower-end portable devices, a number of trade offs are made in order to realize a compact and low-power system. These include the use of folded beam-forming architectures to reduce the number of beam-forming and front-end channels used, as well as reduced transmit voltage in order to limit the power dissipation.

The majority of these devices use proprietary ASICs for the beam-forming and front-end analog functions [76, 85]. Some also use dedicated ASICs for image processing functions. These ASICs realize very compact implementations of the necessary functions while implementing the required tradeoffs in functionality and image quality which are needed to yield a compact low power solution.

As was discussed earlier, every particular exam requires a specific operating frequency and probe form factor, which means that specialized probes must be made available for a broad range of applications. Due to their lack of general applicability and their specialized beam-forming hardware, portable devices are typically sold with a limited number of available probes.

***7.8.1.2. Laptop-Style Units.*** Laptop-style ultrasound scanners realize improvements in image quality and system functionality due to their larger form factor. The larger high-quality screens make it possible for the systems to be used more extensively in clinical practice as they provide much better visualization of diagnostic information as compared to smaller handheld devices.

The additional space available in a laptop form factor may also be used to increase the battery size in order to increase either the available power or the battery life of the device. This can allow for a higher transmit voltage, which in turn can lead to increased penetration on imaging and improved image quality. In addition, more sophisticated beam forming and image processing can be used. These latter features again can contribute further to the general utility of the device in a broader range of clinical areas.

### 7.8.2. Transducer-ASIC Integration Strategies

The ultimate result of system miniaturization in ultrasound will be the migration of all imaging functionality to be integrated directly with the transducer array itself. These highly integrated systems will enable very compact implementations as well as reduced power consumption and increased system functionality due to tightly coupled sensors and electronics. A significant amount of research toward this goal is currently ongoing and will be described in this section.

Advantages of tightly coupling transducers and electronics include:

1. **Reduced capacitance** between the transducer and the front-end electronics. In a typical console system, the cable introduces on the order of 100–200 pF of load capacitance. This additional capacitance must be driven by the high-voltage transmitters and results in a significant amount of excess power dissipation. On receive, the extra capacitance must be driven by the transducer and results in attenuation of the receive signal, which leads to reduced penetration. This is especially problematic for the small elements used in 2D arrays.

2. **Increased transducer channel count.** Tightly coupled systems can lead to an increase in the number of transducers by solving the problem of the bottleneck due to the cable used in traditional systems. This increase is essential to high-quality volume imaging in real time.

3. **Reduced system volume.** Much of the area used in current console systems is devoted to board-level interconnect to route all of the processing channels to and from the beam-forming electronics. This problem is magnified and becomes acute in a high channel count system. Direct integration of the transducer and the front-end electronics greatly reduces the system size by miniaturizing much of the routing.

Each of these advantages may already be seen to a lesser extent in current high channel count systems (such as 3D imagers), where board-level integration of the front-end and first-stage beam formers has already been accomplished. But further improvements are possible and will be essential for the implementation of future handheld systems with high-end image quality.

Two main architectures for integrating transducer arrays directly with ASIC interface electronics have been investigated:

*7.8.2.1. Co-integrated Single-Chip Devices.* Direct integration can be achieved by building the transducers directly on top of the electronics in a secondary fabrication line once the electronics have been completed as shown in Fig. 7.34a [24, 30]. This technique has been used in other areas to realize tightly integrated sensor/ electronics systems [90], although it can be challenging to implement. A number of groups have demonstrated integration prototypes [24, 50]. Figure 7.34b shows one such implementation.

The advantage of building the sensor array directly on top of the electronics is the ability to realize a very dense grid of interconnections between the sensors and their associated electronics. This is especially important in achieving very fine pitch transducers as would be used in high-resolution applications. Another important advantage is that the stray load capacitance due to interconnect is essentially reduced to zero, which results in reduced power dissipation and a larger receive signal.

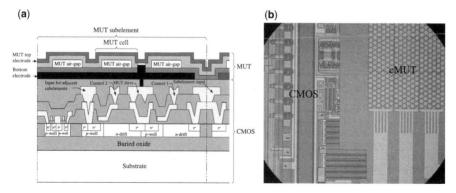

**Fig. 7.34.** Direct integration of cMUT transducers and CMOS electronics. (a) Cross section showing direct integration stack-up. (b) Plan view showing cMUTs on top of CMOS. (Adapted from reference 24.)

The main challenge of the approach is the requirement for process compatibility between the sensor and electronics manufacturers, which is often difficult to achieve and can lead to poor yield and increased cost of implementation.

***7.8.2.2. Highly Integrated Multichip Devices.*** Another way to address the challenge of direct transducer to electronics integration which has attracted a lot of attention in recent years is the integration of separately fabricated die as shown in Fig. 7.35 [80, 91]. Integration of the fabricated die is typically done by flip-chip bump-bonding.

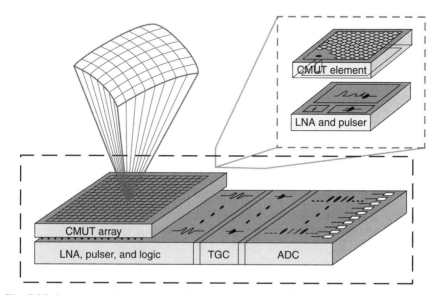

**Fig. 7.35.** Integration of separate transducer and electronics die. (Adapted from reference 92.)

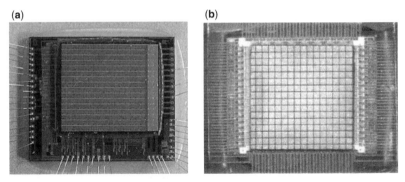

**Fig. 7.36.** Highly integrated multi-chip device prototypes. (a) cMUT and transmit/receive electronics [91]. (b) PZT array cointegrated with dense high-voltage switching matrix [80].

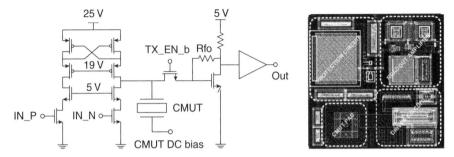

**Fig. 7.37.** Cell-based front-end circuitry for co-integrated transducer/electronics array. (a) Circuit schematic of a single cell with transmit and receive electronics [94]. (b) Layout of a similar circuit that fits into a 200-μm × 200-μm cell [93].

A number of groups have demonstrated working prototypes [80, 92]. Figure 7.36 shows both cMUT and PZT implementations. An important consideration is the ability to design the electronics unit cell to be the same size as the transducer that it interfaces since there is a one to one correspondence in area between the two. This can be a challenging design problem but is being solved as high-voltage processes improve [93]. Figure 7.37 [94] shows some representative cell-based circuits.

### 7.8.3. Challenges to Effective Miniaturization

A number of challenges to effective miniaturization of high-end quality imagers still remain. Each of these must be addressed in turn in order to realize a cost-effective device with clinically useful image quality that will be adopted in the medical community. Some of these include:

*Transmit Voltage.* Higher transmit voltages are needed to improve penetration and also for harmonic imaging. Due to the square law relationship of transmit voltage and power dissipation in the output devices, the dissipated power grows quickly as the transmit voltage is increased. One main solution to this challenge is reduction in load capacitance realized by migrating the transmitters into the probe handle so that they do not have to drive the cable.

*Beam-Former Complexity.* Higher image quality demands a larger number of beam-forming channels. This increased complexity in turn leads to higher power dissipation and requires more volume to implement. The traditional solution to this problem has been to implement dedicated digital beam-former ASICs that benefit from reduction in transistor feature sizes due to Moore's law. A number of groups are also exploring the possibility of software beam formers that make use of DSP chips that are optimized for low-power computations but still retain a degree of flexibility in their operation [69].

*Battery Life.* The increased power draw due to additional processing requirements and higher transmit voltages naturally leads to a reduction in the battery life of the system. For a portable device used in the field, battery life is of course essential. For laptop devices used in a clinical setting where power outlets are available, this may not be as much of an issue. Newer devices that promise highly portable high-quality imaging will be required to have extended battery life. In this respect, portable ultrasound systems can capitalize on the larger trend toward improved battery life that is seen in the portable electronics industry.

*Heat Dissipation.* In a console unit, excess heat can be transferred away from the patient to the system where it is readily dissipated to the ambient using a heat exchanger. In a portable self-contained unit, this is not an option. Therefore heat dissipation will become an increasingly important issue for miniaturized devices as they attempt to improve image quality.

*User Friendliness.* As high-quality ultrasound systems become more affordable, they will be used by a larger segment of the medical profession. This means that doctors who are not specialized in ultrasound will be called upon to glean useful diagnostic information from images that are often difficult to interpret even for career sonographers. One solution to this problem is to improve image quality, but this is especially challenging in a portable system where space and power are at a premium. Console units continue to introduce new software modifications that further automate the interpretation of image data, and this trend will likely continue in newer portable systems and will aid their adoption by the wider medical community.

*Electronic Packaging.* Miniaturized systems will continue to make use of denser and denser packaging technologies in order to increase the amount of processing. Here again, portable ultrasound will benefit from wider trends in the electronics industry.

*Cost.* Highly portable systems must continue to reduce their cost such that they bring increased utility to a broader range of clinicians. Reduction in system cost is expected to lead to the "commoditization" of ultrasound and comes with its own specific issues such as regulatory compliance and modification of well-established service models.

## 7.9. SUMMARY

In this chapter, we have provided an overview of the diagnostic ultrasound field as it stands today with particular emphasis on the key role that electronics plays. The basic principles of ultrasound imaging were reviewed in order to provide a basis for discussions between ultrasound practitioners and electronics engineers. This was followed by a detailed summary of the various constituents of the ultrasound machine including transducers, high-voltage pulsers, and switches as well as the receive chain amplifiers, ADCs and beam-forming electronics. It was seen that each of these component structures stands to gain from electronics miniaturization as the system as a whole continues its inevitable transformation to a completely handheld unit.

The field of ultrasound imaging is currently evolving very rapidly with significant advances in functionality and clinical utility being introduced on a regular basis. Newer, miniaturized devices promise to include many, if not all, of the features that have become expected in high-end units. These advances have been made possible due to the continuous progress that is being experienced in the electronics industry and is a testament to the important role that electronics plays in this field and will continue to play in the coming years.

### ACKNOWLEDGMENTS

We wish to acknowledge the following individuals for their help in reviewing this work: Steven C. Miller, Mirsaid Seyed-Bolorforosh, and Lowell Scott Smith. Special thanks to our editor Kris Iniewski and our publisher, John Wiley & Sons, for their help in bringing this to reality.

Some effort supported by US Army Medical Research Acquisition Activity DAMD17-02-0181, 820 Chandler Street, Fort Detrick, MD 21702-5014. The content of the information does not necessarily reflect the position or the policy of the Government and no official endorsement should be inferred.

Some effort described was supported by Grant Number R01 EB002485 from NIBIB. Contents of this publication are solely the responsibility of the authors and do not necessarily represent the official views of the NIH.

### REFERENCES

1. E. Krestel, *Imaging Systems for Medical Diagnostics*, Siemens AG, Berlin, 1990.
2. F. W. Kremkau, *Diagnostic Ultrasound: Principles and Instruments*, 6th ed., W. B. Saunders, Philadelphia, 2002.
3. K. E. Thomenius, Instrumentation design for ultrasonic imaging, in *Design of Medical Devices and Diagnostic Instumentation*, pp. 25.1–25.18, McGraw-Hill, New York, 2003.
4. K. Raum and W. D. O'Brien, Pulse-echo field distribution measurement technique for high-frequency ultrasound sources, *IEEE Trans. Ultrasonics, Ferroelectrics and Frequency Control* **44**, 810–815, 1997.
5. B. D. Steinberg, *Principles of Aperture and Array System Design*, John Wiley & Sons, New York, 1976.

6. T. L. Szabo, *Diagnostic Ultrasound Imaging*, Elsevier, Burlington, MA, 2004.

7. O. T. Ramm, S. W. Smith, and H. G. Pavy, High-speed ultrasound volumetric imaging system—Part II: Parallel processing and image display, *IEEE Trans. Ultrasonics, Ferroelectrics and Frequency Control*, **38**(2), 109–115, 1991.

8. E. Brunner, Ultrasound system considerations and their impact on front-end components, App-note, Analog Devices, Inc., 2002.

9. T. R. Gururaja and R. K. Panda, Current status and future trends in ultrasonic transducers for medical imaging applications, in *Proceedings of the IEEE International Symposium on Applications of Ferroelectrics*, August 24–27, 1998, pp. 223–228.

10. J. A. Brown, C. E. M. Demore, and G. R. Lockwood, Design and fabrication of annular arrays for high-frequency ultrasound, *IEEE Trans. Ultrasonics, Ferroelectrics and Frequency Control* **51**(8), 1010–1017, 2004.

11. S. Cogan, R. Fisher, K. Thomenius, and R. Wodnicki, Solutions for reconfigurable arrays, in *Proceedings of the IEEE Ultrasonics Symposium*, October 2–6, 2006, pp. 116–119.

12. R. F. Wagner, S. W. Smith, J. M. Sandrik, and H. Lopez, Statistics of speckle in ultrasound B-scans, *IEEE Trans. Sonics and Ultrasonics* **30**(3), 156–163, 1983.

13. AD9271: Octal LNA/VGA/AAF/ADC and Crosspoint Switch, Analog Devices datasheet, Rev. 0, 2007.

14. J. O. Fiering, P. Hultman, W. Lee, E. D. Light, and S. W. Smith, High-density flexible interconnect for two-dimensional ultrasound arrays, *IEEE Trans. Ultrasonics, Ferroelectrics, and Frequency Control* **47**(3), 764–770, 2000.

15. D. G. Wildes, R. Y. Chiao, C. M. W. Daft, K. W. Rigby, L. S. Smith, and K. E. Thomenius, Elevation performance of 1.25D and 1.5D transducer arrays, *IEEE Trans. Ultrasonics, Ferroelectrics and Frequency Control* **44**(5), 1027–1037, 1997.

16. S. W. Smith, W. Lee, E. D. Light, J. T. Yen, P. Wolf, and S. Idriss, Two dimensional arrays for 3-D ultrasound imaging, *Proceedings of the IEEE Ultrasonics Symposium*, Vol. 2, October 8–11, 2002, pp. 1545–1553.

17. D. R. Dietz, S. J. Norton, and M. Linzer, Wideband annular array response, in *Proceedings of the IEEE Ultrasonics Symposium*, 1978, pp. 206–211.

18. PZFlex http://www.pzflex.com/overview/index.html.

19. M. J. Zipparo, C. G. Oakley, D. M. Mills, A. M. Dentinger, and L. S. Sinith, A multirow single crystal phased array for wideband ultrasound imaging, in *Proceedings of the IEEE Ultrasonics Symposium*, Vol. 2, August 23–27, 2004, pp. 1025–1029.

20. J. Chen and R. Panda, Review: Commercialization of piezoelectric single crystals for medical imaging applications, in *Proceedings of the IEEE Ultrasonics Symposium*, September 18–21, 2005, pp. 235–240.

21. D. M. Mills, Medical imaging with capacitive micromachined ultrasound transducer (cMUT) arrays, in *Proceedings of the IEEE Ultrasonics Symposium*, Vol. 1, August 23–27, 2004, pp. 384–390.

22. X. Zhuang, I. O. Wygant, D.T. Yeh, A. Nikoozadeh, O. Oralkan, A. S. Ergun, C.-H. Cheng, Y. Huang, G. G. Yaralioglu, and B. T. Khuri-Yakub, Two-dimensional capacitive micromachined ultrasonic transducer (CMUT) arrays for a miniature integrated volumetric ultrasonic imaging system, in *Proceedings of the SPIE: Medical Imaging 2005: Ultrasonic Imaging and Signal Processing*, Vol. 5750, April 2005, pp. 37–46.

23. O. Oralkan, Jin. Xuecheng, F. L. Degertekin, and B. T. Khuri-Yakub, Simulation and experimental characterization of a 2-D capacitive micromachined ultrasonic transducer array element, *IEEE Trans. Ultrasonics, Ferroelectrics and Frequency Control*, **46**(6), 1337–1340, 1999.

24. C. Daft, S. Calmes, D. da Graca, K. Patel, P. Wagner, and I. Ladabaum, Microfabricated ultrasonic transducers monolithically integrated with high voltage electronics, in *Proceedings of the IEEE Ultrasonics Symposium*, Vol. 1, August 23–27, 2004, pp. 493–496.

25. G. Caliano, R. Carotenuto, E. Cianci, V. Foglietti, A. Caronti, A. Iula, and M. Pappalardo, Design, fabrication and characterization of a capacitive micromachined ultrasonic probe for medical imaging, *IEEE Trans. Ultrasonics, Ferroelectrics and Frequency Control* **52**(12), 2259–2269, 2005.

26. F. Akasheh, J. D. Fraser, S. Bose, and A. Bandyopadhyay, Piezoelectric micromachined ultrasonic transducers: Modeling the influence of structural parameters on device performance, *IEEE Trans. Ultrasonics, Ferroelectrics and Frequency Control* **52**(3), 455–468, 2005.

27. Z. Wang, W. Zhu, J. Mia, H. Zhu, C. Chao, and O. K. Tan, Micromachined thick film piezoelectric ultrasonic transducer array, in *Digest of Technical Papers, 13th International Conference on Solid-State Sensors, Actuators and Microsystems*, Vol. 1, June 5–9, 2005, pp. 883–886.

28. F. Akasheh, J. D. Fraser, S. Bose, and A. Bandyopadhyay, Piezoelectric micromachined ultrasonic transducers: Modeling the influence of structural parameters on device performance, *IEEE Trans. Ultrasonics, Ferroelectrics, and Frequency Control* **52**(3), 455–468, 2005.

29. F. S. Foster, K. A. Harasiewicz, and M. D. Sherar, A history of medical and biological imaging with polyvinylidene fluoride (PVDF) transducers, *IEEE Trans. UFFC* **47**(6), 1363–1371, 2000.

30. Hyun-Joong Kim, and B. Ziaie, Fabrication Techniques for Improving the Performance of PVDF-on-Silicon Ultrasonic Transducer Arrays, in *Proceedings of the IEEE 28th Annual International Conference Engineering in Medicine and Biology Society*, August 2006 pp. 3491–3494.

31. J. Xuecheng, I. Ladabaum, and B. T. Khuri-Yakub, The microfabrication of capacitive ultrasonic transducers, *IEEE J. Microelectromech. Sys.* **7**(3), 295–302, 1998.

32. X. Zhuang, A. S. Ergun, Ö. Oralkan, Y. Huang, I. O. Wygant, G. G. Yaralioglu, D. T. Yeh, and B. T. Khuri-Yakub, Through-wafer trench-isolated electrical interconnects for CMUT arrays, in *Proceedings of the IEEE International Ultrasonics Symposium, Rotterdam, The Netherlands*, September 18–21, 2005.

33. K. A. Wang, S. Panda, and I. Ladabaum, Curved micromachined ultrasonic transducers, in *Proceedings of the IEEE Ultrasonics Symposium*, October 5–8, 2003, pp. 572–576.

34. H. Ballan, and M. Declercq, *High Voltage Devices and Circuits in Standard CMOS Technologies*, Kluwer Academic Publishers, Dordrecht, The Netherlands, 1999.

35. J. D. Plummer and J. D. Meindl, A monolithic 200-V CMOS analog switch, *IEEE J. Solid-State Circuits* **11**(6), 809–817, 1976.

36. F. De Pestel, P. Moens, H. Hakim, H. De Vleeschouwer, K. Reynders, T. Colpaert, P. Colson, P. Coppens, S. Boonen, D. Bolognesi, and M. Tack, Development of a robust 50V 0.35/spl mu/m based Smart Power Technology using trench isolation. in

*Proceedings of the IEEE 15th International Symposium on Power Semiconductor Devices and ICs*, April 14–17, 2003, pp. 182–185.

37. XH035 0.35 μm CMOS Process: 0.35 micron modular analog mixed signal technology with RF capability and HV extensions, XFAB datasheet, rev 3.2 October 2007.

38. www.austriamicrosystems.com.

39. www.amis.com.

40. M. I. Fuller, T. N. Blalock, J. A. Hossack, and W. F. Walker, Novel transmit protection scheme for ultrasound systems, *IEEE Trans. Ultrasonics, Ferroelectrics and Frequency Control* **54**(1), 79–86, 2007.

41. B. Haider, Power drive circuits for Diagnostic Medical Ultrasound, in *Proceedings of the IEEE International Symposium on Power Semiconductor Devices and ICs*, June 4–8, 2006, pp. 1–8.

42. Low Charge Injection 8Channel High Voltage analog switches with bleed resistors, Supertex Datasheet HV230/HV232.

43. Y. Li, R. Wodnicki, N. Chandra, and N. Rao, An integrated 90 V switch array for medical ultrasound applications, in *Proceedings, IEEE Custom Integrated Circuits Conference*, September 10–13, 2006, pp. 269–272.

44. B. Dufort, T. Letavic, and S. Mukherjee, Digitally controlled high-voltage analog switch array for medical ultrasound applications in thin-layer silicon-on-insulator process, in *Proceedings of the IEEE International SOI Conference*, October 7–10, 2002, pp. 78–79.

45. Weir, Basil, US4595847, June 17, 1986.

46. A. S. Sedra and K. C. Smith, *Microelectronic Circuits*, 2nd ed. Holt Rinehart and Winston, Fort Worth, TX, 1987.

47. P. E. Allen and D. R. Holberg, *CMOS Analog Circuit Design*, Oxford: Oxford University Press, 2002.

48. R. L. Goldberg, C. D. Emery, and S. W. Smith, Hybrid multi/single layer array transducers for increased signal-to-noise ratio, *IEEE Trans. Ultrasonics, Ferroelectrics and Frequency Control* **44**(2), 315–325, 1997.

49. B. Stefanelli, J.-P. Bardyn, A. Kaiser, and D. Billet, A very low-noise CMOS preamplifier for capacitive sensors, *IEEE J. Solid-State Circuits* **28**(9), 971–978, 1993.

50. R. A. Noble, R. R. Davies, D. O. King, M. M. Day, A. R. D. Jones, J. S. McIntosh, D. A. Hutchins, and P. Saul, Low-temperature micromachined cMUTs with fully-integrated analogue front-end electronics, in *Proceedings, IEEE Ultrasonics Symposium*, Vol. 2, October 8–11, 2002, pp. 1045–1050.

51. E. Brunner, An ultra-low noise linear-in-dB variable gain amplifier for medical ultrasound applications, *Conference record. Microelectronics Communications Technology Producing Quality Products Mobile and Portable Power Emerging Technologies,* November 7–9, 1995, p. 650.

52. Dual, low-noise variable-gain amplifier with preamp, Burr–Brown datasheet VCA2615, July 2005.

53. R. Oppelt, US5,724,312, March 3, 1998.

54. S.-T. Boaz, M. Kozak, and E. G. Friedman, A 250 MHz delta-sigma modulator for low cost ultrasound/sonar beamforming applications, in *Proceedings of the 11th IEEE International Conference on Electronics, Circuits and Systems*, December 13–15, pp. 113–116.

55. M. Inerfield, G. R. Lockwood, and S. L. Garverick, A sigma-delta-based sparse synthetic aperture beamformer for real-time 3-D ultrasound, *IEEE Trans. Ultrasonics, Ferroelectrics and Frequency Control* **49**(2), 243–254, 2002.

56. S. R. Freeman, M. K. Quick, M. A. Morin, R. C. Anderson, C. S. Desilets, T. E. Linnenbrink, and M. O'Donnell, Delta–sigma oversampled ultrasound beamformer with dynamic delays, *IEEE Trans. Ultrasonics, Ferroelectrics and Frequency Control* **46**(2), 320–332, 1999.

57. Dual, ultralow noise variable gain amplifier, Rev C, 2007. Analog devices datasheet AD604.

58. S. Y. Peng, M. S. Qureshi, A. Basu, P. E. Halser, and F. L. Degertekin, A floating-gate based low-power capacitive sensing interface circuit, in *Proceedings, IEEE 2006 Custom Integrated Circuits Conference*, pp. 257–260.

59. H. L. Chao and D. Ma, CMOS variable-gain wide-bandwidth CMFB-free differential current feedback amplifier for ultrasound diagnostic applications, *Proceedings of IEEE International Symposium on Circuits and Systems*, May 21–24, pp. 649–652.

60. J. Morizio, S. Guhados, J. Castellucci, and O. von Ramm, 64-channel ultrasound transducer amplifier, *Southwest Symposium on Mixed-Signal Design*, February 23–25, 2003, pp. 228–232.

61. L. L. Lay, S. J. Carey, J. V. Hatfield, and C. M. Gregory, Receiving electronics for intra-oral ultrasound probe, in *IEEE International Workshop on Biomedical Circuits and Systems*, December 1–3, 2004, pp. S2/2-9-12.

62. 8-Chanel variable gain amplifier, Burr–Brown datasheet, October 2005.

63. K. Kaviani, O. Oralkan, P. Khuri-Yakub, and B. A. Wooley, A multichannel pipeline analog-to-digital converter for an integrated 3-D ultrasound imaging system, *IEEE J. Solid-State Circuits* **38**(7), 1266–1270, 2003.

64. B. Savord and R. Solomon, Fully sampled matrix transducer for real time 3D ultrasonic imaging, in *Proceedings, IEEE Symposium on Ultrasonics*, Vol. 1, October 5–8, 2003, pp. 945–953.

65. M. O'Donnell, Applications of VLSI circuits to medical imaging, *Proc. IEEE* **76**(9), 1106–1114, 1988.

66. M. Karaman, E. Kolagasioglu, and A. Atalar, A VLSI receive beamformer for digital ultrasound imaging, in *Proceedings, 1992 IEEE International Conference on Acoustics, Speech, and Signal Processing*, Vol. 5, March 23–26, 1992, pp. 657–660.

67. K. E. Thomenius, Evolution of ultrasound beamformers, in *Proceedings, IEEE Ultrasonics Symposium*, Vol. 2, November 3–6, 1996, pp. 1615–1622.

68. Hu Chang-Hong, K. A. Snook, Cao Poi-Jie, and K. Kirk Shung, High-frequency ultrasound annular array imaging. Part II: digital beamformer design and imaging, *IEEE Trans. Ultrasonics, Ferroelectrics and Frequency Control* **53**(2), 309–316, 2006.

69. The Verasonics ultrasound engine, http://www.verasonics.com/pdf/verasonics_ultrasound_eng.pdf.

70. C. R. Hazard and G. R. Lockwood, Developing a high speed beamformer using the TMS320C6201 digital signal processor, in *Proceedings, IEEE Ultrasonics Symposium*, Vol. 2, October 2000, pp. 1755–1758.

71. J. E. Powers, D. J. Phillips, M. A. Brandestini, and R. A. Sigelmann, Ultrasound phased array delay lines based on quadrature sampling techniques, *IEEE Trans. Volume Sonics and Ultrasonics* **27**(6), 1980, pp. 287–294.

72. J. R. Talman, S. L. Garverick, and G. R. Lockwood, Integrated circuit for high-frequency ultrasound annular array, in *Proceedings, IEEE Custom Integrated Circuits Conference*, September 21–24, 2003, pp. 477–480.

73. T. Halvorsrod, W. Luzi, and T. S. Lande, A log-domain μbeamformer for medical ultrasound imaging systems, *IEEE Trans. Circuits and Systems I: Regular Papers* **52**(12), 2563–2575, 2005.

74. M. Yaowu, T. Tanaka, S. Arita, A. Tsuchitani, K. Inoue, and Y. Suzuki, Pipelined delay-sum architecture based on bucket-brigade devices for on-chip ultrasound beamforming, *IEEE J. Solid-State Circuits* **38**(10), 1754–1757, 2003.

75. T. Halvorsrod, L. R. Cenkeramaddi, A. Ronnekleiv, and T. Ytterdal, SCREAM-A discrete time μbeamformer for CMUT arrays-behavioral simulations using system c, in *Proceedings of the IEEE Ultrasonics Symposium*, Vol. 1, September 18–21, 2005, pp. 500–503.

76. A. Chiang, US5,763,785, June 9, 1998.

77. J. D. Fraser, US6,375,617, April 23, 2002.

78. B. Stefanelli, I. O'Connor, L. Quiquerez, A. Kaiser, and D. Billet, An analog beam-forming circuit for ultrasound imaging using switched-current delay lines, *IEEE J. Solid-State Circuits* **35**(2), 202–211, 2000.

79. D. G. Bailey, J. A. Sun, A. Meyyappan, and G. Wade, A computer-controlled transducer for real-time three-dimensional imaging, in *Acoustical Imaging*, Vol. 18, H. Lee and G. Wade, eds., Plenum Press, New York, 1991.

80. R. K. Fisher, K. E. Thomenius, R. Wodnicki, R. Thomas, S. Cogan, C. Hazard, W. Lee, D. M. Mills, B. Khuri-Yakub, A. Ergun, and G. Yaralioglu, Reconfigurable arrays for portable ultrasound, in *Proceedings of the IEEE Ultrasonics Symposium*, pp. 495–499.

81. High Speed $+/-100$V 2A Integrated Ultrasound Pulser, Supertex Datasheet HV732.

82. F. L. Degertekin, R. O. Guldiken, and M. Karaman, Annular-ring CMUT arrays for forward-looking IVUS: Transducer characterization and imaging, *IEEE Trans. Ultrasonics, Ferroelectrics and Frequency Control* **53**(2), 474–482, 2006.

83. W. C. Black, Jr., and D. N. Stephens, CMOS chip for invasive ultrasound imaging, *IEEE J. Solid-State Circuits* **29**(11), 1381–1387, 1994.

84. A. M. Chiang, P. P. Chang, and S. R. Broadstone, PC-Based ultrasound imaging system in a probe, in *Proceedings of the IEEE Ultrasonics Symposium*, 2000, pp. 1255–1260.

85. http://www.sonosite.com.

86. http://www.terason.com.

87. GE Healthcare achieves #1 worldwide market share position in compact ultrasound industry in 2006, GE Press Release, March 26, 2007.

88. B. Casey, New US scanners, CT and MRI upgrades pace Siemens at RSNA 2006, Aunt Minnie.com, November. 26, 2006.

89. U.S. ultrasound market hits all time high, Klein report, 2006 as reported by PRNewswire.

90. S. Ghosh and M. Bayoumi, On integrated CMOS-MEMS system-on-chip, in *The 3rd International IEEE-NEWCAS Conference*, June 19–22, 2005, pp. 31–34.

91. I. O. Wygant, D. T. Yeh, X. Zhuang, A. Nikoozadeh, O. Oralkan, A. S. Ergun, M. Karaman, and B. T. Khuri-Yakub, A miniature real-time volumetric ultrasound imaging system, in *Proceedings of SPIE: Medical Imaging 2005: Ultrasonic Imaging and Signal Processing*, Vol. 5750, April 2005, pp. 26–36.

92. I. O. Wygant, X. Zhuang, D. T. Yeh, A. Nikoozadeh, O. Oralkan, A. S. Ergun, M. Karaman, and B. T. Khuri-Yakub, Integrated Ultrasonic Imaging Systems Based on cMUT Arrays: Recent Progress, in *Proceedings of IEEE International Ultrasonics Symposium*, Montréal, QC, Canada, August 23–27, 2004.

93. I. Cicek, A. Bozkurt, and M. Karaman, Design of a front-end integrated circuit for 3D acoustic imaging using 2D CMUT arrays, *IEEE Trans. Ultrasonics, Ferroelectrics and Frequency Control* **52**(12), 2005, 2235–2241.

94. I. O. Wygant, X. Zhuang, D. T. Yeh, S. Vaithilingram, A. Nikoozadeh, O. Oralkan, A. S. Ergun, M. Karaman, and B. T. Khuri-Yakub, An endoscopic imaging system based on a two-dimensional CMUT array, real-time imaging results, in *Proceedings of the IEEE Ultrasonics Symposium*, 2005, pp. 792–795.

95. G. S. Kino, *Acoustic Waves, Devices, Image and Analog Signal Processing*, Prentice-Hall, Englewood Cliffs, NJ, 1987.

# PART IV
## Magnetic Resonance Imaging

# 8 Magnetic Resonance Imaging

PIOTR KOZLOWSKI

## 8.1. INTRODUCTION

Magnetic resonance imaging (MRI) is a relatively new medical imaging technique primarily used in radiology to generate anatomical and functional images of the human body. Unlike other medical imaging techniques, MRI does not use ionizing radiation and therefore is considered safer than many other techniques. MRI generates images with excellent soft tissue contrast and thus is particularly useful for neurological, musculoskeletal, cardiovascular, and oncological imaging. Although not as sensitive as PET or SPECT, or as fast as CT, MRI is a very versatile technique able to generate great variety of image contrasts for a wide range of clinical and research applications.

MRI detects signals predominantly from hydrogen nuclei (i.e., protons) in water or fat molecules. MRI signal acquisition is based on the phenomenon of nuclear magnetic resonance (NMR), which deals with the interactions between nuclear spins and magnetic fields. NMR was discovered in the 1938 by I. I. Rabi, who received the 1944 Nobel prize in physics for this discovery. It was developed as an experimental technique independently by F. Bloch and E. M. Purcell in 1946, who shared the Nobel prize in physics in 1952 for this work. NMR is a major experimental technique used in solid-state physics and chemistry. Signal localization in MRI is achieved by the application of linear gradients of a magnetic field. It was discovered by P. Lauterbur and P. Mansfield in the early 1970s, who shared the 2003 Nobel prize in medicine for this work. MRI has been in clinical use for almost three decades since commercial MRI scanners became available in the early 1980s. Table 8.1 lists the stepping stones in the discovery and development of NMR and MRI.

MRI is arguably one of the most versatile medical imaging technique currently in use. Its versatility stems from the many types of contrast that can be generated in MRI images with no or little modifications to the imaging equipment. The list of contrast types includes: proton density, $T_1$ and $T_2$ relaxation times, magnetic susceptibility,

*Medical Imaging: Principles, Detectors, and Electronics,* edited by Krzysztof Iniewski
Copyright © 2009 John Wiley & Sons, Inc.

**TABLE 8.1. Stepping Stones in the Discovery and Development of NMR and MRI**

| | | |
|---|---|---|
| 1921 | Stern–Gerlach experiment—first experimental observation of a nuclear spin in a beam of silver atoms. | Gerlach et al. [58] |
| 1938 | Rabi observes resonance absorption of RF energy in a homogenous magnetic field—first NMR observation in an atomic beam. | Rabi et al. [60] |
| 1944 | Isidor Isaac Rabi receives a Nobel prize in physics "for his resonance method for recording the magnetic properties of atomic nuclei." | |
| 1945 | Purcell and co-workers observe resonance effect in a bulk material (paraffin wax)—first NMR observation in bulk material. | Purcell et al. [61] |
| 1946 | Bloch and co-workers observe precession of nuclear magnetization inducing electromotive force in the surrounding RF coil—first NMR observation in bulk material. | Bloch et al. [62] |
| 1949 | First pulse NMR experiment is carried out by Torrey. Hahn discovers spin echoes. | Torrey HC [63] Hahn [64] |
| 1950 | Chemical shift is discovered by a number of researchers. | Arnold et al. [73] |
| 1952 | Edward Mills Purcell and Felix Bloch receive Nobel prize in physics "for their development of new methods for nuclear magnetic precision measurements and discoveries in connection therewith." | |
| 1952 | The first commercial high-resolution NMR spectrometer is built by Varian Associates. | |
| 1964 | Ernst builds the first pulse NMR spectrometer, in collaboration with Varian Inc., and publishes first pulsed NMR spectrum. | Ernst and Anderson [65] |
| 1971 | Jenner discovers 2D NMR, subsequently developed by Ernst. | Mueller et al. [66] |
| 1971 | Damadian discovers that $T_1$ and $T_2$ relaxation times are higher in tumors than in normal tissue. | Damadian [74] |
| 1972 | Lauterbur discovers a spatial localization technique of the NMR signal by application of magnetic field gradients—generates first ever MRI image. | Lauterbur [26] |
| | Mansfield develops an NMR based method to study diffractions in solids, also proposing to use gradients for spatial localization in NMR. | Mansfield and Grannell [67] |
| 1974 | Mansfield develops slice selection technique. First $^{31}$P spectrum from a rat's leg is acquired ex vivo. | Garroway et al. [68] Hoult et al. [69] |
| 1975 | First Fourier MRI method is developed by Ernst and co-workers. | Kumar et al. [70] |
| 1976 | First image of a human body part (finger) is acquired in vivo. | Mansfield and Maudsley [71] |
| 1977 | Mansfield develops echo planar imaging (EPI)—one of the fastest MRI techniques to date. | Mansfield [72] |

*(Continued)*

**TABLE 8.1.** *Continued*

| | |
|---|---|
| 1991 | Richard R. Ernst receives a Nobel prize in chemistry "for his contributions to the development of the methodology of high resolution nuclear magnetic resonance (NMR) spectroscopy." |
| 2002 | Kurt Wüthrich receives a Nobel prize in chemistry "for his development of nuclear magnetic resonance spectroscopy for determining the three-dimensional structure of biological macromolecules in solution." |
| 2003 | Peter C. Lauterbur and Sir Peter Mansfield receive a Nobel prize in medicine "for their discoveries concerning magnetic resonance imaging." |

magnetization transfer, molecular motion and flow, among others, and constantly expands as new techniques are being developed.

$T_1$ and $T_2$ relaxation times of the tissue are the most common sources of contrast in the clinical applications of the MRI. $T_1$ contrast is regularly used to generate anatomical images of various organs and tissues in the human body (e.g., white and gray matter in the brain), while $T_2$ contrast is used to identify pathological tissues, like tumor or edema.

Magnetic Resonance Angiography (MRA) is a technique that allows imaging of flowing blood in veins and arteries. MRA techniques are generally divided into three categories: time of flight (TOF), phase contrast (PC), or contrast-enhanced (CE) angiography. TOF techniques use heavy $T_1$ weighting to saturate signals from the stationary spins in a selected slice, while collecting signals from the fully relaxed spins in the blood flowing into the slice. The maximum intensity projection (MIP) algorithm is used to generate pseudo-three-dimensional images of the blood vessels with the image intensity proportional to the blood flow. PC angiography measures changes in the phase of the MRI signal that results from the coherent motion of the flowing spins. This technique allows three-dimensional quantitative measurements of the blood flow in the arteries and veins. CE angiography uses intravenously injected contrast agents to enhance signal from the flowing blood.

In addition to imaging blood flow in the vessels, MRI is capable of measuring blood flow through capillaries in the tissue—that is, tissue perfusion. Two types of techniques are commonly used for this purpose: arterial spin labeling (ASL) and dynamic contrast enhanced MRI (DCE-MRI). In ASL, spins in major arteries feeding the region of interest (i.e., the imaging slice) are inverted with an inversion RF pulse [1]. Upon entering the imaging slice, the inverted spins exchange with tissue water, effectively altering the total magnetization and lowering the MRI signal from the slice. Images with and without the inversion pulse are collected and the perfusion map is generated by subtracting the two images. DCE MRI is a technique where a bolus of a contrast medium is injected intravenously and, subsequently, a series of fast $T_1$-weighted images are acquired to follow the passage of the bolus through the tissue of interest. Low-molecular-weight contrast agents (e.g., Gd-DTPA), typically used in such studies, leak out of the vasculature, altering $T_1$ of water protons in the surrounding tissue and thus modifying the intensity of $T_1$-weighted images. A two-compartment pharmacokinetic model is commonly used to extract information about tissue perfusion from the $T_1$-weighted images. DCE MRI has been shown to significantly improve tissue characterization and has been used extensively in tumor diagnosis [2–5].

MRI is the only medical imaging technique capable of measuring the molecular diffusion of water in vivo. Molecular diffusion is a process in which molecules move along random paths. When a strong magnetic field gradient is present, this random motion results in irreversible de-phasing of the MRI signal, making diffusion a dominant source of contrast in the MR image [6]. Diffusion of water molecules in tissue is affected by the presence of cellular structures that provide barriers to free movement, and it can therefore characterize these structures and their changes in pathology [7]. Diffusion MRI remains predominantly a research technique; however, its clinical applications steadily expand, and include diagnosis of acute stroke [8] and cancer of the brain [9], liver [10], pancreas [11], and prostate [3].

In 1990, Ogawa introduced a novel contrast in MRI images of the brain called blood oxygenation level dependent (BOLD) contrast [12]. It exploits differences in magnetic properties of the oxygenated and deoxygenated hemoglobin in blood. BOLD became the basis of functional MRI (fMRI), a technique of mapping brain activation in response to a stimulus. Activation of a specific brain region triggers a hemodynamic response—that is, an increase in blood flow to the activation area to satisfy the increase in oxygen consumption by the activated neurons. As a result, activation areas exhibit increased amounts of oxygenated hemoglobin, which can be detected with specific MRI techniques. Since its inception in the early 1990s, fMRI has evolved into an immensely active area of research generating more publications than all the other MRI applications. In addition to being a research tool in neurological sciences, fMRI found its clinical relevance in identifying specific brain areas during the surgery planning process and monitoring recovery from stroke.

Other applications of MRI involve imaging magnetically labeled cells [13], imaging cellular functions and molecular processes using specifically designed contrast agents [14], MRI of hyperpolarized gases [15, 16], studying microstructures of tissue and vasculature with susceptibility weighted imaging [17], and intraoperative and interventional MRI [18]. MRI continues to be a very active area of research with new techniques being developed constantly. The following sections describe the physical basis of NMR and MRI and provide brief account of the most important MRI techniques used in both clinical and research applications of this technology.

## 8.2. NUCLEAR MAGNETIC RESONANCE (NMR)

Nuclear magnetic resonance (NMR) is a phenomenon that provides basis for signal acquisition in magnetic resonance imaging (MRI) techniques. NMR deals with the interactions between nuclear magnetic moments and external magnetic fields. Since atomic nuclei consist of electrically neutral neutrons and positively charged protons, they themselves are positively charged. Both protons and neutrons posses a property called spin. Spin is a quantum mechanical property; however, it can be, somewhat simplistically, understood as the spinning motion of a particle about its own axis. In a nucleus, spins of protons and neutrons add to produce the overall spin of the nucleus, the so-called nuclear spin. The value of the nuclear spin depends on the number of protons and neutrons in the nucleus. In general, nuclei can have (1) no spin, (2) spin $\frac{1}{2}$, or (3) spin larger than $\frac{1}{2}$ [19]. The NMR phenomenon is not observed for the nuclei with no spin. A partial list of NMR-active nuclei that are of interest in biomedical applications of NMR is included in Table 8.2.

Since nuclei are positively charged particles, and the property of spin implies a spinning motion, we can deduce that nuclei with nonzero spin will also posses a magnetic moment. The spinning motion of an electrical charge can be likened to a current in a closed circuit, which possesses a magnetic moment. The nuclear magnetic moment is proportional to the nuclear spin according to the formula

$$\mu = \gamma \cdot I\hbar \tag{8.1}$$

where $I$ is the nuclear spin, $\hbar$ is Plank's constant divided by $2\pi$, and $\gamma$ is the gyromagnetic ratio, which is a characteristic property of nuclei. The fundamentals of the NMR phenomenon are related to the way nuclear magnetic moments interact with magnetic fields.

**TABLE 8.2. List of NMR-Active Nuclei of Interest in Biomedical Applications**

| Nucleus | Larmor Frequency at 1T (MHz) | Natural Abundance (%) | Detection Sensitivity at Constant Field | Spin |
|---|---|---|---|---|
| $^1$H | 42.577 | 99.9844 | 1.00 | $\frac{1}{2}$ |
| $^2$H | 6.536 | 0.0156 | 0.00964 | 1 |
| $^3$H | 45.414 | ~0 | 1.21 | $\frac{1}{2}$ |
| $^3$He | 32.433 | 0.00013 | 0.44 | $\frac{1}{2}$ |
| $^7$Li | 16.547 | 92.57 | 0.294 | $\frac{3}{2}$ |
| $^{13}$C | 10.705 | 1.108 | 0.0159 | $\frac{1}{2}$ |
| $^{14}$N | 3.076 | 99.635 | 0.00101 | 1 |
| $^{15}$N | 4.315 | 0.365 | 0.00104 | $\frac{1}{2}$ |
| $^{17}$O | 5.774 | 0.037 | 0.0000108 | $\frac{5}{2}$ |
| $^{19}$F | 40.055 | 100.0 | 0.834 | $\frac{1}{2}$ |
| $^{23}$Na | 11.262 | 100.0 | 0.0927 | $\frac{3}{2}$ |
| $^{31}$P | 17.235 | 100.0 | 0.0664 | $\frac{1}{2}$ |
| $^{35}$Cl | 4.172 | 75.4 | 0.00471 | $\frac{3}{2}$ |
| $^{39}$K | 1.987 | 93.08 | 0.000508 | $\frac{3}{2}$ |
| $^{87}$Rb | 13.932 | 27.20 | 0.177 | $\frac{3}{2}$ |
| $^{129}$Xe | 11.776 | 26.44 | 0.0212 | $\frac{1}{2}$ |
| $^{133}$Cs | 5.584 | 100 | 0.0474 | $\frac{7}{2}$ |

MRI deals predominantly with the NMR signal detected from hydrogen nuclei (i.e., protons) in water or fat molecules. Thus, we will limit our discussion of the basis of NMR to protons. More complete descriptions of NMR phenomenon, including discussion of other NMR-active nuclei, can be found in many excellent books written on this subject [i.e., references 20–22].

### 8.2.1. Interaction of Protons with Magnetic Fields

When the proton is placed in an external magnetic field $\vec{B}$, its magnetic moment $\vec{\mu}$ tries to align itself with the direction of the field, much the same way the compass needle aligns itself with the earth's magnetic field. However, since a proton possesses spin, its motion will be affected by a torque $\vec{N}$:

$$\vec{N} = \vec{\mu} \times \vec{B} \tag{8.2}$$

forcing the proton's magnetic moment to precess around the direction of $\vec{B}$ (see Fig. 8.1). We can write the equation of motion for the proton's magnetic moment in the external magnetic field as

$$\frac{d\vec{\mu}}{dt} = \gamma \cdot \vec{\mu} \times \vec{B} \tag{8.3}$$

The angular frequency of the precession is dependent on $\vec{B}$ according to the Larmor equation:

$$\vec{\omega} = -\gamma \cdot \vec{B} \tag{8.4}$$

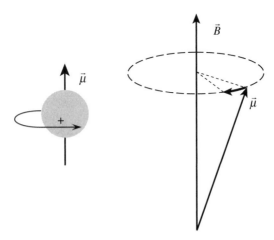

**Fig. 8.1.** Proton, viewed somewhat simplistically as a charged particle spinning about its axis, possesses magnetic moment $\vec{\mu}$ (**left**). When placed in an external magnetic field $\vec{B}$, this magnetic moment will precess about the direction of $\vec{B}$.

The linear dependence between the frequency of the precession and the strength of the magnetic field, as described by the Larmor equation, forms the basis of the signal localization utilized in MRI.

### 8.2.2. Macroscopic Magnetization and $T_1$ Relaxation

Proton's spin is a quantum mechanical property; and thus, when placed in a magnetic field, a proton's magnetic moment can be oriented parallel or anti-parallel to the direction of the field. In real NMR experiments we typically deal with large ensembles of nuclear spins. When such an ensemble of protons is placed in the magnetic field, initially half of the magnetic moments will be oriented parallel and the other half anti-parallel to the direction of the field. It can be shown, using quantum mechanical formalism, that the energy of spins aligned with the field is lower than the energy of the spins anti-parallel to the field (see Fig. 8.2). To minimize their energy, protons will try to align themselves with the direction of the magnetic field by exchanging energy with the surrounding environment (e.g., electrons, other atoms or ions, etc.), the so-called lattice. This process, called spin-lattice relaxation, results in establishing a thermal equilibrium state. The ratio of the number of protons populating the two energy levels in the equilibrium state is governed by the Boltzman distribution:

$$\frac{N_\uparrow}{N_\downarrow} = e^{-(\gamma \cdot \hbar B_0 / kT)} \tag{8.5}$$

where $N_\uparrow$ and $N_\downarrow$ are the numbers of protons parallel and anti-parallel to the direction of the $B_0$ magnetic field, $k$ is Boltzmann's constant, and $T$ is the absolute temperature. NMR or MRI experiments measure the total sum of protons' magnetic moments in the sample. This sum is called the macroscopic magnetization, and in the equilibrium state its value is

$$M_0 = \left(\frac{\rho_0 \gamma^2 \hbar^2}{4kT}\right) B_0 \tag{8.6}$$

where $\rho_0$ is the number of protons per unit volume, or proton density.

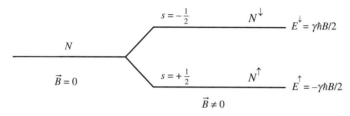

**Fig. 8.2.** Interaction between protons and the external magnetic field $B$ results in splitting the energy levels. Protons aligned parallel to the field have lower energy than the protons anti-parallel to the field. In the equilibrium state, the ratio of the number of protons populating the two energy levels $N^\uparrow / N^\downarrow$ is governed by the Boltzmann distribution.

**TABLE 8.3. Approximate $T_1$ and $T_2$ Values of Various Tissues in the Human Body at $B_0 = 1.5$ T and 37°C**

| Tissue | $T_1$ (ms) | $T_2$ (ms) |
|---|---|---|
| Gray matter | 950 | 100 |
| White matter | 600 | 80 |
| Muscle | 900 | 50 |
| Cerebrospinal fluid | 4500 | 2200 |
| Fat | 250 | 60 |
| Blood | 1200 | 100–200 |

*Source*: Reproduced from Hacke et al. [24].

The population difference between parallel and anti-parallel protons, which determines the macroscopic magnetization, is very small—for example, at the field of 1.5 T, which is the most common field strength of the clinical MRI scanners, $N_\uparrow - N_\downarrow \approx 10$ for every million protons. This small population difference is largely responsible for the low sensitivity of the NMR, as compared to the other spectroscopic techniques. Increasing the strength of the $B_0$ field is one way to increase the sensitivity of both NMR and MRI techniques. With magnet manufacturing technology constantly improving, magnets producing $B_0$ in the excess of 20 T for NMR spectrometers, and 7 T for human MRI scanners, are becoming more common. Another approach to improving sensitivity, recently developed in MRI, is to increase the spin population difference by optical pumping. This technique has been used in MRI of the hyperpolarized gases [23].

The process of establishing equilibrium magnetization is described by the equation

$$M_z(t) = M_z(0) \cdot e^{-t/T_1} + M_0 \left( 1 - e^{-t/T_1} \right) \tag{8.7}$$

where $M_z$ is the component of the macroscopic magnetization $M$ parallel to $B_0$, $M_0$ is the equilibrium magnetization, and $T_1$ is the characteristic time constant of the spin-lattice relaxation process.

Spin-lattice, or $T_1$ relaxation, is a complex process that involves energy exchange between protons and the local magnetic fields fluctuating near Larmor frequency produced by molecular motion (tumbling), and dipole–dipole interactions with other nuclei and electrons (the full description of these processes requires quantum mechanical formalism and is outside the scope of this publication; for more detailed information on this subject see reference 22 and references therein). Hence $T_1$ times will characterize the local environment with which the nuclear spins interact. In MRI, $T_1$ relaxation is a major source of contrast between different tissues. Table 8.3 lists $T_1$ times for various human tissues.

### 8.2.3. Rotating Frame and Resonance Condition

In describing macroscopic magnetization, the commonly used convention defines a reference frame in which the magnetic field $\vec{B}_0$ is aligned with the $z$ axis and $x$ and

$y$ axes are fixed in space. This is the laboratory reference frame. Since the macroscopic magnetization possesses a magnetic moment (being the sum of individual proton magnetic moments), it will interact with $\vec{B}_0$ much the same way individual proton magnetic moments do; that is, it will precess about the direction of $\vec{B}_0$. It is convenient to define another reference frame that rotates with a specific frequency $\vec{\Omega}$ about $\vec{B}_0$ (and thus around the $z$ axis of the laboratory reference frame). The equation of motion for the microscopic magnetization $M(t)$ in this rotating reference frame is

$$\left(\frac{d\vec{M}}{dt}\right)' = \gamma \cdot \vec{M} \times \vec{B}_{eff} \tag{8.8}$$

where the effective field $\vec{B}_{eff}$ is defined as

$$\vec{B}_{eff} = \vec{B}_0 + \frac{\vec{\Omega}}{\gamma} \tag{8.9}$$

Thus, if the rotating reference frame rotates with the Larmor frequency (i.e., $\vec{\Omega} = -\gamma \cdot \vec{B}_0$), the macroscopic magnetization remains stationary in the rotating frame.

Since the equilibrium magnetization is a vector sum of the proton magnetic moments aligned parallel or anti-parallel with $\vec{B}_0$, the equilibrium magnetization itself will always be aligned parallel with $\vec{B}_0$ and therefore will remain stationary. As we will see later on, NMR signal reception relies on the macroscopic magnetization precessing about $\vec{B}_0$. Hence, to generate an NMR signal we have to tip magnetization from its equilibrium state. This is achieved by applying an additional magnetic field, commonly called $\vec{B}_1$, which is perpendicular to $\vec{B}_0$. Let us consider the $\vec{B}_1$ field that rotates with the frequency $\vec{\Omega}$ about $\vec{B}_0$ (see Fig. 8.3a). The equation of motion for the macroscopic magnetization in the rotating frame is described by Eq. (8.8); however, the effective field $\vec{B}_{eff}$ is now defined as

$$\vec{B}_{eff} = \vec{B}_0 + \frac{\vec{\Omega}}{\gamma} + \vec{B}_1 \tag{8.10}$$

When $\vec{\Omega} = -\gamma \cdot \vec{B}_0$ (i.e., the rotating frame rotates with the Larmor frequency) the effective field is equal to $\vec{B}_1$, which is stationary in the rotating frame. In this case, the only field the macroscopic magnetization will "see" in the rotating frame is $\vec{B}_1$ (see Fig. 8.3b). The equation of motion then becomes

$$\left(\frac{d\vec{M}}{dt}\right)' = \gamma \cdot \vec{M} \times \vec{B}_1 \tag{8.11}$$

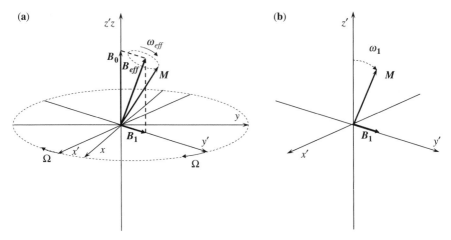

**Fig. 8.3.** (a) Magnetization $M$ in the rotating frame precesses about the effective field $B_{eff}$, which is a sum of $B_0$ and $B_1$ fields. (b) When the rotating frame precesses with the Larmor frequency ($\Omega = \omega_0$), the effective field in the rotating frame is equal to $B_1$, and the magnetization precesses about it.

and thus the magnetization in the rotating frame precesses about $\vec{B}_1$. The requirement that the $\vec{B}_1$ field rotates with the Larmor frequency is called the resonance condition. Typically in NMR or MRI experiments, the strength of the $\vec{B}_1$ field is several orders of magnitude smaller than the strength of the $\vec{B}_0$ field. Therefore, only $\vec{B}_1$ rotating with, or close to, the Larmor frequency can effectively tip magnetization away from its equilibrium position along $\vec{B}_0$, which resembles resonance-like behavior. In quantum mechanical formalism, the resonance condition means that the application of the alternating $\vec{B}_1$ field will result in energy absorption, allowing transition between spin energy levels, only when the $\vec{B}_1$ field alternates with the angular frequency equal to the Larmor frequency (i.e., $\hbar\omega = \Delta E = \gamma \cdot \hbar B_0$ or $\omega = \gamma \cdot B_0$). The names of both NMR (nuclear magnetic resonance) and MRI (magnetic resonance imaging) make reference to this resonance-like behavior.

The $\vec{B}_1$ field is typically applied in the form of short pulses resulting in the magnetization being tipped by a specific angle $\alpha$:

$$\alpha = \gamma \cdot \int_0^\tau B_1(t') \, dt' \tag{8.12}$$

where $\tau$ is the duration of the $\vec{B}_1$ pulse. Of particular interest are two such $\vec{B}_1$ pulses: 90° and 180° pulses, which tip magnetization by 90° and 180°, respectively.

The practical realization of the rotating $\vec{B}_1$ field is achieved by applying an oscillating voltage across a coil (e.g., a loop of wire in the simplest case; see Fig. 8.4a). Such a voltage generates an oscillating current through the coil, which in turn generates an oscillating magnetic field. The oscillating vector of the $\vec{B}_1$ field can be viewed as a sum of two vectors

**(a)**

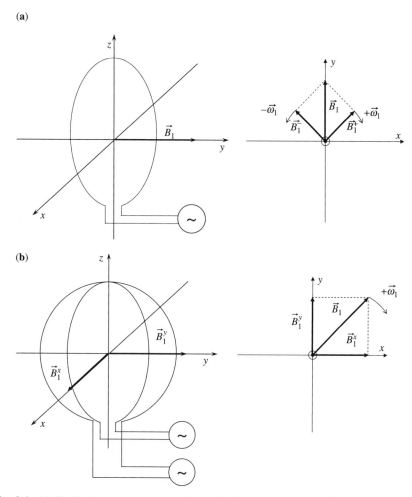

**(b)**

**Fig. 8.4.** (a) Oscillating voltage across a loop of wire generates an oscillating field $B_1$ (**left**), which is a sum of two fields rotating in opposite directions with the same frequency $\pm\omega_1$ (**right**). Only field $B_1^+$ affects magnetization. (b) Two coils, perpendicular to each other, with oscillating currents of the same amplitude and shifted in phase by 90° flowing through each coil (**left**) generate two orthogonal $B_1$ fields, which add to a single field rotating with the frequency $\omega_1$ (**right**).

rotating in opposite directions with the same angular frequency $\pm\vec{\omega}_1$. The field rotating with the $-\vec{\omega}_1$ frequency will not affect the magnetization in any appreciable way, since its rotating frequency is too far from the Larmor frequency (i.e., the resonance condition is fulfilled only for the field rotating with the $+\vec{\omega}_1$ frequency). Hence an oscillating current will generate a rotating $\vec{B}_1$ field, albeit with only half of the available amplitude. The field produced by a single coil is often called a linearly polarized $\vec{B}_1$ field. The more efficient solution is to use two coils perpendicular to each other, with oscillating currents, of the same

same amplitude and shifted in phase by 90°, flowing through each coil (see Fig. 8.4b). The vector sum of the oscillating fields produced by each coil is a field rotating with frequency $+\vec{\omega}_1$. The field produced by such a system of two coils is called a circularly polarized $\vec{B}_1$ field.

### 8.2.4.  $T_2$ Relaxation and Bloch Equations

Once tipped from the equilibrium position parallel to the $\vec{B}_0$ field, the magnetization will precess about $\vec{B}_0$. We can consider the longitudinal (parallel to $\vec{B}_0$) and the transverse (perpendicular to $\vec{B}_0$) components of the magnetization separately (see Fig. 8.5). Let us assume for a moment that the individual spins contributing to the macroscopic magnetization do not interact with one another or the surrounding environment. In this case the longitudinal magnetization will remain stationary, whereas the transverse component will precess about $\vec{B}_0$. We can then write the equation of motion for the macroscopic magnetization as two separate equations:

$$\frac{dM_z(t)}{dt} = 0$$

$$\frac{d\vec{M}_\perp(t)}{dt} = \gamma \cdot \vec{M}_\perp \times \vec{B}_0$$

(8.13)

where $\vec{M}_\perp = M_x \hat{x} + M_y \hat{y}$.

In reality the individual spins interact with one another and with the surrounding environment. The $T_1$ relaxation process, discussed in Section 8.3, represents the spins' interaction with the surrounding environment, or lattice. This process is responsible for generating the equilibrium magnetization along $\vec{B}_0$. The process of spins

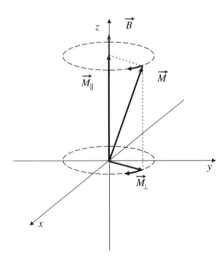

**Fig. 8.5.**  The longitudinal and transverse components of the magnetization.

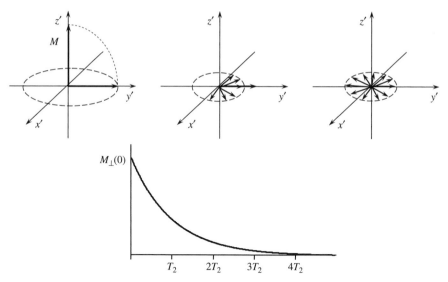

**Fig. 8.6.** Spin–spin relaxation causes the individual spins to lose phase coherence (**top**) resulting in the decay of transverse magnetization (**bottom**).

interacting with one another is called spin–spin relaxation (also called $T_2$ relaxation, where $T_2$ is the characteristic time constant of this process) and is responsible for diminishing the transverse component of the magnetization (see Fig. 8.6).

The decay of the transverse component of the magnetization is caused by the time-dependent variations in local magnetic fields affecting individual spins. Spins experience the applied field ($\vec{B}_0$) as well as the local fields of their neighboring spins. Due to random processes, like diffusion and molecular motion, the local fields vary in time, thereby generating a spread of local precessional frequencies. As a result, the transverse magnetization, which is a vector sum of the magnetic moments of the individual spins, diminishes with time due to spin dephasing (i.e., increasing differences in the precessional frequencies of the individual spins). This is an exponential process with the characteristic time constant $T_2$, which is described by the equation

$$\vec{M}_\perp(t) = \vec{M}_\perp(0) \cdot e^{-t/T_2} \tag{8.14}$$

Unlike $T_1$ relaxation, the $T_2$ relaxation process may occur with or without energy exchange. In general, any process contributing to $T_1$ relaxation will also contribute to $T_2$ relaxation; hence $T_1$ will always be longer than $T_2$. This is intuitively obvious, since before the magnetization can be fully restored by $T_1$ relaxation, it has to first disappear through $T_2$ relaxation. While $T_1$ times characterize the local environment with which the nuclear spins interact, $T_2$ times will characterize the nuclear spins themselves. Figure 8.7 shows the dependence of both $T_1$ and $T_2$ on the rate of molecular "tumbling." Large molecules, solids, and the spins bound to large macromolecules have very short $T_2$ values, while $T_2$ is long for the small molecules, liquids, and "free"

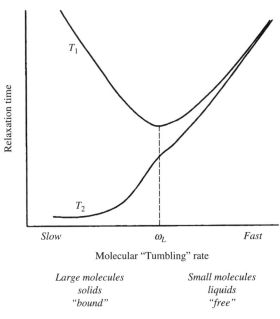

**Fig. 8.7.** The relationship between $T_1$ or $T_2$ and the molecular tumbling rate. $\omega_L$ is the Larmor frequency. (Reproduced from Elster and Burdette JH [59], with permission.)

spins. Like $T_1$, $T_2$ is also a major source of contrast in MRI. Table 8.3 lists $T_2$ for various human tissues.

The equations of motion of the macroscopic magnetization for interacting spins (i.e., taking into account both $T_1$ and $T_2$ relaxation) are called Bloch equations; when the applied field is a constant $B_0$ field along the $z$ axis, the equations are written as

$$\frac{dM_x}{dt} = \omega_0 M_y - \frac{M_x}{T_2}$$
$$\frac{dM_y}{dt} = -\omega_0 M_x - \frac{M_y}{T_2} \qquad (8.15)$$
$$\frac{dM_z}{dt} = \frac{M_0 - M_z}{T_1}$$

where $\omega_0$ is the Larmor frequency and $M_0$ is the equilibrium magnetization. The solutions to Eqs. (8.15) are written as

$$M_x(t) = e^{-t/T_2}(M_x(0) \cos \omega_0 t + M_y(0) \sin \omega_0 t)$$
$$M_y(t) = e^{-t/T_2}(M_y(0) \cos \omega_0 t - M_x(0) \sin \omega_0 t) \qquad (8.16)$$
$$M_z(t) = M_z(0) \cdot e^{-t/T_1} + M_0(1 - e^{-t/T_1})$$

The first two equations in (8.16) describe the precession of the $x$ and $y$ components of the transverse magnetization in the presence of $T_2$ relaxation decay. The last equation in (8.16) describes the recovery of the equilibrium magnetization along the $z$ direction by $T_1$ relaxation.

### 8.2.5. Signal Reception, Free Induction Decay, and Spin-Echo

Detection of the NMR (and MRI) signal is based on Faraday's induction law, which stipulates that a time-varying magnetic flux through a closed circuit (e.g., a loop of wire) will generate an electromotive force across the circuit, and the amplitude of the electromotive force is proportional to the rate of change of the magnetic flux:

$$emf = -\frac{d\Phi}{dt} \tag{8.17}$$

where $\Phi$ is a time-varying magnetic flux defined as

$$\Phi = \iint_\Sigma \vec{B} \cdot d\vec{S} \tag{8.18}$$

where $d\vec{S}$ is an element of surface area of the surface $\Sigma$.

Precessing macroscopic magnetization generates a time-varying magnetic flux. Hence, if we place a coil (e.g., a loop of wire) close to it, according to Faraday's induction law, an electromotive force will be generated across the coil (see Fig. 8.8). It can be shown (see reference 24 for details) that the electromotive force generated by the precessing magnetization is described as

$$emf = -\frac{d}{dt} \int_{sample} d^3 r \vec{M}(\vec{r}, t) \cdot \vec{B}^{rec}(\vec{r}) \tag{8.19}$$

where $\vec{M}$ is the macroscopic magnetization and $\vec{B}^{rec}$ is the magnetic field that would be produced by a unit current in the coil at the point $\vec{r}$. In effect, $\vec{B}^{rec}$ describes the efficiency of the coil in generating current induced by the precessing magnetization—that is, the efficiency of the coil in receiving the NMR signal.

As was discussed in the previous section, only the transverse component of the magnetization precesses about $\vec{B}_0$, thus only the transverse magnetization will contribute to the NMR signal. Hence, we can rearrange the Eq. (8.19) to describe the NMR signal as [24]

$$s(t) \propto \omega_0 \int d^3 r \cdot e^{-t/T_2} e^{-i\omega_0 t + \phi} \vec{M}_\perp(\vec{r}, 0) \cdot \vec{B}^{rec}(\vec{r}) \tag{8.20}$$

where $\vec{M}_\perp(\vec{r}, 0)$ is the transverse magnetization at time $t = 0$, $\phi$ is its initial phase, and the term $e^{-i\omega_0 t} = \cos \omega_0 t - i \sin \omega_0 t$ describes its precession about $\vec{B}_0$.

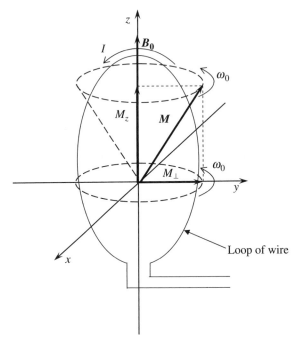

**Fig. 8.8.** Processing transverse magnetization generates a time varying magnetic flux through the coil. As a result, an electromotive force is induced across the coil.

As can be seen from Eq. (8.20) the NMR signal depends linearly on the Larmor frequency $\omega_0$. This dependence is the direct consequence of Faraday's induction law; that is, the faster the magnetization precesses, the larger the changes in the magnetic flux, and thus the larger the induced electromotive force and consequently the larger the signal. Since the Larmor frequency is proportional to $\vec{B}_0$, a higher external magnetic field not only generates larger equilibrium magnetization, but also results in the larger signal due to faster precession.

The NMR signal described by Eq. (8.20) is called the free induction decay (FID), signal and is the signal generated by the freely precessing magnetization (see Fig. 8.9). In modern NMR (or MRI) spectrometers, the acquired signals are digitized and subsequently stored and processed in digital form. Signal detection is carried out through a process called phase-sensitive detection (PSD). The PSD process plays a double role: It splits the NMR signal into two parts shifted in phase by 90°, effectively creating a complex signal, and it "subtracts" the Larmor frequency by demodulating the NMR signal (see Fig. 8.10). The complex signal is needed to distinguish between the positive and negative frequencies (i.e., frequencies larger or smaller than the Larmor frequency), whereas demodulation allows the narrowing of the frequency content in the NMR signal from the radio-frequency range down to the audio-frequency range (in a way, demodulation represents the physical realization of the rotating reference frame during the detection process).

According to Eq. (8.20), the NMR signal decays with the $T_2$ time constant as a result of the spin–spin relaxation process. In reality, any process contributing to

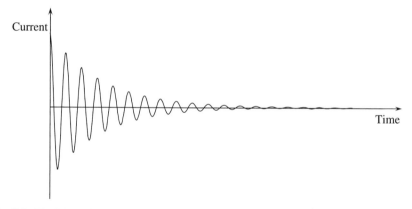

**Fig. 8.9.** The NMR signal, called the free induction decay (FID) signal, generated by the freely precessing transverse magnetization.

spin dephasing will cause the transverse magnetization to decay. Some of these processes are not directly related to the spin–spin interaction, but rather originate with the inhomogeneity of the $\vec{B}_0$ field (e.g., the $\vec{B}_0$ inhomogeneity due to imperfections of the magnet producing $\vec{B}_0$, local inhomogeneities caused by magnetic susceptibility effects, etc.). Hence, in Eq. (8.20), $T_2$ should be replaced with the new time constant, called $T_2^*$:

$$\frac{1}{T_2^*} = \frac{1}{T_2} + \frac{1}{T_2'} \tag{8.21}$$

where $T_2'$ is the characteristic time constant of the spin dephasing processes originated with the $\vec{B}_0$ inhomogeneities.

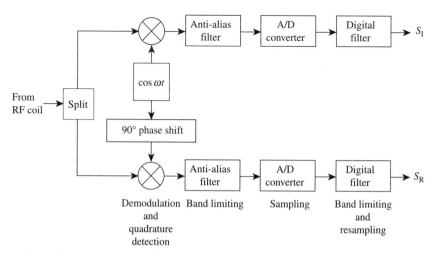

**Fig. 8.10.** Schematic diagram of the phase-sensitive detection (PSD) process. (Reproduced from Bernstein et al. [25], with permission.)

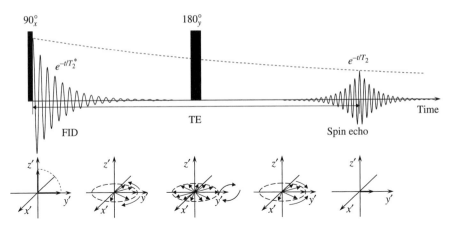

**Fig. 8.11.** Spin-echo technique: A $90°$ pulse flips magnetization to the $x'-y'$ plane. Local field inhomogeneities cause the spins to lose phase coherence resulting in the transverse magnetization decreasing with the $T_2^*$ time constant. The $180°$ refocusing pulse flips the dephased transverse magnetization around the $y'$ axis, reversing the direction of the Larmor precession. As a result, the spins are refocused and form a spin echo. The echo amplitude depends on the $T_2$ relaxation time.

Unlike $T_2$ relaxation, which involves random processes resulting in time-dependent variations of local magnetic fields, the $T_2'$ relaxation processes are static and thus can be reversed. The technique that allows to recover the loss of the magnetization due to $T_2'$ processes is called spin echo. It involves applying a $180°$ pulse, called refocusing pulse, to reverse the direction of the Larmor precession (see Fig. 8.11). As a result, the spins are refocused and generate the so called spin echo signal with the amplitude dependent on the $T_2$ only. The spin-echo technique is ubiquitous in MRI and will be frequently discussed in the sections describing various MRI techniques.

### 8.2.6. Chemical Shift and NMR Spectroscopy

Applying the inverse Fourier transform to the FID signal will generate an NMR spectrum (see Fig. 8.12). The position of the spectral line corresponds to the Larmor frequency, and in the case of liquids the shape is a Lorentzian function (i.e., the inverse Fourier transform of the exponential function, which is the "envelope" of the FID signal).[1] Since the phase-sensitive detection (PSD) process generates a complex FID signal, the NMR spectrum will also be described by a complex function with the real and imaginary parts being the absorption spectrum and the dispersion spectrum, respectively (see Fig. 8.12). The area under the spectral line is equal to the initial amplitude of the FID signal and thus is proportional to the number of spins within the sample.

The Larmor frequency of the nuclear spins is proportional to the total magnetic field acting on the spins. In addition to the external field $\vec{B}_0$, the spins experience the local magnetic fields produced by the electrons surrounding the nuclei. Hence,

---

[1] As shown in previous sections, the FID signal decays with the $T_2$ (or $T_2^*$) time constant: $e^{-t/T_2}$. In the case of solids the FID signal decay is not exponential, and thus the line shape is not a Lorentzian function.

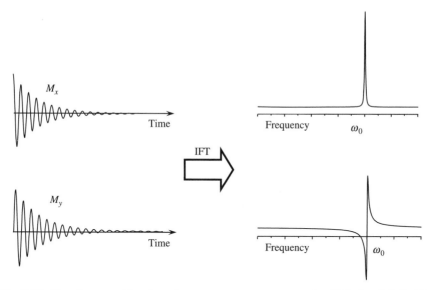

**Fig. 8.12.** The phase-sensitive detection process generates a complex FID signal (**left**). The inverse Fourier transform of the FID signal is a complex spectrum (**right**), where the real part is called the absorption spectrum (**top right**) and the imaginary part is called the dispersion spectrum (**bottom right**). The position of the spectral line $\omega_0$ corresponds to the Larmor frequency.

the effective magnetic field acting on the spins is

$$B_{eff} = B_0 - B_\sigma = B_0(1 - \sigma) \tag{8.22}$$

where $\sigma$ is called the shielding constant and characterizes the chemical environment of the nuclei. For diamagnetic molecules, $\sigma$ is positive and the nuclei experience the effective field larger than $\vec{B}_0$, while for paramagnetic molecules $\sigma$ is negative and the effective field is smaller than $\vec{B}_0$. This change in the local magnetic field results in a shift of the Larmor frequency, called the chemical shift:

$$\Delta\omega = \gamma B_0 \sigma \tag{8.23}$$

Typically, chemical shifts of various spectral lines are normalized to the Larmor frequency (to make it independent of $B_0$) and expressed in parts per million (or ppm). Figure 8.13 shows a theoretical NMR spectrum from lactate, a prominent metabolite in living organisms. The chemical structure of the lactate molecule shows three molecular groups containing protons: $CH_3$, $CH$, and $OH$ exhibiting chemical shifts of 1.33 ppm, 4.12 ppm, and 4.7 ppm, respectively, which correspond to the positions of the three spectral lines in the lactate spectrum.

If the spectral resolution is sufficient, the fine structure of the spectral lines from the molecular groups can be observed. Due to the interactions between protons in neighboring groups (i.e., $CH_3$ and $CH$ in the case of lactate), the spectral lines are split into

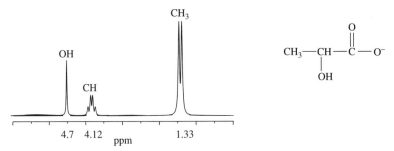

**Fig. 8.13.** Theoretical spectrum of lactate (**left**) and its chemical structure (**right**). Three molecular groups containing protons, $CH_3$, CH, and OH, give rise to three spectral lines with chemical shifts of 1.33 ppm, 4.12 ppm, and 4.7 ppm, respectively. The fine splitting of the $CH_3$ and CH lines is due to interactions between $CH_3$ and CH protons (*J*-coupling).

multiplets (see Fig. 8.13) corresponding to different energy levels.[2] Thus, protons of the $CH_3$ group in the lactate molecule are influenced by the proton of the CH group, which can be orientated parallel ↑ or anti-parallel ↓ to $B_0$. As a result, the $CH_3$ resonance is split into two spectral lines corresponding to the energy levels resulting from ↑ and ↓ orientations of the CH proton. The CH proton, on the other hand, is affected by the $CH_3$ protons, which can have the following orientations: all up (↑↑ ↑), two up one down (↑↑ ↓, ↑↓ ↑, ↓↑ ↑), one up two down (↑↓ ↓, ↓ ↑↓, ↓ ↓↑), and all down (↓↓ ↓). As a result, the CH resonance is split into four spectral lines with the intensity ratios 1 : 3 : 3 : 1. A proton from the OH group is too far from the $CH_3$ and CH groups to be affected by the other protons in the lactate molecule, and thus the OH spectral line does not have any additional fine structure. The interaction between spins in neighboring molecular groups is called spin–spin coupling or J-coupling and plays a significant role in high-resolution NMR spectroscopy, a major tool in analytical chemistry [19].

## 8.3. MAGNETIC RESONANCE IMAGING (MRI)

As mentioned in the NMR section the physics of signal generation and detection in magnetic resonance imaging (MRI) is virtually identical to that of the NMR technique. The main difference between the NMR and MRI techniques is the spatial localization of the signal, which is required to generate MRI images. The basis of spatial localization will be discussed in this section, and will be followed by an overview of the most common MRI techniques.

### 8.3.1. Spatial Localization

The basis for the spatial localization of the NMR signal employed in MRI is the linear relationship between the spins' presessional frequency and the strength of the external

---

[2]Protons in the external magnetic field $B_0$ can be oriented parallel (↑) or anti-parallel (↓) to the field, which will affect the energy level of the proton.

magnetic field $\vec{B}_0$, which is described by the Larmor equation:

$$\omega_0 = \gamma \cdot B_0 \tag{8.24}$$

The Larmor equation stipulates that any spatial dependence of the $B_0$ field will result in the same spatial dependence of the $\omega_0$. In MRI this spatial dependence of the $B_0$ field is achieved by generating a linear gradient of $B$,

$$\vec{G} = \vec{\nabla}B \tag{8.25}$$

making the effective field $B$, and subsequently the precessional frequency $\omega$, a linear function of the position $r$:

$$B(r) = B_0 + \vec{G} \cdot \vec{r}$$
$$\omega(r) = \omega_0 + \gamma\vec{G} \cdot \vec{r} \tag{8.26}$$

A simple way of generating a linear gradient is to use two circular loops of wire that carry currents in opposite directions (called the Maxwell pair; see Fig. 8.14). In practice, elaborate design and manufacturing procedures are employed to produce gradient coils generating strong, linear, fast-switching gradients that are required in modern MRI scanners.

In the process of imaging three-dimensional objects, spatial localization is required in all three dimensions (see Fig. 8.15). In MRI, this is typically achieved by selecting slices in one direction and subsequently localizing the signal in each slice by means of

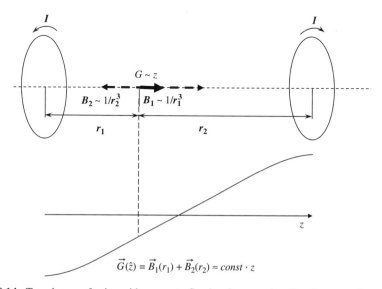

**Fig. 8.14.** Two loops of wire with currents flowing in opposite directions can be used to generate a linear magnetic field gradient along the axis of both coils.

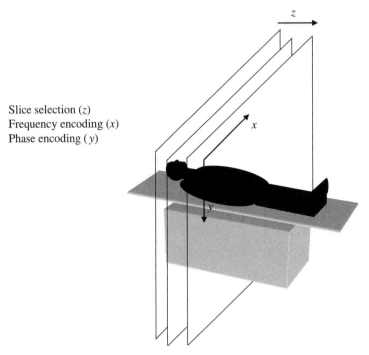

Slice selection (z)
Frequency encoding (x)
Phase encoding (y)

**Fig. 8.15.** Three-dimensional localization in MRI is typically achieved by selecting slices in one direction and subsequently localizing the signal in each slice by means of frequency and phase encoding.

the frequency and phase encoding methods. These localization techniques will be discussed in the following sections.

***8.3.1.1. Slice Selection.*** As was shown in Sections 8.2.4 and 8.2.6, the NMR signal is generated by the transverse magnetization precessing with the Larmor frequency. Typically, a short duration pulse of the $B_1$ field, rotating with the frequency $\omega_1$, is applied to tip magnetization from its equilibrium position to generate the transverse magnetization. Due to the resonance condition, $\omega_1$ has to be close to the Larmor frequency for the $B_1$ field to be effective in tipping the equilibrium magnetization. The nominal flip angle $\alpha$ as a function of the frequency offset $\Delta\omega$ (i.e., the difference between $\omega_1$ and the Larmor frequency) is called the excitation profile of the $B_1$ pulse. It can be shown (see reference 25 for details) that for small flip angles (i.e., $<30°$), the excitation profile can be approximated by the inverse Fourier transform of the shape of the $B_1$ pulse:

$$\alpha(\Delta\omega) \approx \pm\gamma \left| \int_0^t B_1(t')e^{i\Delta\omega t'} dt' \right| \qquad (8.27)$$

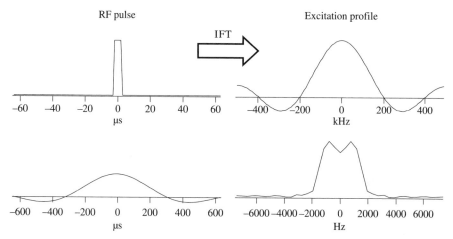

**Fig. 8.16.** For small flip angles ($\alpha < 30°$) the excitation profile of an RF pulse can be approximated by the inverse Fourier transform of its shape. A short-duration high-amplitude "hard" pulse (**top left**) has an excitation profile broad enough to cover a wide range of frequency offsets (**top right**). A long-duration low-amplitude-shaped "soft" pulse (**bottom left**) generates a rectangular narrow excitation profile (**bottom right**) required in slice selection process.

In many applications the $B_1$ pulse is required to excite all the magnetization within a sample. This is achieved by applying a short-duration, high-amplitude pulse (called a hard pulse) with an excitation profile broad enough to cover all frequency offsets present in the sample due to $B_0$ field inhomogeneities (see Fig. 8.16). The slice selection process in MRI, however, often requires the excitation pulses to generate a narrow, rectangular-shape excitation profile.

To select a slice in an MRI experiment, a relatively long $B_1$ pulse (typically several milliseconds) is applied in the presence of a gradient. The $B_1$ pulse will affect only magnetization with Larmor frequency within the bandwidth of the pulse. The magnetic field gradient generates a spatially dependent spread of the precessional frequencies, effectively converting the range of frequencies to a range of positions. As a result, only magnetization within a slice will be excited (see Fig. 8.17). The position of the selected slice is determined by $\omega_1$, while its thickness depends on the bandwidth of the $B_1$ pulse and the strength of the applied slice-selective gradient. These parameters can easily be adapted to select a slice with any position or thickness (within the limits of the gradient's strength) across the sample.

The slice selection process excites a slice of spins perpendicular to the direction of the slice-selective gradient. By changing the direction of the gradient, a slice in any arbitrary orientation, including oblique, can be selected. MRI is unique among the medical imaging techniques in that selecting an arbitrary slice orientation does not require moving the patient or the imaging equipment.[3]

---

[3]Modern PET and CT scanners can generate three-dimensional data sets, from which a slice with an arbitrary orientation can be extracted. However, MRI is unique in acquiring signal from a single slice with an arbitrary orientation without moving the patient or the imaging equipment.

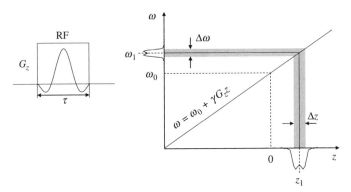

**Fig. 8.17.** Slice selection: The shaped RF pulse of a duration $\tau$ is applied in the presence of gradient $G_z$ (**left**). The gradient generates a spread of the precessional frequencies across a slice. As a result, only magnetization within the slice is excited. The position of the selected slice is determined by $\omega_1$, while its thickness depends on the bandwidth of the RF pulse and the strength of the applied gradient (**right**).

*8.3.1.2. Frequency Encoding.* Frequency encoding is utilized to localize the signal in one of the two perpendicular directions within a selected slice. To explain how it works, let us consider a sample consisting of a sphere filled with water, positioned inside an MRI scanner. Following a $90°$ pulse of the $B_1$ field, which flips the equilibrium magnetization to the transverse plane, the transverse magnetization will precess at the Larmor frequency. Switching on a linear gradient immediately following the pulse will result in a linear variation of the magnetic field along the direction of the gradient, thus generating a spread of Larmor frequencies across the sample. We can divide our sample into a large number of "strips" perpendicular to the direction of the gradient, narrow enough that the magnetic field does not vary across each strip (see Fig. 8.18). Thus the magnetization from each strip will precess with a single Larmor frequency proportional to the magnetic field at the location of the strip. The FID signal generated by each strip will be a single-frequency signal with the amplitude proportional to the number of protons (i.e., the amount of water) in each strip, and the inverse Fourier transform of the FID signal will be a spectrum consisting of a single line at the Larmor frequency with its amplitude proportional to the amount of water in the strip.[4] However, since the gradient converts the spatial location into frequency [see Eq. (26)], the frequency position of the spectral line from each strip will correspond to the physical location of the strip in the sample. The FID signal acquired from the sample will be the sum of signal contributions from all the strips. Therefore, the inverse Fourier transform of the signal will be the sum of all the spectral lines from individual strips, which is the profile of the sample along the direction of the gradient.

---

[4]Strictly speaking, the area under a spectral line is proportional to the number of protons (i.e., the amount of water). We can, however, assume that each spectral line generated by each strip will have roughly the same width, and thus we can approximate the area under the spectral line by its amplitude.

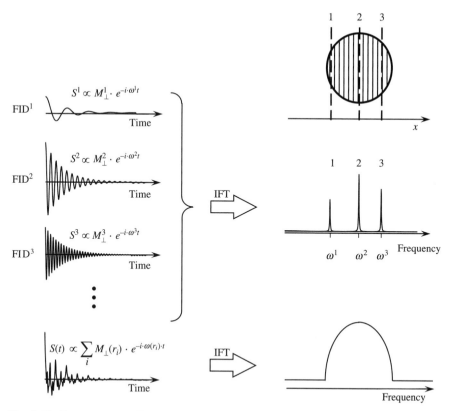

**Fig. 8.18.** Frequency encoding: A linear gradient generates a spread of Larmor frequencies across a spherical sample (**top right**). The FID signal from each "strip" across the sample has a single frequency and its amplitude is proportional to the number of protons in this "strip" (**top left**). The inverse Fourier transform of each FID is a single spectral line with the frequency corresponding to the physical location of the "strip" and its amplitude is proportional to the amount of water in the "strip" (**center right**). The FID from the sample is a sum of contributions from all "strips" (**bottom left**) and its inverse Fourier transform is a profile of the sample along the direction of the gradient (**bottom right**).

A gradient applied during the acquisition of the NMR signal is commonly called the frequency encoding, or read, gradient. The frequency-encoding gradient can be applied in any arbitrary orientation, generating a projection of the sample on any arbitrary direction. This process can be used to generate a series of projections on the large number of directions. By using the back-projection (or filtered back-projection) procedures, commonly used in other medical imaging modalities (e.g., CT, PET), a two-dimensional MRI image can then be reconstructed. This technique was used for the first time by Paul Lauterbur in generating the first-ever MRI image [26]. Although most modern MRI techniques use phase encoding to localize the NMR signal in the second direction within the selected slice, the projection-reconstruction technique is still in use, especially when the duration of the echo time, motion sensitivity, and temporal resolution are of concern [25].

### 8.3.1.3. Phase Encoding.
Although two-dimensional spatial localization can be achieved with the frequency-encoding gradient alone, typically a second gradient, called the phase-encoding gradient, is applied to localize the signal in the direction perpendicular to the frequency-encoding direction. The phase-encoding process relies on the ability to measure the phase of the precessing transverse magnetization through the phase-sensitive detection process described in Section 8.2.6.

Transverse magnetization $\vec{M}_\perp$ is a two-dimensional vector and its position in the transverse plane of the laboratory (or rotating) reference frame can be described either by its components $M_x$ and $M_y$ or by its length and phase: $|\vec{M}_\perp|$ and $\phi$ (see Fig. 8.19). Often the transverse magnetization is described as a complex number in the following form [8]:

$$M_+ \equiv M_x + iM_y$$
$$M_+(t) = e^{-t/T_2}e^{-i\omega_0+\phi_0}M_+(0) \tag{8.28}$$

where $\omega_0$ is the Larmor frequency and $\phi_0$ is the initial phase of the transverse magnetization (see Fig. 8.19).

Phase encoding also uses a linear gradient for spatial localization of the signal. However, unlike frequency encoding, the phase-encoding gradient is applied prior to, rather than during, the signal acquisition (see Fig. 8.20). Let us consider again a

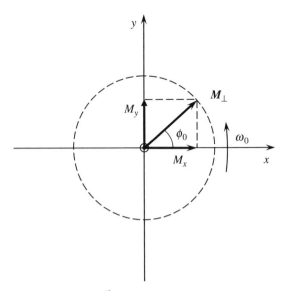

**Fig. 8.19.** Transverse magnetization $\vec{M}_\perp$ is a two-dimensional vector and its position in the transverse plane of the laboratory (or rotating) reference frame can be described either by its components: $M_x$ and $M_y$, or by its length and phase: $|\vec{M}_\perp|$ and $\phi$.

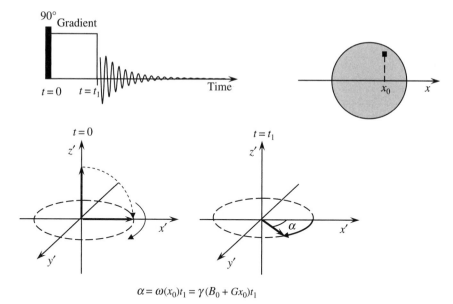

$$\alpha = \omega(x_0)t_1 = \gamma(B_0 + Gx_0)t_1$$

**Fig. 8.20.** Phase encoding: The phase encoding gradient is applied prior to signal acquisition (**top left**). Magnetization from a point within a spherical sample located at $x_0$ (**top right**) will accumulate phase $\alpha$ during the time $t_1$ when the phase-encoding gradient is switched on (**bottom**). The accumulated phase depends on the position of the point within the sample.

spherical sample filled with water positioned inside an MRI scanner. Like before, a linear gradient is switched on immediately following a 90° pulse. Let us consider a single point within the sample, located at $x_0$ along the $x$ axis (see Fig. 8.20). When the phase-encoding gradient is on, the spins located at the $x_0$ position will precess with the Larmor frequency:

$$\omega(x_0) = \gamma B(x_0) = \gamma(B_0 + Gx_0) \tag{8.29}$$

where $G$ is the amplitude of the phase encoding gradient. Since the phase can be defined as

$$\phi = \int \omega\, dt \tag{8.30}$$

at the time when the phase encoding gradient is switched on, just before the signal acquisition, the spins at position $x_0$ will have accumulated the phase

$$\phi(x_0) = \int \omega(x_0)\, dt = \gamma(B_0 + Gx_0)t_1 \tag{8.31}$$

where $t_1$ is the duration of the phase-encoding gradient pulse. Hence, the phase accumulated during precession in the presence of the phase-encoding gradient

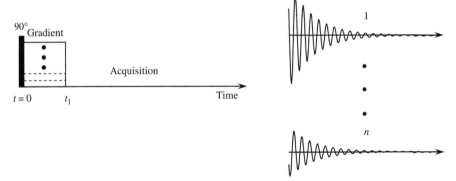

**Fig. 8.21.** Phase encoding: A series of signals is acquired with a differing level of phase encoding gradient, and the spatial information is recovered by applying the inverse Fourier transform.

depends on the spatial location. The NMR signal will of course have contributions from all spins in the sample. To be able to relate their spatial locations to the phase accumulated by these spins due to precession in the presence of the phase encoding gradient, we need to repeat the same experiment many times, but each time with a different amount of phase accumulation. This can be achieved by increasing either the duration or the amplitude of the phase-encoding gradient pulse (the latter is the preferred way). Thus, a series of signals is acquired with a differing level of phase-encoding gradient, and the spatial information is recovered by applying the inverse Fourier transform (see Fig. 8.21).

### 8.3.2. *k*-Space

It has been shown (see Section 8.2.6 and reference 24) that the equation for the NMR signal can be written as

$$s(t) \propto \omega_0 \int d^3r \cdot e^{-t/T_2} e^{-i\omega_0 t + \phi} \vec{M}_\perp(\vec{r}, 0) \cdot \vec{B}^{rec}(\vec{r}) \tag{8.32}$$

We can represent the signal by its magnitude and the time-varying phase [24]:

$$s(t) = \int d^3r \rho(\vec{r}) e^{i\phi(\vec{r}, t)} \tag{8.33}$$

where $\rho(\vec{r})$ is commonly called proton density and includes all the factors that determine the signal's amplitude (e.g., the number of protons per unit volume, relaxation times $T_1$ and $T_2$, the quality of the receiver coil, etc.). As has been shown above (see Section 8.3.1), in the presence of gradient $\vec{G}$ the Larmor frequency becomes spatially dependent [see Eq. (8.26)]. Hence, the time-varying phase of the demodulated signal (i.e., after "subtracting" $\omega_0$ from the signal through the phase-sensitive detection process) can be written as

$$\phi(\vec{r}, t) = -\int_0^t dt'\, \omega(\vec{r}, t') = -\gamma \int_0^t dt'\, \vec{G}(t') \cdot \vec{r} \tag{8.34}$$

We can then rewrite the signal equation [Eq. (8.33)] as

$$s(\vec{k}) = \int d^3r\, \rho(\vec{r}) e^{-2\pi i \vec{k} \cdot \vec{r}} \tag{8.35}$$

where vector $\vec{k}$ is defined as

$$\vec{k}(t) = \frac{\gamma}{2\pi} \int_0^t dt'\, \vec{G}(t') \tag{8.36}$$

Equation (8.35) stipulates that the MRI signal is the Fourier transform of the proton density [as defined in Eq. (8.33)]. Therefore, the proton density, or the MRI image, can be obtained by applying the inverse Fourier transform to the MRI signal. In other words, proton density and the MRI signal form a Fourier pair (see Fig. 8.22):

$$s(\vec{k}) = \int d^3r\, \rho(\vec{r}) e^{-2\pi i \vec{k} \cdot \vec{r}}$$
$$\rho(\vec{r}) = \int d^3k\, s(\vec{k}) e^{+2\pi i \vec{k} \cdot \vec{r}} \tag{8.37}$$

The two- or three-dimensional space defined by vector $\vec{k}$ is called $k$-space, and the MRI signal $s(\vec{k})$ is called the $k$-space data. MRI data acquisition can then be thought of as

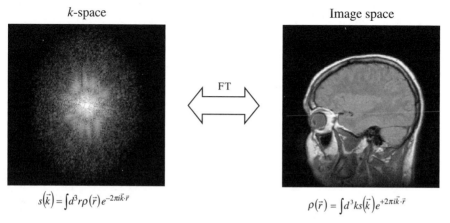

$k$-space    Image space

FT

$$s(\vec{k}) = \int d^3r \rho(\vec{r}) e^{-2\pi i \vec{k} \cdot \vec{r}} \qquad\qquad \rho(\vec{r}) = \int d^3k s(\vec{k}) e^{+2\pi i \vec{k} \cdot \vec{r}}$$

**Fig. 8.22.** MRI signal (**left**) and the proton density (MRI image, **right**) form a Fourier pair.

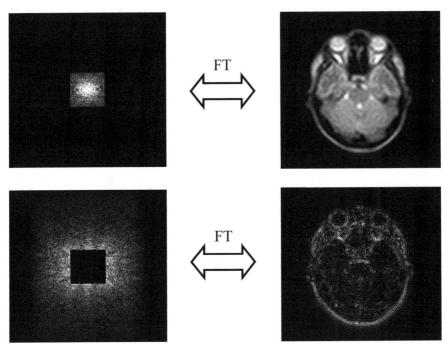

**Fig. 8.23.** The center of $k$-space contains low-frequency components of the image, representing largely the image intensity, i.e. contrast (**top**), while the outskirts of $k$-space contain high-frequency components of the image representing the fine details in the imaged object (**bottom**).

sampling $k$-space—that is, sampling the MRI signal for a wide range of $\vec{k}$ values. Since $\vec{k}$ is fully defined by a gradient $\vec{G}$, traversing $k$-space is equivalent to altering the gradient that affects the transverse magnetization. $k$-space is the Fourier conjugate to spatial image space; thus it contains information about the frequency content of the image. The center of $k$-space contains low-frequency components of the image, representing largely the image intensity (i.e., contrast), while the outskirts of $k$-space contain high-frequency components of the image representing the fine details in the imaged object (see Fig. 8.23).

The concept of $k$-space [27] greatly simplifies understanding of many aspects of MRI technology, including more complicated pulse sequences, image reconstruction and selective excitation. Examples of $k$-space coverage by various MRI techniques will be given below.

### 8.3.3. Basic MRI Techniques

Basic MRI techniques can be broadly divided into two categories: spin-echo and gradient-echo techniques. This distinction is based on the way the NMR signal is generated. In the former case, two excitation pulses, $90°$ and $180°$, are applied to generate a spin echo (as described previously in Section 8.2.6). In the latter case, only one excitation pulse is applied (with the flip angle of 90° or less) and the spins are refocused by a refocusing gradient pulse.

**8.3.3.1. Spin Echo.** The simplest version of the spin-echo MRI technique is a single-slice two-dimensional pulse sequence (see Fig. 8.24a for the pulse sequence diagram). Initially, magnetization from a single slice is flipped to the transverse plane by a slice-selective 90° pulse. The slice-selective gradient is turned on for the duration of the pulse. Since spins at different locations across the slice experience different magnetic fields due to the presence of the gradient, they will precess with different frequencies, which will result in signal dephasing at the end of the pulse[5] (see Fig. 8.25). To compensate for this spin dephasing, a gradient pulse with the opposite polarity, called a refocusing gradient, is applied. Full-signal compensation is achieved if the slice-selective and refocusing gradients are balanced[6] (see Fig. 8.25). The refocusing 180° pulse, applied to generate the spin echo, is slice-selective as well. To ensure that no spin dephasing occurs following the refocusing pulse, the slice-selective gradient needs to be symmetrical around the center of the pulse.[7]

A frequency-encoding gradient is applied during the echo acquisition to provide spatial localization in one direction within the selected slice (as described in Section 8.3.1.2). As is the case with the slice-selective gradient, the frequency-encoding gradient will also result in spin dephasing. To avoid loss of signal, an additional gradient pulse of the opposite polarity is applied prior to the frequency-encoding gradient pulse (the same effect can be achieved by applying a gradient pulse with the same polarity before the 180° refocusing pulse; see Fig. 8.26).

Finally, a phase-encoding gradient is applied prior to the echo acquisition to provide spatial localization along the second direction within the slice. The duration of the phase-encoding gradient pulse is kept constant, and its amplitude varies in equal steps while a series of echoes is acquired. Between each echo acquisition, which has a different phase-encoding gradient amplitude, a time delay (called repetition time or TR) is required to allow magnetization to recover along the direction of the $\vec{B}_0$ field.

$k$-space coverage by the spin-echo sequence is shown in Fig. 8.24b. $k_x$ and $k_y$ are frequency-encoding and phase-encoding directions, respectively. We start from the center of $k$-space. The negative frequency-encoding gradient lobe and the first (negative) phase encoding gradient pulse shifts the position to the lower left corner of $k$-space. As the spin-echo signal is acquired in the presence of the positive frequency-encoding gradient, the first (most lower) line in $k$-space is acquired. The positive frequency-encoding gradient shifts the position toward the positive $k_x$

---

[5]The dephasing process starts at the center of the 90° pulse, since at this moment magnetization reaches the transverse plane.

[6]Balancing the gradients in this case means that the area under the refocusing lobe is equal to the area under the slice-selective gradient pulse, calculated from the center of the radio-frequency pulse to the end of the slice-selective gradient pulse (see Fig. 8.25).

[7]Since the refocusing pulse reverses the direction of the spin precession (see Section 8.2.6), there is no need to reverse the gradient to avoid the signal loss.

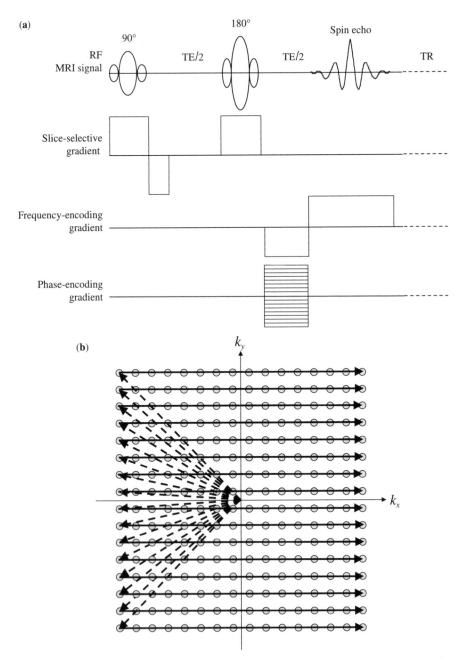

**Fig. 8.24.** (a) Diagram of a two-dimensional spin-echo MRI pulse sequence. TE is the echo time and TR is the repetition time. (b) $k$-space coverage by the pulse sequence shown in Fig. 8.24a. $k_x$ and $k_y$ are frequency-encoding and phase-encoding directions, respectively. The gray dots represent sampled points of the spin-echo signal.

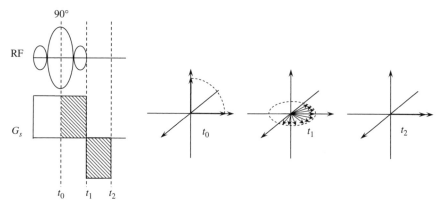

**Fig. 8.25.** Slice selection: At $t_0$, magnetization is flipped to the transverse plane. Slice-selective gradient causes signal dephasing at the end of the RF pulse ($t_1$). A negative refocusing gradient is applied to recover the signal at $t_2$. The full recovery is achieved if the two gradients are balanced, that is, the shaded areas are equal.

values; at the center of the echo, $k_x$ equals zero. After the TR delay we start again from the center of $k$-space. The negative frequency-encoding lobe and the second, less negative, phase-encoding gradient pulse shifts the position to the most negative $k_x$ and second most negative $k_y$ values. The second line in the $k$-space is then acquired. The process continues until the $n$th (most positive in $k_y$) line is acquired. The image

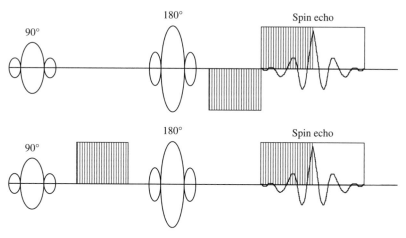

**Fig. 8.26.** The frequency-encoding gradient results in spin dephasing. To avoid loss of signal, an additional gradient pulse of the opposite polarity is applied prior to the frequency-encoding gradient pulse (**top**). The same effect can be achieved by applying a gradient pulse with the same polarity before the $180°$ refocusing pulse (**bottom**). Full signal recovery is achieved when the gradient pulses are balanced, that is, the shaded areas are equal.

is obtained by calculating the two-dimensional inverse Fourier transform of the k-space data.

The MRI signal generated in the spin-echo sequence is a function of proton density (i.e., the water content in the sample) and the relaxation times $T_1$ and $T_2$. The relative contribution of these factors to the signal amplitude, and subsequently to the MRI image contrast, can be affected by altering two parameters of the pulse sequence: the echo time TE, and the repetition time TR. To generate the proton density weighting in the image (i.e., make the water content a dominant factor affecting the image contrast), the contributions from $T_1$ and $T_2$ relaxation processes have to be minimized. This is achieved when TR $\sim 5T_1$[8] and TE $\ll T_2$. The image is $T_2$-weighted when TR $\sim 5T_1$ and TE $> T_2$, while $T_1$ weighting is realized when TR $< T_1$ and TE $\ll T_2$. All three types of contrast are routinely used in clinical applications of MRI, as will be discussed later (see Section 8.4). The fact that the same spin-echo technique can generate these different image contrasts by a simple adjustment of two pulse sequence parameters is an example of the great versatility of MRI technology.

Three-dimensional information about the sample can be obtained by acquiring images from multiple slices. As discussed in Section 8.3.1.1, slice position is determined by the frequency of the slice-selective pulse. Hence, multiple slices are selected by the application of the multiple slice-selective pulses with appropriate frequencies (see Fig. 8.27a). Since the slice-selective pulse does not affect magnetization outside the selected slice, multiple slices can be selected within the same repetition time TR; hence the total imaging time is the same for both the single-slice and multiple-slice acquisition.[9]

An alternative way of collecting three-dimensional information from the sample is to apply a second phase-encoding gradient rather than selecting multiple slices. Typically, a thick slab is selected in the slice direction and a phase-encoding gradient is then used to localize the signal across the slab (see Fig. 8.27b). The resulting k-space data matrix is three-dimensional, and therefore the image is reconstructed by applying a three-dimensional inverse Fourier transform. A disadvantage of the 3D versus multi-slice acquisition is the longer total imaging time of 3D acquisitions, because only one phase-encoding step per TR is acquired. However, 3D acquisitions provide better signal-to-noise ratio (SNR) per unit time than the multi-slice technique.[10]

***8.3.3.2. Gradient Echo.*** The gradient-echo technique is very similar to the spin-echo technique, except that no refocusing pulse is applied. Instead, the transverse magnetization generated by the slice-selective excitation pulse is refocused only by the negative lobes of the slice-selective and frequency-encoding gradients to form the so-called gradient echo (see Fig. 8.28 for the pulse sequence diagram). Unlike

---

[8]At the time equal $5T_1$ magnetization will reach approximately 99% of its equilibrium value.

[9]Typically, there is a limit on the number of slices that can be selected within one TR. If this limit is exceeded, TR value has to be increased or the selected slices are divided into smaller batches acquired in separate experiments.

[10]3D acquisition is preferred when $T_1$-weighted images are generated with very short TR (i.e., only one slice per TR can be acquired).

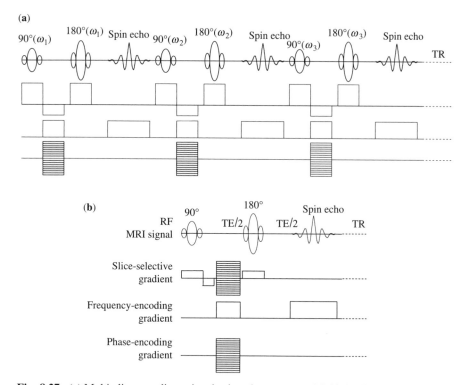

**Fig. 8.27.** (a) Multi-slice two-dimensional spin-echo sequence. Multiple slices are selected by the application of multiple slice-selective pulses with the frequencies corresponding to the slice positions. All slices are acquired within the same repetition time TR. (b) Three-dimensional spin-echo pulse sequence. A thick slab is selected in the slice direction, and a phase-encoding gradient is used to localize the signal across the slab.

with the spin-echo technique, in the gradient-echo technique spin dephasing due to local field inhomogeneities is not refocused (since no refocusing radio-frequency pulse is applied), and thus the echo amplitude will depend on the $T_2^*$ rather than $T_2$: $s \sim e^{-TE/T_2^*}$ (see Section 8.2.6).

Similarly to the spin-echo technique, three-dimensional images of the sample can be generated either with a multi-slice or a 3D version of the gradient-echo pulse sequence. As will be discussed later, the gradient-echo technique is frequently used with very short repetition times to minimize the total imaging time. Hence, the 3D version of this technique is far more common than the multi-slice version.

### 8.3.4. Signal and Noise in MRI

As discussed in Section 8.2.3, NMR, and thus MRI, is a relatively insensitive technique; thus SNR is often a factor limiting the performance of both technologies. In MRI, SNR can be defined as a ratio of the signal amplitude in a single image pixel to

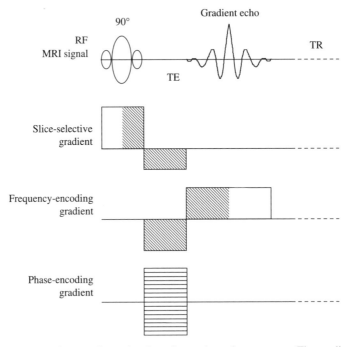

**Fig. 8.28.** Diagram of a two-dimensional gradient echo pulse sequence. The gradient echo is formed by the negative lobes of the slice-selective and frequency-encoding gradients. To avoid signal loss due to spin dephasing, slice-selective and phase-encoding gradients need to be balanced (i.e., the shaded areas have to be equal). TE is the echo time and TR is the repetition time.

the standard deviation of the noise in the image. The signal amplitude can be approximated by [28]

$$S \sim \omega_0 B_\perp M_0 V_{pixel} \tag{8.38}$$

where $\omega_0$ is the Larmor frequency, $B_\perp$ is the transverse component of the $B_1$ field produced by a unit current in the receive coil (i.e., the measure of the efficiency of the coil receiving the MRI signal), $M_0$ is the equilibrium magnetization, and $V_{pixel}$ is the volume of the image pixel.

The source of noise in MRI is random Brownian motion of the electrons in the sample, the receiver coil, and the electronic equipment used to convert the MRI signal into digital data. The root mean square of the random voltage generated by randomly moving electrons is given by the Nyquist equation:

$$N = \sqrt{4\kappa \cdot T \cdot R \cdot BW} \tag{8.39}$$

where $T$ is the absolute temperature, $R$ is the resistance, $BW$ is the bandwidth of the noise voltage detecting system (i.e., the receive system in MRI), and $\kappa$ is

Boltzmann's constant. Typically, noise from the electronic equipment (e.g., preamplifier, phase-sensitive detector, etc.) is small compared to the noise produced by the receive coil and the sample itself. Since, in the clinical setting, the noise recorded from the sample cannot be influenced, other than by lowering the bandwidth of the receive system, the only strategy to minimize the overall noise level is to ensure that the noise produced by the receive coil is smaller than the noise from the sample (i.e., the human body). It can be shown that for the Larmor frequencies and the sizes of the subjects (e.g., different parts of the body) typical for the clinical MRI scanners the noise generated by the subject will be much larger than the noise generated by the coil itself [28]. One way of reducing the subject noise is to reduce the *effective* size of the sample (i.e., the size of the sample that is "seen" by the coil). This can be done by using a smaller coil to receive the MRI signal[11] (e.g., a small circular coil placed on the surface of the sample—the so-called surface coil [29]). The drawback of this approach is that for the large regions of interest positioned far away from the surface, the $B_\perp$ field of the coil is weak and results in poor SNR. This problem is remedied by using an array of surface coils (called the phased array [30]) which covers a larger volume of interest. Signals detected by the individual coils are combined to yield a signal from the entire volume. Unlike the signals, noise detected by the individual coils is not correlated; thus the overall SNR is improved. In research applications on smaller animal MRI scanners, other strategies—involving the use of superconducting receive coils and lowering the temperature of the preamplifier by immersing it in liquid nitrogen—are being developed.

One simple way of improving SNR is averaging the signal. As a result of averaging, the signal will add coherently, while, due to its random nature, the noise will add incoherently; hence the SNR will increase. It can be shown that SNR increases with the square root of the number of averages. Signal averaging will inevitably increase the total imaging time, which cannot always be afforded (especially that increasing the SNR by a factor of $N$ would require lengthening the total acquisition time by a factor of $N^2$).

As is evident from the Nyquist equation [Eq. (8.39).], lowering the bandwidth of the receive system will lower the noise level in the acquired signal and thus increase the SNR. However, lowering the bandwidth will lengthen the acquisition of the MRI signal, which in turn will lengthen the echo time in both spin-echo and gradient-echo techniques. In addition, it may result in geometric distortions in the image and increased chemical shift artifacts.[12] Hence necessary tradeoffs have to be made. Excellent analysis of the SNR dependence on various imaging parameters can be found in reference 24.

---

[11]Reducing the effective sample size will reduce the total amount of noise received by the coil, and subsequently improve the SNR within a specific region of interest (ROI), provided that the smaller coil is equally effective in receiving the signal from the ROI.

[12]Chemical shift artifact results from the difference in the Larmor frequency between two chemical species present in the sample (e.g., water and fat). Even though both are physically located in the same place within the sample, they appear shifted in the image. Chemical shift artifacts typically occur in the frequency encoding direction.

### 8.3.5. Fast MRI Techniques

Total imaging time is a very important parameter in MRI because it has significant influence on issues such as patient comfort, motion artifacts in the images, temporal resolution of dynamic scans, and so on. In general, the total imaging time depends on three parameters: the repetition time TR, the number of phase encoding steps $N_{pe}$, and the number of averages $N_{ave}$:

$$T_{im}^{total} \sim TR \cdot N_{pe} \cdot N_{ave} \tag{8.40}$$

Changing any of these parameters will influence the resulting image: TR will affect the SNR and the image contrast, the number of phase encoding steps will affect the spatial resolution and SNR, and the number of averages will also affect the SNR. Depending on the specific requirements of a particular application, these parameters may be adjusted to minimize the total imaging time. Figure 8.29 presents a partial list of pulse sequences designed specifically to shorten the total imaging time by minimizing one of these parameters. The most common fast-imaging techniques are briefly discussed in the following sections.

***8.3.5.1. RARE Imaging.*** *Rapid acquisition with relaxation enhancement* (RARE) imaging is a fast MRI technique that minimizes the total imaging time by applying multiple phase encoding steps (i.e., acquiring multiple $k$-space lines) per TR. It was first described by Hennig et al. [31] and quickly became one of the most commonly used techniques in the clinical applications of MRI [it is also known as *fast spin echo* (FSE) or *turbo spin echo* (TSE)]. In the RARE experiment the initial $90°$ pulse is followed by a train of the $180°$ refocusing pulses, generating a train of spin echoes (see Fig. 8.30a). Each echo is encoded with a different value of the phase-encoding gradient. As a result, the number of $k$-space lines acquired during one TR is equal to the number of different phase encoding steps applied during one TR, which in turn is equal to the echo train length (ETL), which is the number of echoes acquired during one TR (see Fig. 8.30b). The total imaging time is reduced

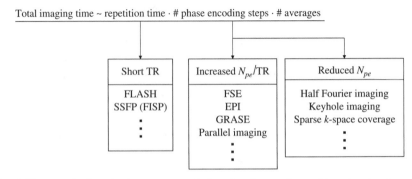

**Fig. 8.29.** A partial list of pulse sequences designed to shorten the total imaging time by minimizing repetition time TR or the number of phase-encoding steps $N_{pe}$.

**(a)**

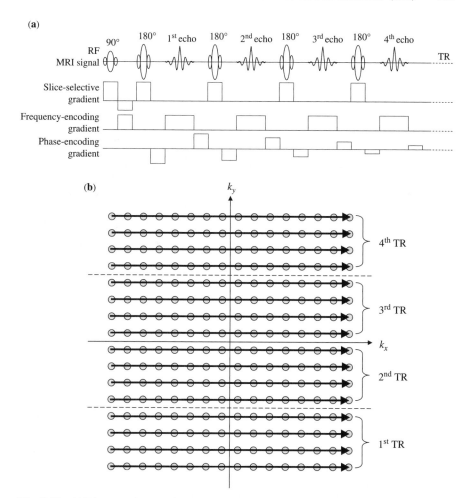

**Fig. 8.30.** (a) Diagram of a two-dimensional RARE pulse sequence. (b) $k$-space coverage by the RARE pulse sequence shown in Fig. 8.30a. Four $k$-space lines per TR are covered, which reduces the total imaging time by a factor of four. $k_x$ and $k_y$ are frequency-encoding and phase-encoding directions respectively. The grey dots represent sampled points of the spin-echo signal.

by a factor of $N_{pe}/\text{ETL}$; that is, the longer the echo train, the shorter the total imaging time. In practice, the maximum ETL is determined by the echo spacing (i.e., time distance between consecutive echoes) and the $T_2$ value of the sample.[13]

As was discussed in Section 8.3.2, the center of the $k$-space contains information about the image's contrast. Since the order of the phase-encoding gradient in the RARE acquisition can be chosen arbitrarily, the minimum value of the phase encoding gradient may be applied to any echo within the echo train (i.e., any echo can

---

[13]As discussed in Section 8.2.6, amplitude of the spin echo is dependent on $e^{-t/T_2}$; thus the echo train will decay with the time constant equal $T_2$ of the sample.

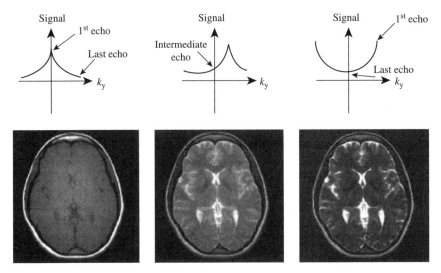

**Fig. 8.31.** Examples of three phase encoding schemes in the RARE sequence (**top**), which result in a short (**left**) intermediate (**center**) and long (**right**) effective echo time. Images acquired with such phase-encoding schemes (**bottom**) will have three different types of contrast: $T_1$-weighted (**left**), moderately $T_2$-weighted (**center**) and heavily $T_2$-weighted (**right**). (Reproduced from Bernstein et al. [25], with permission.)

correspond to the central $k$-space line). The echo time of the echo acquired with the smallest phase-encoding gradient is called the effective echo time $T_{eff}$, and the contrast in the RARE image is determined by the value of $T_{eff}$.[14] Figure 8.31 shows three different phase-encoding schemes corresponding to three different values of $T_{eff}$ along with the corresponding RARE images with three different types of contrast.

One of the problems frequently encountered in RARE imaging is image blurring in the phase-encoding direction caused by the $T_2$ decay of the echo train [25]. This decay produces a weighting function on the $k$-space data, which, after applying inverse Fourier transform, will result in the convolution of a Lorentzian function[15] with the proton density. This convolution manifests itself as blurring in the reconstructed image. RARE images may also suffer from the artificial edge enhancement and ghosting arising from the step-like behavior of the $T_2$ weighting function mentioned above. Various techniques designed to minimize these image artifacts are discussed in reference 25.

***8.3.5.2. Steady-State Magnetization Imaging.*** Steady-state magnetization imaging techniques are gradient-echo techniques that afford short total imaging time by applying short, or very short, repetition time TR (unlike with the RARE imaging, in the steady-state magnetization techniques only one $k$-space line is acquired per TR).

---

[14]Since the echo time of an echo within the echo train is a multiple of the echo spacing, $T_{eff}$ will also be a multiple of the echo spacing.
[15]The weighting function in this case is an exponential function $e^{-t/T_2}$. The inverse Fourier transform of the exponential function is a Lorentzian function, and the inverse Fourier transform of a product of two functions is a convolution of their inverse Fourier transforms.

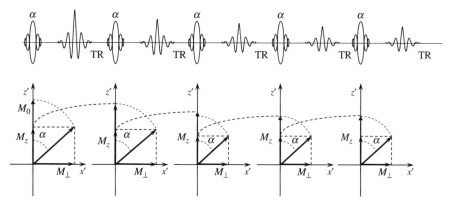

**Fig. 8.32.** A slice-selective pulse with the flip angle $\alpha$ is repeatedly applied to tip magnetization from its position along the $z$ axis. If TR $\ll T_1$ and TR $\gg T_2$, a steady state of the longitudinal magnetization is achieved; that is, the same amount of the longitudinal magnetization recovers in between each two $\alpha$ pulses, and $M_z$ has the same value at the time of each $\alpha$ pulse.

Let us consider an experiment where a slice selective pulse with a flip angle $\alpha$ is repeatedly applied to tip magnetization from its position along the $z$ axis (see Fig. 8.32). The first pulse $\alpha$ will generate the following transverse and longitudinal components of magnetization immediately following the pulse:

$$M_\perp(0) = M_0 \sin \alpha$$
$$M_z(0) = M_0 \cos \alpha \qquad (8.41)$$

where $M_0$ is the initial equilibrium magnetization. Let us assume that TR is much shorter than $T_1$ but much longer than $T_2$. The latter will ensure that the transverse magnetization decayed to zero before the second $\alpha$ pulse is applied. The former means that the longitudinal magnetization will have recovered to a value smaller than the equilibrium magnetization before the application of the second $\alpha$ pulse. Repetitive application of the $\alpha$ pulse will eventually result in a steady state of the longitudinal magnetization; that is, the same amount of the longitudinal magnetization will recover in between each two $\alpha$ pulses, and $M_z$ will have the same value at the time of each $\alpha$ pulse (see Fig. 8.32). Such longitudinal magnetization is called steady-state incoherent magnetization and is dependent on the $T_1$, TR, and $\alpha$ [24]:

$$M_{ze} = \frac{M_0(1 - e^{-TR/T_1})}{(1 - e^{-TR/T_1} \cos \alpha)} \qquad (8.42)$$

It can be shown that the maximum value of the steady-state incoherent magnetization will be reached, for specific TR and $T_1$, when the flip angle satisfies the condition

$$\cos \alpha_E = e^{-TR/T_1} \qquad (8.43)$$

where $\alpha_E$ is called the Ernst angle.

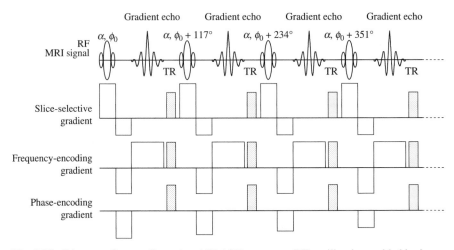

**Fig. 8.33.** Diagram of a two-dimensional FLASH sequence. RF spoiling is provided by incrementing the phase of the $\alpha$ pulses by $117°$. Gradient spoiling is achieved by the application of additional gradient pulses (*shaded*), which destroy transverse magnetization before each $\alpha$ pulse.

An imaging technique that employs the steady-state incoherent magnetization is called fast low-angle shot (FLASH) imaging and was first introduced by Haase et al. [32]. The pulse sequence diagram of the FLASH technique is shown in Fig. 8.33. As discussed above, the steady-state incoherent magnetization can be achieved when $TR \ll T_1$ and $TR \gg T_2$. While the first condition is easily achievable, the second condition is often difficult to realize in practice.[16] This can be rectified by spoiling the transverse magnetization after the signal acquisition and before the application of the next $\alpha$ pulse. Two such spoiling mechanisms are typically applied: gradient spoiling and RF spoiling. Gradient spoiling involves the application of the gradient pulses, which results in dephasing the residual magnetization before each $\alpha$ pulse (see Fig. 8.33). A more effective way of destroying the transverse magnetization is to increase the phase of each consecutive $\alpha$ pulse by $117°$ [24]. This mechanism is called the RF spoiling (see Fig. 8.33).

FLASH (known also as spoiled gradient echo (SPGR) or fast field echo (FFE)) is one of the most commonly used MRI pulse sequences. Like the spin echo sequence discussed in Section 8.3.3.1, different contrast in FLASH images can be achieved by altering pulse sequence parameters. When the flip angle is much lower than the Ernst angle $\alpha \ll \alpha_E$ and $TE \ll T_2^*$, proton density (PD) becomes the dominant source of contrast in FLASH images; when $TE \sim T_2^*$ the images are $T_2^*$-weighted, and when $\alpha \geq \alpha_E$ the images are $T_1$ weighted. Since FLASH can generate images much faster than the spin-echo technique, PD- and $T_1$-weighted images are commonly acquired using FLASH.

---

[16]The typical range of $T_2$ values in the human body varies from tens of milliseconds for most types of tissue to over 2 seconds for cerebrospinal fluid. Typical TR values for the FLASH pulse sequence are on the order of tens of milliseconds. Thus steady-state incoherent magnetization is not achieved for many types of tissue.

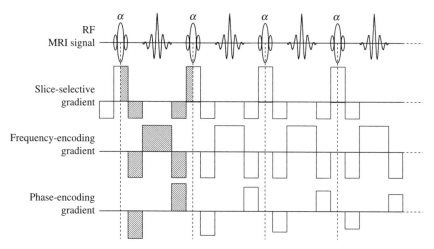

**Fig. 8.34.** Diagram of a two-dimensional balanced SSFP sequence. To achieve the SSFP state, all imaging gradients have to be fully balanced within each TR period (i.e., the area under every gradient within each TR has to be zero).

When TR $\sim T_2$ and no spoiling is applied, another type of steady-state magnetization, called the steady-state coherent magnetization, can be achieved. In this case, residual transverse magnetization, mixed with longitudinal magnetization, will also contribute to the acquired signal. To assure that steady state is achieved for both longitudinal and transverse magnetization, the total phase accumulation for the transverse magnetization has to be same within each TR period (i.e., in between each two $\alpha$ pulses). The phase of the transverse magnetization is affected by $B_0$ field inhomogeneities, as well as imaging gradients. If the phase accumulation due to imaging gradients is zero, the transverse magnetization is in the so-called steady-state free precession[17] (SSFP) equilibrium [24]. To achieve the SSFP state, all imaging gradients have to be fully balanced within each TR period (i.e., the area under every gradient within each TR has to be zero—see Fig. 8.34).

The imaging technique that employs steady-state coherent magnetization is called balanced SSFP; other names used for this technique include true fast imaging with steady-state precession (True FISP), balanced fast field echo (balanced FFE), and fast imaging employing steady-state acquisition (FIESTA). The signal in the balanced SSFP sequence is a complicated function of $T_1$, $T_2$, and the flip angle $\alpha$. It can be shown [24] that for the optimal flip angle the transverse magnetization is

$$M_\perp \cong \frac{1}{2} M_0 \sqrt{\frac{T_2}{T_1}} \tag{8.44}$$

---

[17]Some authors use the term balanced SSFP for this type of steady state equilibrium, while reserve the SSFP term for the weaker condition of the phase accumulation being the same, but not necessarily zero, within each TR period [25].

It is interesting to note that the signal in Eq. (8.44) does not depend on TR; thus it enables extremely short TR values to be used without the penalty of the low SNR (typical TR values used in the balanced SSFP are less than 10 ms, which results in the total imaging time of 2 seconds or less). The balanced SSFP technique provides the most signal per total imaging time of all known MRI techniques [33], which makes it a very popular method whenever SNR and/or the total acquisition time are of concern.

***8.3.5.3. Echo Planar Imaging.*** Echo planar imaging (EPI) is a fast MRI technique that minimizes the total imaging time by acquiring multiple $k$-space lines per TR. It was originally introduced by Mansfield et al. [34] in 1976; however, it was not widely used until the late 1990s when very fast, actively shielded gradient coils became commercially available. EPI is considered one of the fastest MRI techniques available, capable of generating an image in as little as 60 ms.

EPI is a multi-echo gradient echo sequence, with each echo being encoded with a different value of the phase-encoding gradient. A 90° slice-selective pulse tips the equilibrium magnetization into the transverse plane and multiple gradient echoes are generated by a series of frequency encoding gradient pulses with alternating polarity (see Fig. 8.35a). Phase encoding is achieved by a series of short gradient pulses applied just before the acquisition of consecutive echoes. $k$-space is covered in a zigzag pattern (see Fig. 8.35b), and with a sufficient number of echoes in the echo train the entire $k$-space can be covered (i.e. the entire image generated) in a single acquisition.

The echo amplitudes in the EPI echo train will decay with the time constant $T_2^*$. To provide adequate SNR across the entire $k$-space, the signal sampling rate, or the acquisition bandwidth, needs to be very high (typical sampling rates are $5-10\,\mu s$ per data point, or bandwidths of $100-200\,kHz$). Such high-bandwidth places very stringent requirements on the strength and the speed of the gradient coils producing the frequency and phase-encoding gradients.[18] To speed up the acquisition process, the data are commonly acquired continuously throughout the echo train. Since the frequency-encoding gradient is not switched instantaneously (i.e., each gradient pulse has a ramp of a finite duration), signal sampling that is uniform in time results in nonuniform sampling of the $k$-space, which leads to artifacts in the reconstructed images. This problem can be remedied by varying the signal sampling or applying re-gridding algorithms prior to the image reconstruction.

The exact positioning of the echoes in the EPI echo train depends on proper balancing of frequency-encoding gradient pulses. However, the echo position is determined by refocusing all the gradients affecting the spins; thus any residual gradient (e.g., generated by the local field inhomogeneity, eddy currents, or other sources) will potentially shift the echo position (see Fig. 8.36a). Since echoes are formed by alternating the polarity of the frequency-encoding gradient, the residual gradient will shift the

---

[18]This is the main reason why EPI was not routinely in use until over 20 years after it was originally introduced.

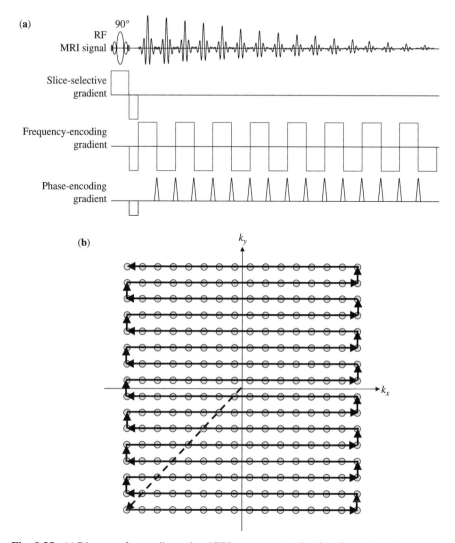

**Fig. 8.35.** (a) Diagram of a two-dimensional EPI sequence. A train of gradient echoes is generated by changing the polarity of successive frequency encoding gradient pulses. Phase encoding is achieved by a series of short gradient pulses applied just before the acquisition of consecutive echoes. Entire $k$-space can be covered in a single TR. (b) $k$-space coverage by the EPI sequence shown in Fig. 8.35a. Alternating $k$-space lines are sampled in reverse directions due to the alternating sign of the frequency-encoding gradient. Entire $k$-space is sampled in a single TR.

even and odd echoes in opposite directions. In addition, the error in the echo position caused by the residual gradient accumulates throughout the echo train, resulting in the increasingly larger shifts of the later echoes (see Fig. 8.36b). These echo shifts cause an oscillating phase error in $k$-space data, which results in a ghosting artifact (called Nyquist or $N/2$ ghost), characteristic for the EPI images. $N/2$ ghost can be minimized

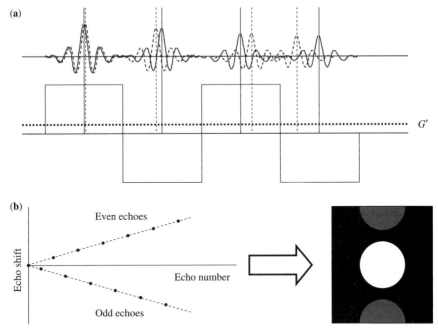

**Fig. 8.36.** (a) The echo position in EPI is determined by refocusing all the gradients affecting the spins. The residual gradient $G'$, generated by the local field inhomogeneity, eddy currents, or other sources, shifts the echo position. Since echoes are formed by alternating the polarity of the frequency encoding gradient, even and odd echoes are shifted in opposite directions. Solid lines represent the properly positioned echoes, and dashed lines represent the shifted echoes. (b) Error in the echo position caused by the residual gradient accumulates throughout the echo train and results in the increasingly larger shifts of the later echoes (**left**). These echo shifts cause an oscillating phase error in $k$-space data that results in a ghosting artifact (called Nyquist or $N/2$ ghost), characteristic for the EPI images (**right**).

by applying frequency-dependent phase correction to the $k$-space data prior to the image reconstruction.[19]

EPI, like many other gradient echo sequences, is particularly sensitive to magnetic susceptibility variations and local field inhomogeneities that produce various artifacts in EPI images. Many of these artifacts can be minimized by using a spin-echo version of the EPI technique (see Fig. 8.37). In the spin-echo EPI both 90° and 180° slice-selective pulses are applied to generate the spin-echo signal. Similar to the gradient-echo EPI, a series of the frequency-encoding gradient pulses with alternating polarity generate a series of gradient echoes; however, due to the application of the refocusing pulse, the echo amplitudes will depend on $T_2$ rather then $T_2^*$. If the echo time of the spin echo coincides with the effective echo time of the EPI sequence (i.e., the echo time of the gradient echo within the echo train that corresponds to the center of the

---

[19]The amount of phase correction can be estimated from the echo shifts in the data.

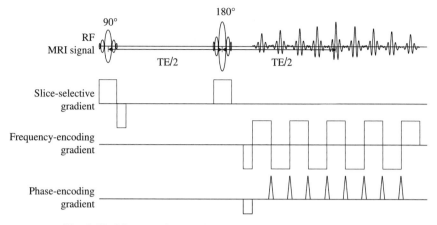

**Fig. 8.37.** Diagram of a two-dimensional spin-echo EPI sequence.

$k$-space), the sensitivity of the pulse sequence to magnetic susceptibility variations and local field inhomogeneities is significantly reduced and the resulting image is largely $T_2$-weighted [25].

EPI allows all of $k$-space to be covered in a single acquisition. However, due to $T_2^*$ decay of the echo train, the number of $k$-space lines with sufficient SNR is often limited, which compromises the spatial resolution of the reconstructed image. One solution to this problem is to cover $k$-space using multiple EPI acquisitions, with each acquisition covering only part of $k$-space (see Fig. 8.38). This so-called multi-shot EPI technique can produce higher resolution and higher quality (i.e., better SNR, less geometric distortions, reduced blurring, and reduced ghosting) images than single-shot EPI, albeit at the expense of longer total imaging time. Multi-shot EPI can employ either sequential or interleaved acquisition (see Fig. 8.38b). The latter is often preferred, because the interleaved acquisition scheme reduces phase errors and amplitude modulations reducing ghosting artifacts in the image [25]. One drawback of multi-shot EPI, apart from increased total imaging time, is its sensitivity to motion. Any movement of the subject between the acquisition of the individual shots will result in sharp phase discontinuities in the $k$-space, resulting in multiple ghosts in the image. This can be remedied by using navigator echoes [35]. In the simplest case, the navigator echo is an additional echo included in the echo train, which is not phase-encoded. Hence, the navigator echo can be used to measure the motion-induced phase, which can subsequently be removed from the EPI data acquired with the individual shots, minimizing the phase discontinuities in the $k$-space [36].

***8.3.5.4. Other Fast Imaging Techniques.*** Another way of reducing the total imaging time is to reduce the number of phase-encoding steps $N_{pe}$ (i.e., the number of $k$-space lines acquired in the phase-encoding direction). Undersampling $k$-space typically results in either lower spatial resolution or aliasing artifacts in the reconstructed images. A number of image reconstruction techniques have recently been developed to deal with these problems.

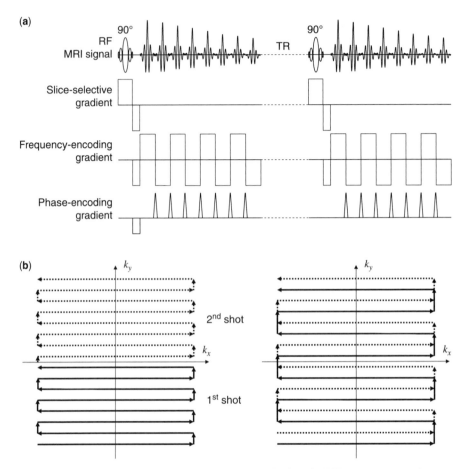

**Fig. 8.38.** (a) Diagram of a multi-shot two-dimensional spin-echo EPI sequence (two shots are shown). (b) $k$-space coverage by a multi-shot EPI sequence. Multi-shot EPI can employ either sequential (**left**) or interleaved (**right**) acquisition. The solid line represents the first shot and the dotted line represents the second shot.

Parallel imaging techniques combine the use of phased array coils (see Section 8.3.4) to record the MRI signal with undersampling of the $k$-space. The distance between the $k$-space lines in the phase encoding direction is increased by a factor $R$ (called the reduction or acceleration factor) effectively reducing the field of view (FOV) while keeping the spatial resolution in the image unchanged.[20] This normally results in signal aliasing, since the Nyquist criterion is not fulfilled. In parallel imaging, the $B_1$ sensitivity profiles of the individual coils of the phased array are used to either prevent or correct for the signal aliasing. One approach is employed by a

---

[20]The FOV is inversely proportional to the distance between consecutive lines in $k$-space, while the image spatial resolution is inversely proportional to the extent of the $k$-space coverage (i.e., the distance between the lines corresponding to the largest and the smallest $k$ value).

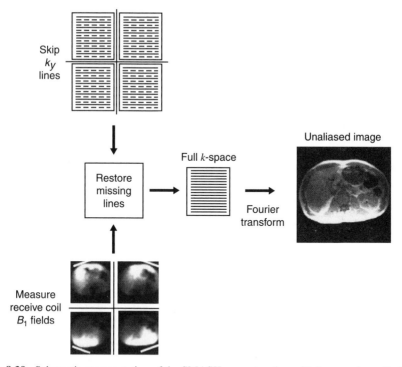

**Fig. 8.39.** Schematic representation of the SMASH reconstruction with four receive coils. The dashed lines represent the skipped $k$-space lines. (Reproduced from Bernstein et al. [25], with permission.)

technique called simultaneous acquisition of spatial harmonics (SMASH) [37]. In SMASH, combinations of the $B_1$ sensitivity profiles are used to synthesize the missing $k$-space lines and restore the full $k$-space coverage prior to the image reconstruction through the inverse Fourier transform (see Fig. 8.39). The technique called sensitivity encoding (SENSE) [38] employs the second approach; that is, the aliased images are reconstructed from the undersampled $k$-space, and the $B_1$ sensitivity profiles are then used to unwrap the aliasing (see Fig. 8.40). Because of the shorter acquisition time, both techniques suffer from reduced SNR; however, in many clinical applications, when the SNR is sufficient, these techniques are routinely used to reduce the total imaging time.

Partial Fourier reconstruction is another way of generating images from the undersampled $k$-space. Unlike with fully sampled $k$-space (or the parallel imaging techniques discussed above), in partial Fourier acquisition $k$-space is covered asymmetrically, with only one-half of $k$-space and a small portion of the other half fully sampled[21] (see Fig. 8.41). Zero-filling is a simple technique of restoring full

---

[21] In partial Fourier acquisition, $k$-space can be undersampled in either the phase or the frequency encoding direction. The latter is called the partial echo data and is typically acquired in a gradient-echo experiment to minimize the echo time; that is, the echo is shifted toward the excitation pulse and thus is not centered in the acquisition window.

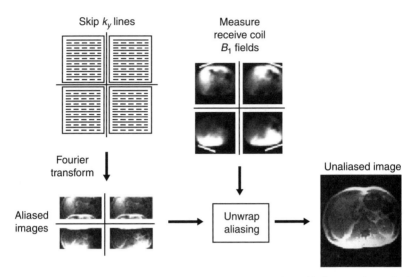

**Fig. 8.40.** Schematic representation of the SENSE reconstruction with four receive coils. The dashed lines represent the skipped $k$-space lines. (Reproduced from Bernstein et al. [25], with permission.)

$k$-space coverage by replacing missing data points with zeros. It is equivalent to the image interpolation (i.e., the missing lines in the image are equal to the average of the neighboring lines) and may reduce partial volume artifacts, if enough $k$-space coverage is available [25]. Homodyne processing is a more involved partial Fourier reconstruction technique, which takes advantage of the symmetries of $k$-space when the reconstructed image is real.[22] In homodyne processing, missing $k$-space data points are replaced by the complex conjugate of the measured data points symmetrically positioned in the $k$-space; that is, $s(k) = \mathrm{Re}(s(k) + \mathrm{Im}(s(k))$ is replaced by $s^*(-k) = \mathrm{Re}(s(-k) - \mathrm{Im}(s(-k))$. In addition, phase errors resulting from the discontinuities of the homodyne filter are corrected by estimating the phase of the missing high-frequency $k$-space lines using the phase of the measured low-frequency $k$-space lines.

*Keyhole* imaging [39] is a technique that is frequently used to accelerate data acquisition when imaging a dynamic process (e.g., passage of a contrast agent through tissue). It relies on the fact that the image contrast is determined by the central region of the $k$-space. Thus to monitor changes in the image contrast in a dynamic process, a full-resolution reference image is acquired prior to dynamic scans, followed by the repetitive acquisition of a small number of phase encoding steps covering only the central part of the $k$-space. Images are reconstructed by combining the high-frequency $k$-space lines from the reference data with the low-frequency $k$-space lines acquired during the dynamic scans. The image reconstruction methods involve direct substitution, weighted

---

[22]The Fourier transform of a real function is a Hermitian function (i.e., $f(x) = f^*(-x)$). Since MRI images represent physical entities (i.e., proton density), they can be described by real functions, and thus $k$-space data can be described by a Hermitian function.

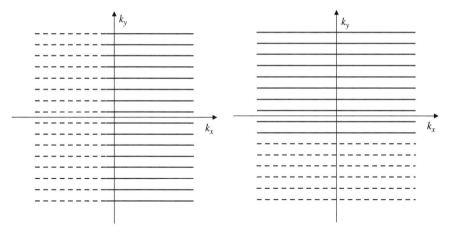

**Fig. 8.41.** $k$-space coverage with partial Fourier acquisition. $k$-space is covered asymmetrically, with only one-half of $k$-space and a small portion of the other half fully sampled either in the frequency encoding (**left**) or the phase encoding (**right**) direction. The dashed lines represent missing $k$-space data.

substitution, or more complicated techniques (e.g., generalized series method [40]). Similar techniques based on the *keyhole* imaging idea include BRISK [41], designed specifically for cardiac imaging, and TRICKS [42], used for 3D contrast-enhanced scans.

Recently, even larger undersampling of $k$-space has been proposed to further reduce the total imaging time using a technique called sparse MRI or compressed SENSE [43]. Sparse MRI reduces $k$-space sampling by as much as a factor of 8. To avoid aliasing, sampling is carried out in a pseudorandom fashion (see Fig. 8.42). Unlike with regular undersampling, which results in the coherent aliasing that cannot be easily recovered, pseudorandom undersampling results in incoherent aliasing, where strong signals can be distinguished from the interference and detected by thresholding. The interference of strong signal components is subsequently calculated and subtracted allowing the weaker signals to be recovered (see Fig. 8.42). Random $k$-space sampling has recently become a subject of major research effort in MRI; and undoubtedly new, even faster MRI techniques will be soon developed.

### 8.3.6. Magnetic Resonance Spectroscopy (MRS)

As discussed in Section 8.2.7, molecules present in a sample can be identified in an NMR spectrum based on their frequency position—the chemical shift (i.e., the shift in the Larmor frequency due to variations in the local magnetic fields produced by the electrons surrounding the nuclei). Although NMR spectroscopy is routinely applied to study the molecular content of various chemicals, the same technique can be used to identify a range of metabolites in human (or animal) bodies in vivo by acquiring spectra of a wide range of NMR-active nuclei (see Table 8.4 for a partial list of nuclei of interest in biomedical applications and the biochemical parameters

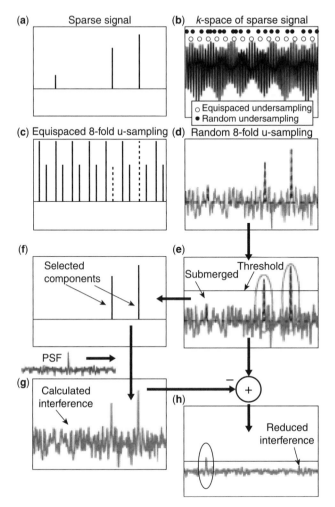

**Fig. 8.42.** An intuitive reconstruction of a sparse signal from pseudorandom $k$-space undersampling. A sparse signal (a) is 8-fold undersampled in $k$-space (b). Equispaced undersampling results in coherent signal aliasing (c) that cannot be recovered. Pseudorandom undersampling results in incoherent aliasing (d). Strong signal components stick above the interference, are detected (e) and recovered (f) by thresholding. The interference of these components is computed (g) and subtracted (h), lowering the total interference level and enabling recovery of weaker components. (Reproduced from Lustig et al., [43], with permission.)

they measure). Figure 8.43 shows an example of proton ($^1$H) and phosphorus ($^{31}$P) spectra acquired from a rat brain in vivo—a number of metabolites present in the brain are identified in each spectrum based on the chemical shift values. The relative concentrations of the metabolites are estimated from the areas under the spectral lines. Measurements of the absolute molar concentrations of the metabolites require a concentration reference—that is, a substance with a known concentration producing a

**TABLE 8.4. Partial List of NMR-Active Nuclei and Biochemical Parameters They Measure**

| | |
|---|---|
| $^{31}P$ | ATP, inorganic phosphate, phosphocreatine, phosphomonoesters, intra- and extracellular pH |
| $^{1}H$ | Lactate, lipids, N-acetyl aspartate, glutamine/glutamate, GABA, citrate |
| $^{13}C$ | Metabolic products of $^{13}C$-labeled compounds |
| $^{23}Na$ | Intra- and extracellular $Na^+$ levels |
| $^{39}K$ | Intra- and extracellular $K^+$ levels, potassium equilibrium potential |
| $^{87}Rb$ | $Rb^+$ influx and efflux (congener for $K^+$) |
| $^{7}Li$ | $Li^+$ influx and efflux (congener for $Na^+$) |
| $^{19}F$ | Metabolic products of $^{19}F$-labeled drugs, intracellular free $Ca^{2+}$ levels |
| $^{14}N$, $^{15}N$ | Ammonium ion, amino acids |
| $^{35}Cl$ | Extracellular $Cl^-$ levels, extracellular volumes |
| $^{2}H$ | Water flow, metabolic products of $^{2}H$-labeled compounds |
| $^{3}H$ | Metabolic products of $^{3}H$-labeled compounds |
| $^{17}O$ | Water flow |

strong NMR signal that can be easily identified in the spectrum. Both internal and external concentration references are commonly used, with each approach posing specific technical challenges [44, 45]. Due to relatively low NMR sensitivity (see Section 8.2.3), typically concentrations in excess of 1 mM are needed for the metabolites to be detected in the in vivo applications.

The utility of magnetic resonance spectroscopy (MRS) in providing biochemical information in vivo relies on the ability to localize the MRS signal within a specific region in the body. MRS localization techniques can be divided into two categories: single-voxel and multi-voxel techniques. The most common approach to select a three-dimensional volume (i.e., voxel) in MRS is the application of three slice-selective pulses, selecting three slices in mutually orthogonal directions—hence the selected

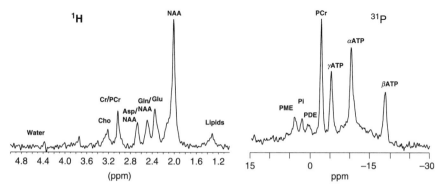

**Fig. 8.43.** Proton (**left**) and phosphorus (**right**) spectra acquired from a rat brain in vivo at 9.4 T. The voxel size for the proton spectrum is $2.5 \times 2.5 \times 3$ mm$^3$ and for the phosphorus spectrum $5 \times 5 \times 3$ mm$^3$. Spectra are extracted from the two-dimensional spectroscopic imaging data sets.

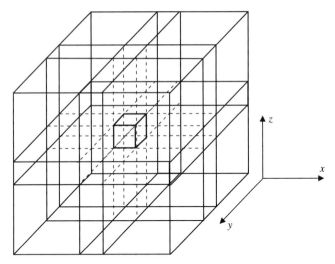

**Fig. 8.44.** The most common approach to select a three-dimensional volume (i.e., voxel) in MRS is the application of three slice-selective pulses, which select three mutually orthogonal slices. The selected voxel is the intersection of the selected slices.

voxel is the intersection of the selected slices (see Fig. 8.44). In a single-voxel technique called stimulated echo acquisition mode (STEAM) [47], three 90° slice selective pulses are used to generate the so called stimulated echo[23] from a voxel (see Fig. 8.45a). The second half of the stimulated echo is acquired without the presence of any gradient to preserve the spectral information. The signal is then Fourier transformed to generate the spectrum from the localized volume. The advantage of the STEAM method is that a short echo time can be used to minimize the signal loss to $T_2$ relaxation.[23] The drawback is, however, that only half of the signal is recovered. Another single-voxel technique called point resolved spectroscopy (PRESS) [46] is more commonly used to acquire localized MR spectra in vivo. PRESS is a double-echo spin-echo technique, where the slice-selective excitation and both refocusing pulses are used to localize the volume of interest (see Fig. 8.45b).

Many metabolites detected with the STEAM or PRESS technique exhibit multiple spectral lines with fine structures resulting from $J$-coupling (see Section 8.2.7). Thus the timing of both STEAM and PRESS sequences is typically adjusted to maximize the signal from these metabolites. In general, the shortest possible echo time allows detection of most metabolites by minimizing both $T_2$ and $J$-coupling effects. However, other timing schemes (or other pulse sequences) are often designed to specifically detect certain metabolites. This is especially the case in $^1$H spectroscopy, where spectral lines from multiple metabolites often overlap. Optimizing the pulse

---

[23]The first pulse flips the magnetization into the transverse plane, whereas the second pulse partly returns it to the position along the $B_0$ direction. The magnetization is then stored during the so-called mixing time, when it is affected by the $T_1$, rather than $T_2$ relaxation. The third pulse then returns the magnetization to the transverse plane and the stimulated echo is formed. The amplitude of the stimulated echo is half of the amplitude of the spin echo for the same echo time [52].

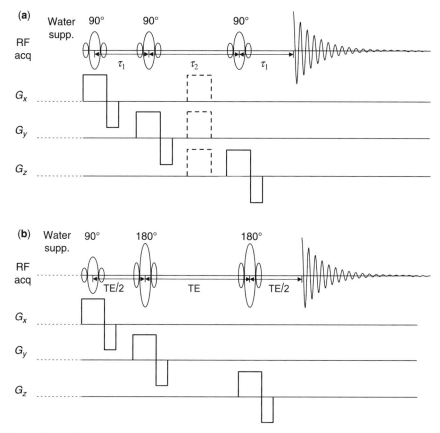

**Fig. 8.45.** (a) Stimulated echo acquisition mode (STEAM). Three 90° slice-selective pulses are used to generate stimulated echo from a voxel. To preserve the spectral information, the second half of the stimulated echo is acquired without the presence of any gradient. (b) Point resolved spectroscopy (PRESS) is a double-echo spin-echo technique, where the slice-selective excitation and both refocusing pulses are used to localize the volume of interest. To preserve the spectral information, the second half of the stimulated echo is acquired without the presence of any gradient.

sequence for the detection of a specific metabolite is called spectral editing; examples of metabolites that are often detected through spectral editing techniques include lactate [48], GABA [49], glutamine/glutamate [50], and citrate [51].

Multi-voxel MRS techniques apply phase-encoding gradients to localize the spectral information in two or three spatial directions. These can be applied either as part of the PRESS or STEAM sequence or in a spin-echo sequence (see Fig. 8.46). Multi-voxel techniques are called two-dimensional or three-dimensional spectroscopic imaging (SI) techniques (often also called, somewhat incorrectly, chemical shift imaging (CSI) techniques). As with the single-voxel techniques, no frequency-encoding gradient is applied to preserve the spectral information. 2D (or 3D) SI

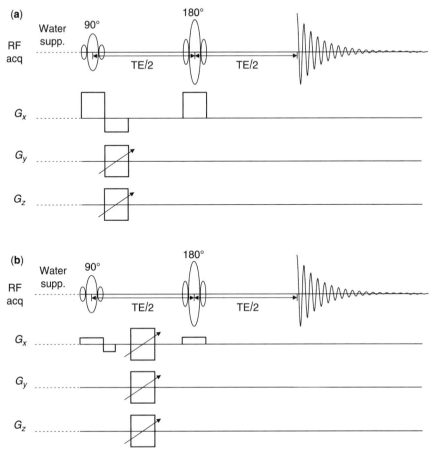

**Fig. 8.46.** (a) Two-dimensional spin-echo spectroscopic imaging (2D SI). Spatial localization is achieved by slice selection and two phase-encoding gradients. 2D SI generates three-dimensional data with two spatial one spectral dimensions. (b) Three-dimensional spin-echo spectroscopic imaging (3D SI). Spatial localization is achieved by three phase-encoding gradients. 3D SI generates four-dimensional data with three spatial one spectral dimensions.

experiments generate three (or four)-dimensional data sets with two (or three) spatial dimensions and one time, or spectral dimension. Reconstruction thus involves 3D (or 4D) Fourier transform and generates a 2D (or 3D) matrix of spectra from individual voxels. Metabolic maps representing spatial distributions of specific spectral lines corresponding to different metabolites (also called chemical shift images) can then be generated from the reconstructed data (see Fig. 8.47). The advantage of the multi-voxel approach over the single voxel techniques is that spectral information regarding the entire volume (or a large region of it) can be acquired in a single experiment. However, the total acquisition time of the SI data is much longer than that of a single-voxel spectrum (typically over 20 minutes for the former as opposed to 5 minutes for the latter). Spectral quality critically depends on the homogeneity of the $B_0$ field at

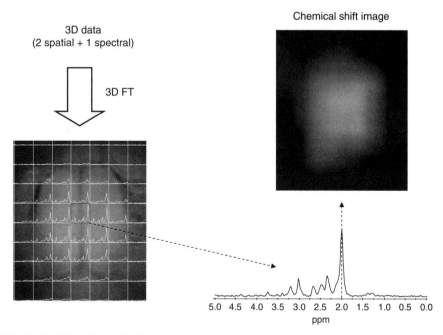

Fig. 8.47. Three-dimensional data acquired in a 2D SI experiment are processed by the three-dimensional Fourier transform. Spectra from multiple voxels are overlaid on a pilot image (**bottom left**). A metabolic map representing spatial distribution of a specific spectral line (also called chemical shift image: **top right**) is generated from the reconstructed data. Data acquired from a rat brain in vivo at 9.4 T.

the location from which the spectra are collected. Since it is easier to improve the $B_0$ homogeneity through shimming across a small voxel, rather than a large volume, single voxel spectra in general have superior quality compared to the multi-voxel ones.

$^1$H MRS is complicated by the ubiquitous presence of water in the tissue. Typical concentrations of the metabolites detected with MRS are within the range of 1–10 mM, while

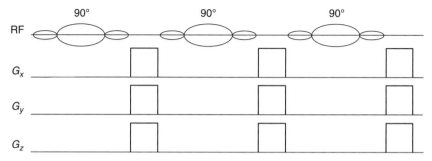

Fig. 8.48. Chemical shift selective saturation (CHESS)—the most often used water suppression technique. Long frequency-selective pulses excite only a narrow frequency range centered on the Larmor frequency of water protons. Thus only the water magnetization is flipped into the transverse plane and is subsequently destroyed by the gradient pulses.

the water concentration is estimated at $\sim$55 M. Hence, a strong water signal may obscure the weak metabolites' signals in the MR spectrum. Commonly, this problem is rectified by selectively suppressing the water signal without affecting the metabolites' signals. The most often used water suppression technique, called chemical shift selective saturation (CHESS), is based on the application of long-frequency selective pulses exciting only a narrow frequency range centered on the Larmor frequency of water protons. Thus only the water magnetization is flipped into the transverse plane and is subsequently destroyed by the gradient pulses (see Fig. 8.48). Recent progress in the detection systems of the modern MRI scanners has improved the dynamic range such that sufficient quality spectra can be acquired without suppressing the water signal. Unsuppressed water resonance can then be used as an internal concentration reference allowing absolute quantitative measurements of the metabolites' concentrations [53].

MRS is predominantly a research tool, but advances in hardware development and pulse sequence designs by the commercial MRI scanner manufacturers have led to increased interest in using MRS techniques in clinical applications in the brain [53], prostate [54], heart [55], and other organs [45, 56, 57].

## REFERENCES

1. D. S. Williams, J. A. Detre, J. S. Leigh, and A. P. Koretsky, Magnetic resonance imaging of perfusion using spin inversion of arterial water, *Proc. Natl. Acad. Sci. USA* **89**, 212–216, 1992.

2. P. Gibbs, G. P. Liney, M. Lowry, P. J. Kneeshaw, and L. W. Turnbull, Differentiation of benign and malignant sub-1 cm breast lesions using dynamic contrast enhanced MRI, *Breast* **13**, 115–121, 2004.

3. P. Kozlowski, S. D. Chang, E. C. Jones, K. W. Berean, H. Chen, and S. L. Goldenberg, Combined diffusion-weighted and dynamic contrast-enhanced MRI for prostate cancer diagnosis—correlation with biopsy and histopathology, *J. Magn. Reson. Imaging* **24**, 108–113, 2006.

4. F. Montemurro, F. Russo, L. Martincich, S. Cirillo, M. Gatti, M. Aglietta, and D. Regge, Dynamic contrast enhanced magnetic resonance imaging in monitoring bone metastases in breast cancer patients receiving bisphosphonates and endocrine therapy, *Acta Radiol.* **45**, 71–74, 2004.

5. A. Tekes, I. Kamel, K. Imam, G. Szarf, M. Schoenberg, K. Nasir, R. Thompson, and D. Bluemke, Dynamic MRI of bladder cancer: Evaluation of staging accuracy, *AJR Am. J. Roentgenol.* **184**, 121–127, 2005.

6. D. Le Bihan, E. Breton, D. Lallemand, P. Grenier, E. Cabanis, and M. Laval-Jeantet, MR imaging of intravoxel incoherent motions: Application to diffusion and perfusion in neurologic disorders, *Radiology* **161**, 401–407, 1986.

7. D. Le Bihan, Molecular diffusion nuclear magnetic resonance imaging, *Magn. Reson. Quaterly* **7**, 1–30, 1991.

8. W. Hacke and S. Warach, Diffusion-weighted MRI as an evolving standard of care in acute stroke, *Neurology* **54**, 1548–1549, 2000.

9. K. Yanaka, S. Shirai, H. Kimura, T. Kamezaki, A. Matsumura, and T. Nose, Clinical application of diffusion-weighted magnetic resonance imaging to intracranial disorders, *Neurol. Med. Chir. (Tokyo)* **35**, 648–654, 1995.

10. T. Namimoto, Y. Yamashita, S. Sumi, Y. Tang, and M. Takahashi, Focal liver masses: Characterization with diffusion-weighted echo-planar MR imaging, *Radiology* **204**, 739–744, 1997.

11. Y. Yamashita, T. Namimoto, K. Mitsuzaki, J. Urata, T. Tsuchigame, M. Takahashi, and M. Ogawa, Mucin-producing tumour of the pancreas: Diagnostic value of diffusion-weighted echo-planar MR imaging, *Radiology* **208**, 605–609, 1998.

12. S. Ogawa, T. M. Lee, A. R. Kay, and D. W. Tank, Brain magnetic resonance imaging with contrast dependent on blood oxygenation, *Proc. Natl. Acad. Sci. USA* **87**, 9868–9872, 1990.

13. J. W. Bulte, T. Douglas, B. Witwer, S. C. Zhang, E. Strable, B. K. Lewis, H. Zywicke, B. Miller, P. van Gelderen, B. M. Moskowitz, I. D. Duncan, and J. A. Frank, Magnetodendrimers allow endosomal magnetic labeling and in vivo tracking of stem cells, *Nat. Biotechnol.* **19**, 1141–1147, 2003.

14. P. M. Winter, S. D. Caruthers, S. A. Wickline, and G. M. Lanza, Molecular imaging by MRI, *Curr. Cardiol. Rep.* **8**, 65–69, 2006.

15. K. Golman and J. S. Petersson, Metabolic imaging and other applications of hyperpolarized 13C1, *Acad Radiol* **13**, 932–942, 2006.

16. S. R. Hopkins, D. L. Levin, K. Emami, S. Kadlecek, J. Yu, M. Ishii, and R. R. Rizi, Advances in magnetic resonance imaging of lung physiology, *J. Appl. Physiol.* **102**, 1244–1254, 2007.

17. N. Raz, K. M. Rodrigue, and E. M. Haacke, Brain aging and its modifiers: insights from in vivo neuromorphometry and susceptibility weighted imaging, *Ann. N.Y. Acad. Sci.* **1097**, 84–93, 2007.

18. T. Schulz, S. Puccini, J. P. Schneider, and T. Kahn, Interventional and intraoperative MR: Review and update of techniques and clinical experience, *Eur. Radiol.* **14**, 2212–2227, 2004.

19. J. H. Nelson, *Nuclear Magnetic Resonance Spectroscopy*, Pearson Education, Inc., Upper Saddle River, NJ, 2003.

20. A. Abragam, *The Principles of Nuclear Magnetism*, Oxford University Press, Oxford, 1983.

21. R. R. Ernst, G. Bodenhausen, and A. Wokaun, *Principles of Magnetic Resonance in One and Two Dimensions*, Clarendon Press, Oxford, 1987.

22. C. P. Slichter, *Principles of Magnetic Resonance*, Springer-Verlag, Berlin, Heidelberg, 1990.

23. H. Kauczor, R. Surkau, and T. Roberts, MRI using hyperpolarized noble gases, *Eur. Radiol.* **8**, 820–827, 1998.

24. E. M. Haacke, R. W. Brown, M. R. Thompson, and R. Venkatesan, *Magnetic Resonance Imaging. Physical Principles and Sequence Design*, John Wiley & Sons, New York, 1999.

25. M. A. Bernstein, K. F. King, and X. J. Zhou, *Handbook of MRI Pulse Sequences*, Elsevier, New York, 2004.

26. P. C. Lauterbur, Image formation by induced local interactions: Examples employing nuclear magnetic resonance, *Nature* **242**, 190–191, 1973.

27. S. Ljunggren, A simple graphical representation of Fourier-based imaging methods, *J. Magn. Reson.* **54**, 338–343, 1983.

28. C. N. Chen and D. I. Hoult, *Biomedical Magnetic Resonance Technology*, 1989.

29. J. J. Ackerman, T. H. Grove, G. G. Wong, D. G. Gadian, and G. K. Radda, Mapping of metabolites in whole animals by 31P NMR using surface coils, *Nature* **283**, 167–170, 1980.

30. P. B. Roemer, W. A. Edelstein, C. E. Hayes, S. P. Souza, and O. M. Mueller, The NMR phased array, *Magn. Reson. Med.* **16**, 192–225, 1990.

31. J. Hennig, A. Nauerth, and H. Friedburg, RARE imaging: A fast imaging method for clinical MR, *Magn. Reson. Med.* **3**, 823–833, 1986.

32. A. Haase, J. Frahm, D. Matthei, W. Hannicke, and K. D. Merboldt, FLASH imaging: Rapid NMR imaging using low flip-angle pulses, *J. Magn. Reson.* **67**, 258–266. 1986.

33. K. Scheffler and S. Lehnhardt, Principles and applications of balanced SSFP techniques, *Eur. Radiol.* **13**, 2409–2418, 2003.

34. P. Mansfield, A. A. Maudsley, and T. Baines, Fast scan proton density imaging by NMR, *J. Phys. E: Sci. Instrum.* **9**, 271–278, 1976.

35. R. L. Ehman and J. P. Felmlee, Adaptive technique for high-definition MR imaging of moving structures, *Radiology* **173**, 255–263, 1989.

36. R. J. Ordidge, J. A. Helpern, Z. X. Qing, R. A. Knight, and V. Nagesh, Correction of motional artifacts in diffusion-weighted MR images using navigator echoes, *Magn. Reson. Imaging* **12**, 455–460, 1994.

37. D. K. Sodickson and W. J. Manning, Simultaneous acquisition of spatial harmonics (SMASH): Fast imaging with radiofrequency coil arrays, *Magn. Reson. Med.* **38**, 591–603, 1997.

38. K. P. Pruessmann, M. Weiger, M. B. Scheidegger, and P. Boesiger, SENSE: Sensitivity encoding for fast MRI, *Magn. Reson. Med.* **42**, 952–962, 1999.

39. J. J. van Vaals, M. E. Brummer, W. T. Dixon, H. H. Tuithof, H. Engels, R. C. Nelson, B. M. Gerety, J. L. Chezmar, and J. A. den Boer, "Keyhole" method for accelerating imaging of contrast agent uptake, *J. Magn. Reson. Imaging* **3**, 671–675, 1993.

40. Z. P. Liang and P. C. Lauterbur, An efficient method for dynamic magnetic resonance imaging, *IEEE Trans. Med. Imaging* **13**, 677–686, 1994.

41. M. Doyle, E. G. Walsh, G. G. Blackwell, and G. M. Pohost, Block regional interpolation scheme for *k*-space (BRISK): A rapid cardiac imaging technique, *Magn. Reson. Med.* **33**, 163–170, 1995.

42. F. R. Korosec, R. Frayne, T. M. Grist, and C. A. Mistretta, Time-resolved contrast-enhanced 3D MR angiography, *Magn. Reson. Med.* **36**, 345–351, 1996.

43. M. Lustig, D. Donoho, and J. M. Pauly, Sparse MRI: The application of compressed sensing for rapid MR imaging, *Magn. Reson. Med.* **58**, 1182–1195, 2007.

44. J. F. Jansen, W. H. Backes, K. Nicolay, and M. E. Kooi, 1H MR spectroscopy of the brain: Absolute quantification of metabolites, *Radiology* **240**, 318–332, 2006.

45. G. J. Kemp, M. Meyerspeer, and E. Moser, Absolute quantification of phosphorus metabolite concentrations in human muscle in vivo by 31P MRS: A quantitative review, *NMR Biomed.* **20**, 555–565, 2007.

46. P. A. Bottomley, Spatial localization in NMR spectroscopy in vivo, *Ann. N.Y. Acad. Sci.* **508**, 348, 1987.

47. J. Frahm, K. D. Merboldt, and W. Hannicke, Localized proton spectroscopy using stimulated echoes, *J. Magn. Reson* **72**, 502–508, 1987.

48. D. Bourgeois and P. Kozlowski, A highly sensitive lactate editing technique for surface coil spectroscopic imaging *in-vivo*, *Magn. Reson. Med.* **29**, 402–406, 1993.

49. A. H. Wilman and P. S. Allen, *In vivo* NMR detection strategies for γ-aminobutyric acid, utilizing proton spectroscopy and coherence-pathway filtering with gradients, *J. Magn. Reson. B* **101**, 165–171, 1993.

50. C. Choi, N. J. Coupland, P. P. Bhardwaj, N. Malykhin, D. Gheorghiu, and P. S. Allen, Measurement of brain glutamate and glutamine by spectrally-selective refocusing at 3 Tesla, *Magn. Reson. Med.* **55**, 997–1005, 2006.

51. A. H. Wilman and P. S. Allen, Double-quantum filtering of citrate for in vivo observation, *J. Magn. Reson. B* **105**, 58–60, 1994.

52. J. Hennig, Multiecho imaging sequences with low refocusing flip angles, *J. Magn. Reson.* **78**, 397–407, 1988.

53. Y. Rosen and R. E. Lenkinski, Recent advances in magnetic resonance neurospectroscopy, *Neurotherapeutics* **4**, 330–345, 2007.

54. U. G. Mueller-Lisse and M. K. Scherr, Proton MR spectroscopy of the prostate, *Eur. J. Radiol.* **63**, 351–360, 2007.

55. M. Ten Hove and S. Neubauer, MR spectroscopy in heart failure—Clinical and experimental findings, *Heart Fail. Rev.* **12**, 48–57, 2007.

56. F. Fischbach and H. Bruhn, Assessment of in vivo 1H magnetic resonance spectroscopy in the liver: A review. *Liver Int.* **28**, 297–307, 2008.

57. G. M. Tse, D. K. Yeung, A. D. King, H. S. Cheung, and W. T. Yang, In vivo proton magnetic resonance spectroscopy of breast lesions: An update, *Breast Cancer Res. Treat.* **104**, 249–255, 2007.

58. W. Gerlach and O. Stern, Der experimentelle Nachweis des magnetischen Moments des Silberatoms, *Z. Phys.* **8**, 110–111, 1921.

59. A. D. Elster and J. H. Burdette, *Questions and Answeres in Magnetic Resonance Imaging*, Mosby, St. Louis, 2001.

60. I. I. Rabi, J. R. Zacharias, S. Millman, and P. Kusch, A new method of measuring nuclear magnetic moment [Letter], *Phys. Rev.* **53**, 318, 1938.

61. E. M. Purcell, H. C. Torrey, and R. V. Pound, Resonance absorption by nuclear magnetic moments in a solid, *Phys. Rev.* **69**, 37–38, 1946.

62. F. Bloch, W. W. Hansen, and M. Packard, Nuclear induction, *Phys. Rev.* **69**, 127, 1946.

63. H. C. Torrey, Transient nutations in nuclear magnetic resonance, *Phys. Rev.* **76**, 1059–1068, 1949.

64. E. L. Hahn, Spin echoes, *Phys. Rev.* **80**, 594, (1950).

65. R. R. Ernst and W. A. Anderson, Application of Fourier transform spectroscopy to magnetic resonance, *Rev. Sci. Instrum.* **37**, 93–102, 1966.

66. L. Mueller, A. Kumar, and R. R. Ernst, Two dimensional carbon-13 NMR spectroscopy, *J. Chem. Phys.* **63**, 5490–5491, 1975.

67. P. Mansfield and P. K. Grannell, NMR diffraction in solids, *J. Phys. C* **6**, L422–L426, 1973.

68. A. N. Garroway, P. K. Grannell, and P. Mansfield, Image formation in NMR by a selective irradiative process, *J. Phys. C* **7**, L457–L462, 1974.

69. D. I. Hoult, S. J. W. Busby, D. G. Gadian, G. K. Radda, R. E. Richards, and P. J. Seeley, Observation of tissue metabolites using [31]P nuclear magnetic resonance, *Nature* **252**, 285–287, 1974.

70. A. Kumar, D. Welti, and R. R. Ernst, NMR Fourier zeugmatography, *J. Magn. Reson.* **18**, 69–83, 1975.

71. P. Mansfield and A. A. Maudsley, Planar and line-scan imaging by NMR, in *Proceedings of the XIXth Congress Ampere*, Heidelberg, 1976, p. 247.

72. P. Mansfield, Multi-planar image formation using NMR spin echoes, *J. Phys. C* **10**, L55–L58, 1977.

73. J. T. Arnold, S. S. Dharmatti, and M. Packard, Chemical effects on nuclear induction signals from organic compounds, *J. Chem. Phys.* **19**, 507, 1951.

74. R. Damadian, Tumor detection by nuclear magnetic resonance, *Science* **171**, 1151–1153, 1971.

# 9 MRI Technology: Circuits and Challenges for Receiver Coil Hardware

NICOLA de ZANCHE

## 9.1. INTRODUCTION

### 9.1.1. The MRI System

Modern nuclear magnetic resonance imaging ((N)MRI) systems are based on superconducting magnets that produce a highly uniform static field ($B_0$) inside a cylindrical bore (Fig. 9.1). Within this field the abundant hydrogen nuclei (i.e., protons) found in biological samples (mostly in water molecules) tend to align with the field, resulting in a small bulk magnetization, $M$. The magnetic resonance phenomenon consists of the precession of the individual nuclear moments around the $B_0$ vector (Fig. 9.2) with a frequency (known as the Larmor frequency) that is the product of the field strength and the proton's magnetogyric constant (42.5 MHz/T). Typical field strengths of clinical systems are in the range of $1-3$ tesla, yielding Larmor frequencies in the 42.5- to 128-MHz range. The magnetic resonance phenomenon is observable when the bulk magnetization is tilted from this equilibrium position of alignment with $B_0$, and precession gives rise to radio-frequency (RF) magnetic fields that can be detected with pickup loop coils. The rates at which the longitudinal and transverse components of the magnetization return to their respective equilibrium values form a powerful method of creating contrast between different biological tissues.

The energy necessary to produce the tilting or excitation of the magnetization is provided by a short, intense pulse of RF magnetic field at the Larmor frequency produced by a large "body" transmission coil. Such pulses can reach peak power levels of several kilowatts, and they consequently deposit energy into the anatomy in the form of heat. Patient safety is ensured by monitoring the specific absorption rate (SAR, in W/kg) so that it does not exceed values found in international standards.

*Medical Imaging: Principles, Detectors, and Electronics*, edited by Krzysztof Iniewski
Copyright © 2009 John Wiley & Sons, Inc.

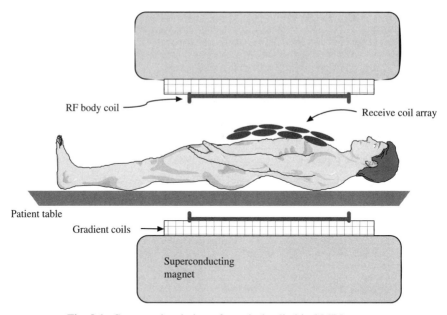

**Fig. 9.1.** Cross-sectional view of a typical cylindrical MRI system.

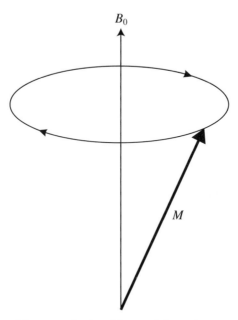

**Fig. 9.2.** Precession of the magnetization, $M$, around the static magnetic field ($B_0$) vector.

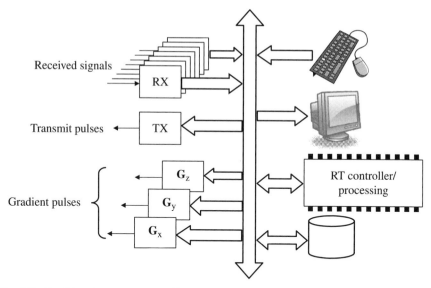

**Fig. 9.3.** Traditional spectrometer architecture. All communication and data transfer takes place over a single bus.

Spatial encoding occurs after the excitation pulse by driving time-variant currents (frequency content <100 kHz) in the gradient field coils, during which the RF signal induced by the magnetization in receiver coils is acquired. This measurement procedure, or "pulse squence" [1], can be repeated anywhere between one and several hundreds of times in order to acquire enough data to produce an image.

The pulse sequence is generated by the spectrometer (with reference to the traditional application of NMR in physics and chemistry), a computer system which contains cards for the synthesis of gradient waveforms and RF pulses, as well as acquisition boards for the signal detected by the receiver coils [2]. A common spectrometer architecture is shown in Fig. 9.3.

### 9.1.2. Typical RF Receive Coil Array

A receive coil array consists of multiple conductive loops arranged to cover the anatomical region of interest [3]. Each loop is made resonant at the Larmor frequency and connected to a low-noise preamplifier (LNA) through a matching network (Fig. 9.4). Signal-to-noise ratio (SNR) is optimized by ensuring that the LNA is noise-matched to the coil [4]. To prevent coupling among the resonant coil loops, the preamp decoupling technique [3] is employed, whereby a highly reflective preamp input impedance is used to present a high impedance between the loop's terminals, thus blocking the currents required for inductive coupling to take place. This technique allows arrays with a large number of independent channels to be constructed.

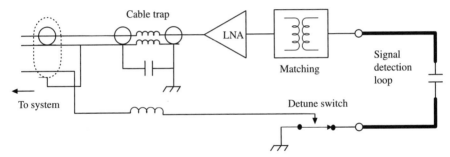

**Fig. 9.4.** Schematic of the typical components of one channel of a receive-only RF coil for MRI. An array is assembled by placing several such elements around the imaging region.

The presence of tuned receiver coils near the patient's body within the body transmit coil is a potential safety hazard due to the high peak transmit powers (often >10 kW). Currents induced in the receiver coils can not only distort the uniform transmit field (causing a degradation of image quality) but also increase the local power deposition (SAR) in the tissue. This is prevented by a detuning switch that is synchronized with the transmit RF pulses to ensure that no dangerous currents flow on the coils. Common-mode currents (see, e.g., Fig. 9.1 of reference 5) on the necessary cabling can present an additional hazard that is mitigated by the use of baluns and cable traps. The coil of Fig. 9.4 includes a cable trap immediately following the preamplifier; others may be needed at additional locations [4].

## 9.2. CONDUCTORLESS SIGNAL TRANSMISSION

The desire to send the MR signal from the receiver coil to the spectrometer by means other than coaxial cables arises from multiple issues. From the RF electronics standpoint, the presence of coaxial (or, in general, conductive) cables is responsible for introducing coupling among the coils due to the interaction of the electric fields they produce with cabling which is practically at ground potential. This phenomenon is more easily managed when numbers are small by routing the cables along symmetry planes where electric fields are minimal [3]; but as the number of coils increases, this routing strategy is more difficult or impossible to exploit, especially at higher frequencies. Ensuring safety during transmission also becomes difficult because of the complicated (and often unpredictable) interactions and resonances that can exist among the cables, making the traditional trap circuits [6, 7] less effective. In general, these effects also limit the scalability of the modular approach to array design [4].

As the number of channels is increased, mechanical cabling issues must also be considered. The dimensions of composite cables used to connect coils to the system are limited by the amount of DC power required and the minimal dimensions of coaxial cables. Both DC lines and coaxes cannot be miniaturized arbitrarily due to losses and isolation constraints, leading to cables that can be heavy, stiff, and thick—and consequently too clumsy for comfortable daily use.

The ideal RF coil array therefore should be untethered from the MRI system and transmit its signals without galvanic connections to the system or among array elements [5].

### 9.2.1. Possible Implementations

Sending MR signals without traditional cables is an area of ongoing research and development, and several methods have been considered. In most the transmission is performed either using a wireless radiofrequency link or with optical signals guided by dielectric fibers. Although technically still a cabled system, optical fibers can be packed more tightly than coaxial cables and are not conductive, thereby eliminating the problems due to coupling with RF electric fields. We can also make the distinction between (a) a system that sends signals in the analog domain and (b) one that performs an analog-to-digital conversion beforehand thus sending digital data. A summary of features and limitations of the resulting methods is provided in Table 9.1.

*9.2.1.1. Analog Transmission over Optical Fiber.* Perhaps the simplest implementation transmits analog RF signals over optical fiber by directly modulating the current in a laser or light-emitting diode or by modulation of a continuous optical signal using a Mach–Zender interferometer or other optical modulator [8, 9]. This technology is commonly known as RF-over-fiber and is relatively commonplace [10] therefore readily adapted for MRI applications. Preliminary studies [8, 11] have shown that the optical source's relative intensity noise (RIN) is among the limitations

**TABLE 9.1. Summary of Possible Methods of Sending MR Signals from the RF Coil to the Spectrometer; Advantages ( + ) and Disadvantages ( − )**

| Encoding Method | Transmission Method | |
|---|---|---|
| | Optical (Fiber) | RF (Wireless) |
| Analog | + Simple implementation<br>+ No interference with measurement<br>− Performance limited by laser noise (RIN) → tradeoff NF and dynamic range<br>− Laser uses/dissipates much power | + No need for digital electronics<br>− RF transmit pulses potential for interference/damage<br>− Phase stability/channel fading (e.g., patient motion)<br>− Requires analog frequency multiplexing |
| Digital | + Use standard optical links (e.g., Gb ethernet)<br>+ No RF spectrum/bandwidth limits<br>+ No fading due to patient motion<br>− AD conversion, processing requires several chips, power | + Most flexible option<br>+ Bandwidth efficiency using diversity/ MIMO<br>− Data rate pushes current wireless technology (e.g., WiFi) → 60 GHz Si technology? (5 Gb/s, 5 m) |

that bound noise, distortion, and dynamic range peformance of such links. A heterodyne approach [12] may be required to overcome these limitations. Although the direct-modulation approach can be implemented at lower cost than with indirect modulation it suffers from the large power required to operate the laser diode at the coil. A dedicated preamplifier/modulator unit with limited dimensions and power consumption is needed for RF over fiber to be practical for MR signal transmission in large arrays.

***9.2.1.2. Wireless Analog Transmission.*** A truly wireless approach can be implemented by modulating the MR signal with an analog carrier of much larger frequency [13]. Providing this reference frequency to a number of coils in a stable and synchronous manner has proven to be a technical challenge due to phase shifts and signal cancelations that can easily occur by slight motion of dielectric tissue within the electromagnetic boundary conditions imposed by the MRI system. Furthermore, transmission of multiple channels requires frequency multiplexing whereby each channel is allocated a precise region of RF spectrum, requiring individual fine-tuning and perhaps filtering to avoid interference from adjacent channels. The number of channels that can be accommodated within a given band is thus limited, and the RF modulation and amplification functionalities require several components that occupy space and use power.

The RF electronics (especially the stages connected directly to the antenna) must also be able to withstand the multi-kilowatt transmit pulses generated by the system for excitation of the imaging signal. Such pulses produce RF magnetic fields up to $20\,\mu T$ in strength, along with electric fields that can locally exceed $1\,kV/m$. Preliminary studies using a Bluetooth module [14] have shown that the transmit pulses can cause temporary interruptions of the wireless link without, however, causing permanent damage.

***9.2.1.3. Digital Transmission over Optical Fiber.*** The performance limitations (in terms of noise figure, dynamic range, and linearity) of analog links can be overcome by performing the analog-to-digital (AD) conversion at the coil *before* transmission out of the system's bore. A successful implementation of on-coil AD conversion will need to directly (under)sample the preamplified signal at a sufficient rate (50–100 MHz for typical MRI systems) and resolution (12 bits or more) without generating digital noise that could cause image artifacts. Such data rates could be supported directly by Gigabit Ethernet (IEEE 802.3z) to be sent to the data acquisition computer or spectrometer for further processing. Similarly to RF-over-fiber, each channel would require a dedicated fiber and would benefit over wireless systems from the lack of RF spectrum limits and easily managed interaction with transmit pulses (e.g., by shielding since no antennas are required).

***9.2.1.4. Wireless Digital Transmission.*** Transmitting gigabit-rate data using current wireless technology is not feasible; and even considering the anticipated 60-GHz WLAN standards (IEEE 802.15.3c and Ecma TC32-TG20), it is reasonable to believe that RF spectrum limits would not allow the use of the large number of channels (32–128+) that are expected to become commonplace in MRI coil arrays [15, 16]. Fortunately, MR signals have bandwidths rarely exceeding 1 MHz, thus suggesting the possibility of performing a lossless data compression down to rates of the order of 20 Mb/s or less. While the implementation of reference 17 performs analog demodulation before digitization to achieve the same goal, the ability to perform processing digitally introduces a flexibility that can be exploited by techniques that will be discussed in Section 9.3. Regardless of the specific data compression method employed, the required processing hardware will require valuable space and power.

### 9.2.2. General Issues

The implementations described above will require solution of some problems common to all. Operation in fields of several tesla requires that minimal amounts of ferromagnetic materials be used in all components to prevent the generation of mechanical forces and distortion of the highly uniform main magnetic field. Unfortunately, many off-the-shelf semiconductor circuits do not meet this requirement and may need to be obtained in die form or custom manufactured. The strong magnetic fields can also influence the operation of semiconductor devices by disrupting current flow and generating Hall voltages [18]. This phenomenon is especially evident in high-mobility semiconductors. Consequently, all circuits must be tested for sensitivity to high magnetic fields in all orientations to fully assess MRI compatibility.

In addition to standard system-to-coil communications to synchronize coil transmit and receive states, all implementations will profit from a means for the individual coils to communicate to the system other information such as fault conditions like low supply voltage or unreliable clock. As a largely independent unit, the wireless MRI array will benefit greatly from an ability to perform self-diagnostics, either within or outside of the actual system bore (e.g., on the test bench).

The processing electronics in all but the analog RF-over-fiber system will need to be well-designed and thoroughly tested against potential interference with the MR signal which could lead to image artifacts such as lines or ghosting. Digital electronics notoriously produce broadband interference at the clock frequency and its harmonics, while leakage of an analog carrier frequency can cause itermodulation and other problems.

### 9.2.3. Power Use and Delivery

The elimination of copper conductors creates a challenge for delivering sufficient power to the indvidual coil modules. The presence of components for signal transmission in addition to those for preamplification could easily bring power consumption to 0.5 W/coil or more. For large arrays, simply dissipating tens of watts of heat could be a significant challenge. In fact, the array's housing must be made of insulating material for RF safety reasons, thus precluding the use of metallic heat dissipators.

The housing's temperature also must not exceed $41°C$ at any time to avoid burns [19]. Power delivery, management, and dissipation are therefore key engineering challenges.

As with many portable devices, battery power is an obvious choice for unwired MRI coils. However, the typical continuous workflow of an MRI system would not allow recharging for eight or more hours daily, thus requiring battery packs holding tens of watt-hours of energy. Such battery packs add considerable mass to the array, reducing comfort and ease of handling. Furthermore, most cells for portable electronics are housed in ferromagnetic metal casings and are hence incompatible with the high magnetic fields. Nonmagnetic alternatives include sealed lead-acid packs and some small Li-ion cells. Ultra- (or super-) capacitors [20] are especially suited to hold a charge temporarily during periods of high power consumption (i.e., actual scanning) when a continuous power delivery to the array may not be practical.

For tethered coils (fiber-optic signal connections) DC power can be delivered through conductive connections in which parallel resonant LC traps are placed at regular intervals to block RF currents. A solution offering a purely dielectric connection would be to send optical power through fibers and convert it to electrical power using high-efficiency photovoltaic cells [21]. Several manufacturers (e.g., JDSU, Photonic Power) offer power-over-fiber systems that can readily deliver between $\frac{1}{2}$ and $1\,W$.

For untethered coils recent advances in wireless resonant power transfer [22] may hold promise. This technique is based on the ability to send power efficiently between magnetically coupled circuits that resonate at the same frequency. This frequency must be chosen with care to maximize the efficiency of power transfer while avoiding interference with the MR experiment or coil electronics. A related technique uses power scavenged from the RF excitation pulses [23], although extracting sufficient power without distorting the RF field homogeneity will likely be challenging.

### 9.2.4. Low-Power Alternatives to PIN Diodes

Efforts to reduce power consumption in wireless coils will need to address power drawn by all circuits, including preamplifiers and detuning switches. In a standard receive-only surface coil the detuning switch in Fig. 9.4 is typically implemented (Fig. 9.5a) using a resonant trap that is switched using a PIN diode, a three-layer diode in which the traditional P and N regions are separated by a layer of intrinsic semiconductor [24]. This construction allows the diode to switch a large RF current with a small DC current while presenting low RF resistance (typically less than $1\,\Omega$ for MRI coil applications). Such traps are robust but require a DC power source with an output of $\frac{1}{4}\,W$ or more (e.g., 50 mA from 5 V). The DC current can also create undesirable magnetic field distortions in its vicinity as well as mechanical forces. Consequently, it is preferable to replace the PIN diode switch with a device that is not controlled by such a large current.

One such device is the MOSFET (Fig. 9.5b), which has been used in RF switching applications in both its silicon (including integrated into CMOS) and more recent GaAs implementations [25]. MOSFETs require negligible power to control the switch and furthermore do not require a decoupling inductor due to the high isolation

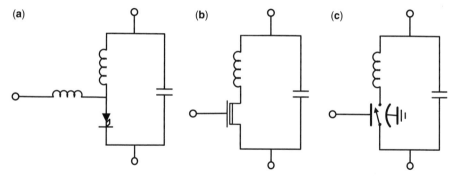

**Fig. 9.5.** Coil detuning traps. (a) Traditional PIN diode, (b) MOSFET switch, and (c) MEMS switch.

provided by the gate. In the closed state the device presents a resistance between source and drain that can be as high as several ohms. Another potential candidate is the MEMS switch (Fig. 9.5c), which allows large currents to flow with small resistance but requires large voltages (e.g., 40–120 V) to actuate the electrostatic contact. Although the energy required for switching is very small, producing such large voltages further complicates the design of wireless MRI coils. Low-voltage MEMS switches have been reported in the literature but are not yet available commercially. Reliability may continue to be a concern for MRI applications since coil detuning is an important safety feature that cannot allow failures.

Although there are no reports on the performance of these alternative devices for MRI coil detuning switches, any potential candidate may be evaluated by considering the series resistance of the device while in the closed state as well as maximal allowed current and dissipation. The series resistance is related to the trap's minimum blocking impedance required during transmit, which may be calculated using Faraday's law of induction. Depending on whether the limiting factor is the current induced in the coil [26], or the distortion field produced at the center of a circular loop [27], for 1.5-T operation (64 MHz) the minimum trap impedance is 2.8 $\Omega/\text{cm}^2$ and 80 $\Omega/\text{cm}$ (normalized to the coil's area, or radius, respectively).

The detuning switch is a parallel resonant LC trap in which all of the losses are assumed to be in series with the inductor (including those of the inductor itself and those of the RF switch). The maximum allowed resistance, $R_L$, is then given by

$$R_L \approx \frac{X_L^2}{R_{block}} \tag{9.1}$$

where $R_{block}$ is the desired blocking impedance and $X_L$ is the reactance of the inductor at resonance. For a typical blocking impedance of 1 k$\Omega$ and a reactance of 100 $\Omega$ the maximum $R_L$ is hence 10 $\Omega$, a value that is readily met by PIN diodes used in power applications, but may require a careful choice of components for MOSFET or MEMS solutions.

## 9.3. ON-BOARD DATA COMPRESSION: THE SCALEABLE, DISTRIBUTED SPECTROMETER

The acquisition of signals from the elements of a coil array is performed by the spectrometer, whose function is also to generate the precisely timed gradient and RF pulses that are required to excite and encode spatial information into the nuclear magnetization. Due to the strict timing constraints (tolerance for pulse timing is well below 1 μs and phase stability must be maintained for several seconds for some sequences), a clinical spectrometer is typically a real-time computer system able to adapt and synchronize its measurement sequence to physiological inputs such as heart rate. Cards for generating waveforms and acquiring signals are connected to a single bus (Fig. 9.3). A limitation of this architecture is the lack of scalability, an important restriction that is incompatible with the expected increase in the number of acquisition channels to 64 or 128 in the near future [15, 16]. The feasibility of connecting multiple such spectrometers in a parallel configuration has been demonstrated [28], but ultimately newer approaches will be required in future systems to deal with the large amount of data produced.

Traditionally, the spectrometer is located in an equipment room that also houses amplifiers and control equipment, and it is adjacent to the MRI suite but outside of the main magnet's fringe field. The wireless digital coil approach of Section 9.2.1.4 modifies the data acquisition task of the spectrometer by performing it in a distributed manner on hardware located next to each coil element within the system bore. This distributed architecture introduces opportunities to perform additional processing operations before the data are stored in the system's main memory.

### 9.3.1. On-Coil Detection and Demodulation

Typically, a modern spectrometer uses a field-programmable gate array (FPGA) or similar numerical processing devices to perform a quadrature amplitude demodulation (QAM) on the undersampled digitized signal from each coil (Fig. 9.6). The demodulation frequency is generated by direct digital synthesis (DDS) such that the resulting baseband spectrum is centered on 0 Hz, which is equivalent to an analog QAM demodulation at the Larmor frequency of the system. The result is a complex baseband signal that is normally low-pass filtered and decimated to reduce the data rate and reject noise (see, e.g., Section 3.4 of reference 1; also see reference 29). All measurement channels must perform sampling and demodulation in synch with the spectrometer's excitation and gradient pulses in order for the images to be reconstructed without artifacts.

Reductions in the amount of data could be beneficial to the systems of Sections 9.2.1.3 and especially 9.2.1.4, where RF spectrum limitations restrict the number of channels that may be transmitted simultaneously. The current trend of building arrays with larger numbers of smaller coil elements means that each coil sees a smaller portion of the imaging region, which translates into a smaller frequency band in the presence of encoding field gradients. This reduced frequency band cannot be immediately taken advantage of because gradients vary in amplitude and sign during the acquisition. Consequently, coils at the edges of the field of view will still see a signal with a full frequency excursion. However, the instantaneous

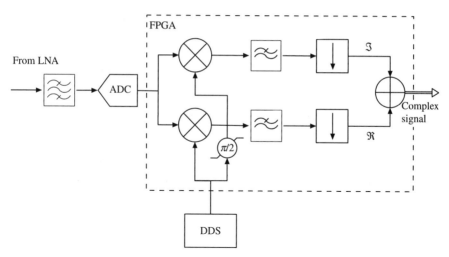

**Fig. 9.6.** Typical directly sampled MR receiver architecture (single channel).

frequency content *is* limited; and in order to take advantage of this fact, it is sufficient to demodulate with a time-varying carrier whose frequency is determined uniquely by the instantaneous strength of the gradient at the region of the coil's maximum sensitivity.

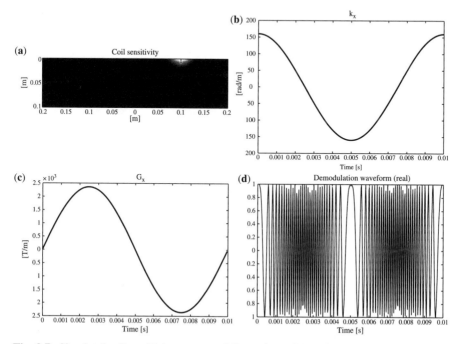

**Fig. 9.7.** Simulated coil sensitivity pattern and dimensions of the uniform rectangular phantom (a), *k*-space position (b), gradient strength (c), and demodulation waveform (d).

To illustrate this method, simulations of the signal from a surface coil with limited field of view were performed using typical values (Fig. 9.7) and a sinusoidally oscillating gradient along the horizontal ($x$) direction. Results (Fig. 9.8) show that most of the signal energy is confined to a bandwidth that is a factor 2 smaller than that required in the traditional case. This bandwidth reduction is not associated with an increase in dynamic range; therefore a true factor 2 reduction in data can be achieved. Similar results are obtained for other common gradient excitation waveforms.

Hardware implementations of this acquisition strategy have yet to be demonstrated, but would likely replace the constant-frequency DDS with a digitally synthesized frequency-modulated signal similar to those used for the production of RF pulses. The ability to demodulate with a dynamic carrier frequency is also necessary to implement techniques such as off-center spiral imaging [30]. A further, more adaptive variant may be to allow each receiver to perform a homodyne demodulation [31] on the acquired signal.

A distributed system of synchronized, but functionally independent acquisition units could also be exploited advantageously by assigning parameters such as the center (demodulation) frequency on a per-channel basis. This would enable the simultaneous acquisition of proton imaging data and spatiotemporal field evolution data used for advanced artifact-free reconstruction [32] using NMR magnetometry probes

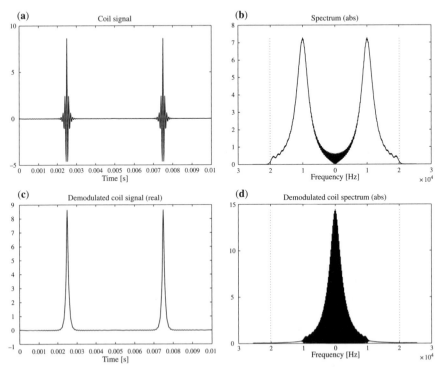

**Fig. 9.8.** Simulated signals and respective spectra acquired from the coil (a, b) and after demodulation (c, d). A bandwidth reduction of approximately a factor 2 is achieved.

[33] that use nuclei other than protons (e.g., $^{19}$F). Current spectrometers typically do not allow the acquisition of data from different nuclei simultaneously, but in this application it would result in a reduction in interference between the main imaging experiment and magnetometry. A per-channel adjustment of gain is also readily implemented [17] and would be beneficial to maximize the available ADC resolution in situations where large differences in signal strength are present (e.g., array elements that are far from the imaging region compared to those that are close).

Implementation of the on-coil distributed spectrometer will face similar challenges to those in Sections 9.2.1.3 and 9.2.1.4. The number of components and power they dissipate will limit practical dimensions, with the ultimate solution being the integration of an ADC with at least 12-bit resolution sampling at tens of megasamples/s, an FPGA, and a digital frequency synthesizer in a single package. Synchronization of all the modules with the system's operation will also be a challenge, perhaps requiring standard 10-MHz reference signals to be sent to each spectrometer module using optical fiber.

### 9.3.2. Online Data Pre-processing: Array Compression, Virtual Arrays, and Preconditioning

A distributed approach to spectrometer design naturally allows for linear combinations of data from multiple channels to be performed directly on the demodulated data rather than in a time-consuming step following the storage of data in computer memory. A major advantage of this approach would be the reduction of RAM required to store and subsequently process the data, as well as the speed with which this operation could be performed using dedicated hardware. A reduction in data can also accelerate image reconstruction (in which data from all channels is fused into a single image) while retaining the flexibility in range of application, high SNR, and field of view that large arrays provide.

Performing phased summations of signals from all coils (i.e., the equivalent of beamforming performed in antenna arrays) rarely leads to images with acceptable quality since the spatial distribution of signal phase in a surface coil is dominated by the geometrical orientation of its RF magnetic field sensitivity pattern. The relative signal phases from different points in the field of view can therefore vary widely between coils positioned in different regions, leading to cancelations since acquired signals (which are in spatial frequency space) can be directly combined only with the same constant coefficients throughout the field of view. Most MR images are therefore reconstructed from the individual coil images *after* Fourier transformations since in image space the coefficients can be assigned on a pixel-by-pixel basis [3]. A typical image-production workflow is shown in Fig. 9.9, where it is also clear that the amount of data storage and operations required scales at least linearly with the number of coils. While a large number of coils is desirable for maximal SNR or cutting-edge techniques that aim, for example, to reduce acquisition time [34–36], other imaging tasks (e.g., acquiring preparatory scout images) should not be burdened by the large amounts of data produced. A reduction in the number of channels is therefore computationally beneficial.

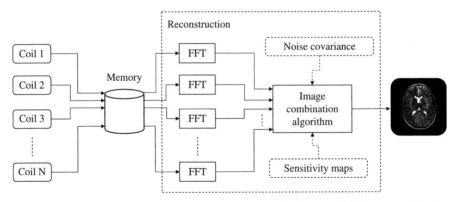

**Fig. 9.9.** Typical traditional workflow for MR image production; sensitivity maps and noise covariance information are required by advanced image reconstruction algorithms [34–36]. With a standard system architecture, access to memory can become a bottleneck when a large number of coils are used.

Several methods have been proposed to effectively reduce the amount of data while preserving image quality. One is to select a subset of channels using a switching matrix [37] based on proximity of the coil to the imaging region or signal intensity [38]. In a more general approach, channels are linearly combined, thus producing data from a (sub)set of "virtual" coils. Typical criteria for choosing effective combination coefficients are maximal signal-to-noise ratio [39] and preconditioning to diagonalize the noise covariance matrix [40]. All such methods can be implemented online by dedicated hardware *before* storage in memory as shown in Fig. 9.10.

The primary difference between this architecture and that of the traditional spectrometer is the elimination of the bottleneck imposed by the bus by providing

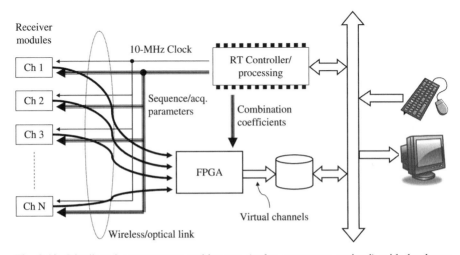

**Fig. 9.10.** Distributed spectrometer architecture (pulse generators omitted) with hardware channel combination that is performed in real time before data storage.

an independent path to memory. Data from each receiver module (which could be at the coil as in Section 9.3.1 or, more traditionally, in the equipment room) is first fed to a processing unit whose function is to perform linear combinations of the received data based on a predetermined matrix of coefficients. Similarly to ultrasound receive beam formers, this function could be implemented using standard FPGAs. The combined data are then stored in memory from which they can be used to perform the desired image reconstruction.

The on-coil detection approach, along with wireless transmission, naturally lends itself to this new architecture since data are acquired on boards that are physically separate from each other and the rest of the spectrometer. The transmission of the data to the central processing unit in which they will be reconstructed into images provides the opportunity to perform the additional hardware channel combination step that would be impossible to carry out efficiently with the traditional spectrometer of Fig. 9.3.

## 9.4. CONCLUSION

Current trends in MRI using large receiver coil arrays will require new approaches to constructing and receiving data from such arrays. The main issues are RF interactions and quantity of data, which can be addressed by the use of (a) wireless or fiber-optic communication and (b) dedicated pre-processing hardware, respectively. The technological solutions that have been put forward provide opportunities to apply recent advances in semiconductor technology including photonics, wireless high-speed communications, large-scale integration, and MEMS. Due to the highly magnetic environment, new devices will need to be tested against the effects of the field as well as against potential interference with the MR measurement.

## REFERENCES

1. Z.-P. Liang and P. C. Lauterbur, *Principles of Magnetic Resonance Imaging: A Signal Processing Perspective*, M. Akay, ed., IEEE Press, New York, 2000.

2. C. N. Chen and D. I. Hoult, *Biomedical Magnetic Resonance Technology*, Adam Hilger, Bristol, UK, 1989.

3. P. B. Roemer, W. A. Edelstein, C. E. Hayes, S. P. Souza, and O. M. Mueller, The NMR Phased Array, *Magn. Resonan. Med.* **16**(2), 192–225, 1990.

4. N. De Zanche, J. Massner, C. Leussler, and K. Pruessmann, 16-Channel interface boxes for adaptable MRI array systems, in *Proceedings, 14th Scientific Meeting, International Society for Magnetic Resonance in Medicine*, May 2006, p. 2030.

5. H. Fujita, New horizons in MR technology: RF coil designs and trends, *Magn. Reson. Med. Sci.* **6**(1), 29–42, 2007.

6. D. M. Peterson, B. L. Beck, G. R. Duensing, and J. R. Fitzsimmons, Common mode signal rejection methods for MRI: Reduction of cable shield currents for high static magnetic field systems, *Concepts Magn. Reson. Part B: Magn. Reson. Eng.* **19B**(1), 1–8, 2003.

7. D. A. Seeber, I. Jevtic, and A. Menon, Floating shield current suppression trap, *Concepts Magn. Reson. Part B: Magn. Reson. Eng.* **21B**(1), 26–31, 2004.

8. G. Koste, M. Nielsen, T. Tolliver, R. Frey, and R. Watkins, Optical MR receive coil array interconnect, in *Proceedings, 13th Scientific Meeting, International Society for Magnetic Resonance in Medicine*, May 2005, p. 411.

9. I. Yuan, G. Shen, and J. Wei, Comparison of optical modulation methods for RF coil interlinks, in *Proceedings, 14th Scientific Meeting, International Society for Magnetic Resonance in Medicine*, May 2006, p. 784.

10. C. H. Cox III, *Analog Optical Links, Theory and Practice*, Cambridge University Press, New York, 2004.

11. J. Yuan, P. Qu, J. Wei, and G. Shen, Noise figure and dynamic range optimization in optical links for MRI applications, in *Proceedings, 14th Scientific Meeting, International Society for Magnetic Resonance in Medicine*, May 2006, p. 2031.

12. S. Datta, S. Agashe, and S. R. Forrest, A high bandwidth analog heterodyne RF optical link with high dynamic range and low noise figure, *Photonics Technol. Lett. IEEE* **16**(7), 1733–1735, 2004.

13. G. Scott and K. Yu, Wireless transponders for RF coils: Systems issues, in *Proceedings, 13th Scientific Meeting, International Society for Magnetic Resonance in Medicine*, Miami, 2005, p. 330.

14. N. De Zanche, R. Luechinger, D. Blaettler, S. Jaeger, R. Mudra, E. Keller, and K. P. Pruessmann, Compatibility study for concurrent bluetooth wireless transmission and MR imaging, in *Proceedings of the European Society for Magnetic Resonance in Medicine and Biology, 21st Annual Meeting*, Copenhagen, 2004, p. 449.

15. C. J. Hardy, R. O. Giaquinto, J. E. Piel, K. W. Rohling, L. Marinelli, E. W. Fiveland, C. J. Rossi, K. J. Park, R. D. Darrow, R. D. Watkins, and T. K. Foo, 128-Channel body MRI with a flexible high-density receiver-coil array, in *Proceedings, 15th Scientific Meeting, International Society for Magnetic Resonance in Medicine*, Berlin, 2007, p. 244.

16. M. Schmitt, A. Potthast, D. E. Sosvonik, G. C. Wiggins, C. Triantafyllou, and L. L. Wald, A 128 channel receive-only cardiac Coil for 3T, in *Proceedings, 15th Scientific Meeting, International Society for Magnetic Resonance in Medicine*, Berlin, 2007, p. 245.

17. J. Wei, Z. Liu, Z. Chai, J. Yuan, J. Lian, and G. X. Shen, A realization of digital wireless transmission for MRI signals based on 802.11b, *J. Magn. Reson.* **186**(2), 358–363, 2007.

18. J. R. Bodart, B. M. Garcia, L. Phelps, N. S. Sullivan, W. G. Moulton, and P. Kuhns, The effect of high magnetic fields on junction field effect transistor device performance. *Rev. Sci. Instrum.* **69**(1), 319–320, 1998.

19. IEC 60601-1. Geneva, Switzerland: International Electrotechnical Commission, 1988. Report no. CEI/IEC 60601-1: 1988.

20. J. Schindall, The charge of the ultra-capacitors, *IEEE Spectrum* **44**(11), 38–42, 2007.

21. J. G. Werthen, M. J. Cohen, T.-C. Wu, and S. Widjaja, Electrically isolated power delivery for MRI applications, in *Proceedings, 14th Scientific Meeting, International Society for Magnetic Resonance in Medicine*, Seattle, May 2006, p. 1353.

22. A. Kurs, A. Karalis, R. Moffatt, J. D. Joannopoulos, P. Fisher, and M. Soljacic, Wireless power transfer via strongly coupled magnetic resonances, *Science* 1143254, 2007.

23. M. J. Riffe, J. A. Heilman, and M. A. Griswold, Power scavenging circuit for wireless DC power, in *Proceedings, 15th Scientific Meeting, International Society for Magnetic Resonance in Medicine*, 2007, p. 3273.

24. W. E. Doherty, Jr., and R. D. Joos, *The Pin Diode Circuit Designers' Handbook*, Microsemi Corp., Watertown, MA, 1998.

25. D. Gotch, A review of technological advances in solid-state switches, *Microwave J.* **50**(11), 24–34, 2007.

26. A. Kocharian, P. J. Rossman, T. C. Hulshizer, J. P. Felmlee, and S. J. Riederer, Determination of appropriate RF blocking impedance for MRI surface coils and arrays, *Magn. Reson. Mater. Biol. Physics Med.* **10**(2), 80–83, 2000.

27. B. L. Beck and G. R. Duensing, Design of decoupling circuits for patient safety, in *Proceedings, 5th Scientific Meeting, International Society for Magnetic Resonance in Medicine*, Vancouver, 1997, p. 1526.

28. Y. Zhu, C. J. Hardy, D. K. Sodickson, R. O. Giaquinto, C. L. Dumoulin, G. Kenwood, T. Niendorf, H. Lejay, C. A. McKenzie, M. A. Ohliger, and N. M. Rofsky, Highly parallel volumetric imaging with a 32-element RF coil array, *Magn. Reson. Med.* **52**(4), 869–877, 2004.

29. P. Perez, A. Santos, and J. J. Vaquero, Potential use of the undersampling technique in the acquisition of nuclear magnetic resonance signals. *Magn. Reson. Mater. Biol. Physics Med.* **13**(2), 109–117, 2001.

30. J. H. Lee, G. C. Scott, J. M. Pauly, and D. G. Nishimura, Broadband multicoil imaging using multiple demodulation hardware: A feasibility study, *Magn. Reson. Med.* **54**(3), 669–676, 2005.

31. D. C. Noll, D. G. Nishimura, and A. Macovski, Homodyne detection in magnetic-resonance-imaging, *IEEE Trans. Med. Imaging* **10**(2), 154–163, 1991.

32. C. Barmet, N. De Zanche, and K. P. Pruessmann, Spatiotemporal magnetic field monitoring for MR. *Magn. Reson. Med.* **60**(1), 187–197, 2008.

33. N. De Zanche, C. Barmet, J. A. Nordmeyer-Massner, and K. P. Pruessmann, NMR probes for measuring magnetic fields and field dynamics in MR systems, *Magn Reson. Med.* **60**(1), 176–186, 2008.

34. M. A. Griswold, P. M. Jakob, R. M. Heidemann, M. Nittka, V. Jellus, J. Wang, B. Kiefer, and A. Haase, Generalized autocalibrating partially parallel acquisitions (GRAPPA). *Magn. Reson. Med.* **47**(6), 1202–1210, 2002.

35. K. P. Pruessmann, M. Weiger, M. B. Scheidegger, and P. Boesiger, SENSE: Sensitivity encoding for fast MRI. *Magn. Reson. Med.* **42**(5), 952–962, 1999.

36. D. K. Sodickson and W. J. Manning, Simultaneous acquisition of spatial harmonics (SMASH); Fast imaging with radiofrequency coil arrays, *Magn. Reson. Med.* **38**(4), 591–603, 1997.

37. A. Reykowski, *Design of Dedicated MRI Systems for Parallel Imaging. Parallel Imaging in Clinical MR Applications, Medical Radiology*. Springer, Berlin, 2007, pp. 155–159.

38. S. Müller, R. Umathum, P. Speier, S. Zühlsdorff, S. Ley, W. Semmler, and M. Bock, Dynamic coil selection for real-time imaging in interventional MRI. *Magn. Reson. Med.* **56**(5), 1156–1162, 2006.

39. M. Buehrer, K. P. Pruessmann, P. Boesiger, and S. Kozerke, Array compression for MRI with large coil arrays. *Magn. Reson. Med.* **57**(6), 1131–1139, 2007.

40. S. B. King, S. M. Varosi, and G. R. Duensing, Eigenmode analysis for understanding phased array coils and their limits. *Concepts Magn. Reson. Part B: Magn. Reson. Eng.* **29B**(1), 42–49, 2006.

# INDEX

*Medical Imaging: Principles, Detectors, and Electronics*, edited by Krzysztof Iniewski
Copyright © 2009 John Wiley & Sons, Inc.